Enhancing Forest Resilience through Plant Functional Traits

This book explores the relationship between plant functional traits and the resilience of forest ecosystems. Understanding how these traits influence forest ecosystems has significant implications for conservation, forestry, and ecosystem management. This book explores the intersection of ecology, botany, and forestry, providing a comprehensive guide to using plant functional traits as a tool to advance sustainable forest development as sustainable forest management becomes increasingly important in the face of global environmental challenges.

This book also discusses the role that plant functional traits play in ecosystem dynamics and how they support forest resilience. It emphasizes how crucial these characteristics are for dealing with environmental stresses like drought and climate change, as well as how they affect soil interactions, carbon sequestration, and nutrient cycling. By integrating contemporary developments in trait-based research and analytical instruments, this book provides useful tactics for improving forest resilience via sustainable management techniques.

Key Features:

- A comprehensive overview of the latest advancements in plant functional traits and their role in forest resilience.
- Detailed explanations of the anatomical, physiological, and genetic aspects of plant functional traits, emphasizing their importance in tree growth and adaptation.
- Insights into how functional traits influence nutrient cycling, carbon sequestration, soil interactions, and responses to environmental stressors like drought and climate change.
- Discussion of modern tools and methodologies in trait-based research, along with future directions for integrating functional traits into forest management strategies.
- Practical applications of functional traits in conservation, biodiversity management, and sustainable forest management to enhance ecosystem resilience.

This book serves as a valuable reference for researchers and academicians studying species composition within forest ecosystems and exploring ways to enhance forest resilience in the context of climate change. It is also suitable as a textbook for graduate, postgraduate, and Ph.D. students in the fields such as agriculture, botany, plant physiology, forestry, biotechnology, environmental science, and agroforestry.

Enhancing Forest Resilience through Plant Functional Traits

A Practical Approach

Edited by
Narendra Kumar, Amit Kumar,
and Pramod Kumar Pandey

CRC Press
Taylor & Francis Group
Boca Raton London New York

CRC Press is an imprint of the
Taylor & Francis Group, an **informa** business

Designed cover image: Shutterstock

First edition published 2026
by CRC Press
2385 NW Executive Center Drive, Suite 320, Boca Raton FL 33431

and by CRC Press
4 Park Square, Milton Park, Abingdon, Oxon, OX14 4RN

CRC Press is an imprint of Taylor & Francis Group, LLC

ISBN: 9781032899060 (hbk)
ISBN: 9781032899084 (pbk)
ISBN: 9781003545231 (ebk)

DOI: 10.1201/9781003545231

Typeset in Times
by codeMantra

Contents

About the Editors

Narendra Kumar obtained his Ph.D. in Plant Physiology from G.B. Pant University of Agriculture and Technology, Pantnagar, India. His research focused on the response of plants to abiotic stresses. Later, he served as an Assistant Professor in various reputed academic institutes, especially Surajmal Aggarwal (Pvt.) Kanya Mahavidyalaya, Kichha, US Nagar, India; Dev Bhoomi Group of Institutions, Dehradun, India; and Uttaranchal University, Dehradun, India. He also worked as Research Associate at Forest Research Institute, Dehradun. His research interests primarily include plant physiology, ecophysiology, plant response to abiotic stresses, climate change adaptation and mitigation, functional traits, and climate change modeling. He has supervised many M.Sc. students, with many more currently under supervision. He has published various papers/chapters/articles in national and international peer-reviewed journals with high impact factors. He has edited many books with international publishers, and has served as an editor for many journals on current subjects. He is currently employed as Assistant Professor in the College of Agriculture at Central Agriculture University (Imphal), Kyrdemkulai, Meghalaya, India.

Amit Kumar is a distinguished researcher, a renowned author, and a leading expert in the fields of plant physiology, forestry, and climate change. Driven by an insatiable curiosity about the natural world and a deep-rooted passion for environmental conservation, he has dedicated his life to understanding the intricate relationships between plants, ecosystems, and the ever-changing climate. He earned his Ph.D. in Forestry from Forest Research Institute, Dehradun, India. His research is primarily focused on plant functional traits of forestry ecosystems. Earlier, he worked as an Assistant Professor in many organizations, i.e., Dolphin P.G. Institute of Biomedical and Natural Sciences, Dehradun, and Alpine Group of Institutes, Dehradun, Uttarakhand. He has published several research articles, books, and book chapters in peer-reviewed journals with high impact factors. He has edited many books with international publishers. He has received several awards from prestigious organizations, i.e., Young Scientist Award from UCOST, Dehradun, Uttarakhand; Emerging Scientist Award from VDGOOD Foundation, Kolkata; and Best Research Paper Award from Indian Association of Soil and Water Conservationists, Dehradun. He has participated in more than 30 national and international conferences, symposia, and seminars. He has published more than 25 research papers in well-reputed national and international journals. Currently, he is working as Scientist in CSIR-Institute of Himalayan Bioresource Technology, Palampur, Himachal Pradesh.

Pramod Kumar Pandey is presently working as an Assistant Maize Breeder in All India Coordinated Research Project (AICRP) on Maize, Genetics and Plant Breeding, College of Agriculture, Central Agriculture University (Imphal), Kyrdemkulai, Meghalaya. He completed his postgraduate degree in Biotechnology at KIIT University, Bhubaneswar, Odisha. He obtained his Ph.D. in Biotechnology from Dibrugarh University, Dibrugarh, Assam. He has authored more than 50 scholarly research papers and articles and contributed to various book chapters. He has served as an editor for many journals on current subjects. He has supervised many M.Sc. and Ph.D. students, and many more are under supervision. He has around 10 years of academic and professional experience in his area of expertise.

Contributors

Arushi Arora
Department of Plant Breeding and Genetics
Punjab Agricultural University
Ludhiana, Punjab, India

Kousik Atta
Faculty of Agricultural Sciences
GLA University
Mathura, Uttar Pradesh, India

R. S. Bali
Veer Chandra Singh Garhwali Uttarakhand University of Horticulture & Forestry
Bharsar, Pauri Garhwal, Uttarakhand, India

Rajneesh Bhardwaj
Graphic Era Hill University
Dehradun, Uttarakhand, India

Amit Bijlwan
Department of Agrometeorology
G.B. Pant University of Agriculture and Technology
Pantnagar, Uttarakhand, India
and
Birsa Agricultural University
Kanke, Ranchi, Jharkhand, India

Charu Bisht
Teerthanker Mahaveer University
Moradabad, Uttar Pradesh, India

Deepa Bisht
Wood Anatomy Discipline
Forest Research Institute
Dehradun, Uttarakhand, India

Sarita Bisht
Forest Research Institute
Dehradun, Uttarakhand, India

Jitendra Singh Butola
Department of Forestry and Natural Resources
Hemvati Nandan Bahuguna Garhwal University (A Central University)
Srinagar Garhwal, Uttarakhand, India

Divya Chaudhary
ICAR-National Bureau of Plant Genetic Resources
New Delhi, India

Subrata Debnath
Division of Plant Physiology
ICAR-Indian Agricultural Research Institute
New Delhi, India

Garima
College of Horticulture and Forestry, Thunag, Mandi
Dr YS Parmar University of Horticulture and Forestry
Nauni, Himachal Pradesh, India

Tuhina Ghosh
Division of Seed Science and Technology
ICAR-Indian Agricultural Research Institute
New Delhi, India

Ram Gopal
Department of Forestry
North Eastern Hill University
Tura Campus, Meghalaya, India

Himshikha Gusain
Department of Forestry and Natural Resources
Hemvati Nandan Bahuguna Garhwal University (A Central University)
Srinagar Garhwal, Uttarakhand, India

Ankita Jha
ICAR-Indian Institute of Water Management
Bhubaneswar, Odisha, India

Amit Joshi
Department of Biochemistry
Kalinga University
Naya Raipur, Chhattisgarh, India

Anjali Joshi
ICFRE-Arid Forest Research Institute
Jodhpur, Rajasthan, India

Bokado Kangujam
Lovely Professional University
Phagwara, Punjab, India

Rajesh Kaushal
ICAR-Indian Institute of Soil and Water Conservation
Dehradun, Uttarakhand, India

Rowndel Khwairakpam
Graphic Era Hill University
Dehradun, Uttarakhand, India

Shivani Kothiyal
Department of Agrometeorology
G.B. Pant University of Agriculture and Technology
Pantnagar, Uttarakhand, India

Amit Kumar
CSIR-Institute of Himalayan Bioresource Technology
Palampur, Himachal Pradesh, India

Anand Kumar
Faculty of Agricultural Sciences
GLA University
Mathura, Uttar Pradesh, India

Narendra Kumar
College of Agriculture
CAU (Imphal)
Kyrdemkulai, Meghalaya, India

Pankaj Kumar
Veer Chandra Singh Garhwali Uttarakhand University of Horticulture & Forestry
Bharsar, Pauri Garhwal, Uttarakhand, India

Roomesh Kumar Jena
ICAR-Indian Institute of Water Management
Bhubaneswar, Odisha, India

Yourmilla Kumari
College of Horticulture and Forestry, Thunag, Mandi
Dr YS Parmar University of Horticulture and Forestry
Nauni, Himachal Pradesh, India

Pankaj Lal
KVK, Jehanabad
Bihar Agricultural University
Sabour, Bhagalpur, Bihar, India

Jharna Maiti
Department of Biochemistry
Kalinga University
Naya Raipur, Chhattisgarh, India

Aruna Mehta
College of Horticulture and Forestry, Thunag, Mandi
Dr YS Parmar University of Horticulture and Forestry
Nauni, Himachal Pradesh, India

Saptarshi Mondal
Department of Crop and Soil Sciences
University of Georgia
Griffin, Georgia

Neha
Genetics and Tree Improvement Division
Forest Research Institute
Dehradun, Uttarakhand, India

Pragya
Forest Research Institute
Dehradun, Uttarakhand, India

Aditya Pratap Singh
Department of Genetics and Plant Breeding
Bidhan Chandra Krishi Viswavidyalaya
Mohanpur, West Bengal, India

Rajeev Ranjan
Department of Agrometeorology
G.B. Pant University of Agriculture and Technology
Pantnagar, Uttarakhand, India
and
Senior Scientist, ICAR-Central Marine Fisheries Research Institute
Kochi, India

Meenakshi Sati
Department of Environmental Science
College of Agricultural Science and Applied Research, Bharatiya Engineering Science and Technology Innovation University
Gownivaripalli, Andhra Pradesh, India

Upasna Sharma
Wood Anatomy Discipline
Forest Research Institute
Dehradun, Uttarakhand, India

Naresh Singh
Department of Tree Improvement
Veer Chandra Singh Garhwali Uttarakhand University of Horticulture & Forestry
Bharsar, Pauri Garhwal, Uttarakhand, India

Kalyani Supriya
Himachal Pradesh University
Shimla, Himachal Pradesh, India

Nada Tadi
Silviculture and Forest Management Division
ICFRE-Rain Forest Research Institute
Jorhat, Assam, India

Priyanka Thakur
Veer Chandra Singh Garhwali Uttarakhand University of Horticulture & Forestry
Bharsar, Pauri Garhwal, Uttarakhand, India

Victor Thingujam
Farming System Research Centre for Hill and Plateau Region (ICAR-Research Complex for Eastern Region)
Ranchi, Jharkhand, India

Laxmidas Verma
Department of Genetics and Plant Breeding
Acharya Narendra Deva University of Agriculture & Technology
Kumarganj, Ayodhya, Uttar Pradesh, India

Kashish Walia
College of Horticulture and Forestry, Thunag, Mandi
Dr YS Parmar University of Horticulture and Forestry
Nauni, Himachal Pradesh, India

Anamika Yadav
Indian Institute of Technology
Guwahati, Assam, India

1 Introduction to the Concept of Plant Functional Traits

Himshikha Gusain

1.1 INTRODUCTION

1.1.1 Functional Traits (FTs)

There are numerous ways to view FTs. On the basis of functional characteristics, plant communities have been categorized for a very long time. Generally, the term "functional trait" denotes, a well-defined, measurable property at the individual level used across the species (Nock et al., 2016). FT can also be recognized as the most significant kind of trait, which strongly influences the performance of any organism (McGill et al., 2006). Standard definition describes FTs as any such characteristics (structural, biochemical, morphological, physiological, etc.) that exert a significant impact on the behavior of organisms toward their environment (Nock et al., 2016; Violle et al., 2007; Kumar et al., 2020a,b). FTs thus include all those ecological strategies that are being applied by plants to control the assemblage of their communities, their biodiversity, and their functioning (Li and Prentice, 2024). Due to these properties, FTs are also important to quantify the impact of community migration on their surroundings and to have an understanding of plant communities in response to any sudden or periodic change.

In higher plant forms, FTs are those characteristics that affect individual performance and response based on interactions within the environment and, thus, further impact the fitness and functionality of communities. The species' characteristics and functional composition heavily (directly or indirectly) influence different processes in most ecosystem types. Assessment of FTs also provides an accurate measurement of system processes and responses directly from the individual plants to complex ecosystem levels. Since plants have been considered the foundation and functional units of any ecosystem, changes in plant traits with the due course of time act as essential drivers of various ecological processes and their interaction with locality factors significantly influences multiple ecosystem operations.

1.1.1.1 Plant Functional Traits (PFTs)

Plant functional traits or PFTs are defined as morphological, physiological or phenological features measurable at the individual level, which directly or indirectly affect

DOI: 10.1201/9781003545231-1

plant performance through growth, reproduction and survival (Violle et al., 2027). Violle et al (2007) explains trait functional only if impacts on performance and ecological outcome. Furthermore, Diaz et al (2004) assert that such functional traits or FTs also help to predict communities composition, resilience abilities against disturbance and their contribution in shaping ecosystem services.

From above definitions, it is clear that in the plant kingdom, certain groups of species share similar traits in nature, known as plant functional traits or PFTs. The explanations of PFT are based on many traits that exert specific responses to multiple challenges, such as ecological disturbance occurring within the environment. PFTs are the physical manifestations holding unique features in the life history of a plant species and some PFTs, especially those derived based on single (or few) traits addressing a particular ecological challenge; for example, fire (Tsakalos et al., 2019), nutrient availability, and moisture stress. Generally, it is much easier to define and implement such PFTs to explore plant assemblages.

1.1.2 Conceptual frame work

Plant functions can be understood mostly as the leaf economics spectrum or LES represented by specific leaf area and the wood economics spectrum or WES represented by wood density (Bu et al., 2019), while responding to the external environment. For example, in plants, the leaf characteristics are determined by the availability of nutrients in soil, followed by the topographic factors (Butterfield and Suding, 2013), whereas the wood density is mostly affected by the fertility of soil. Apart from this, the plant's characteristics can also be classified under manageable PFTs (Prentice and Cowling, 2013).

Ecologists have elaborated FTs in numerous ways. The theory of FTs has contributed to gaining multiple ecological insights by concentrating on characteristics that indicate organizational interaction with surroundings. Certainly, ecologists have applied methods focusing on species diversity with unique FTs, but the traits that co-relate ecosystem functioning with existing plant or species diversity are still highly controversial. While functional diversity is explored in a few cases only, species diversity is an inadequate surrogate to utilize multiple (Diaz and Cabido, 2001; Kumar et al., 2020b, 2021a,b). On the other side, ecosystem functioning is primarily determined by the type, range, and values of ecological diversity. In plant ecology, recent years have witnessed a transition from species centric approaches to trait based functional frameworks. Such methods utilize functional characteristics which influence the dynamics of ecosystem and performance of plants therein. This shift, rooted in the growing need to understand plant-environment interactions at multiple scales, has led to the development and application of PFT paradigm.

In the long run, any coordinated variation in plant traits forms PFTs (Snapp, 2017; Kumar et al., 2021c,d). In biosystematics, the species concept aims to summarize the history of its evolution, while the PFT concept is applied to recapitulate species similarity in the context of defined challenges such as stress faced by it. Here, the species concept varies in biosystematics, the PFT explores certain other primary challenges such as nutrient stress (nutrition), competition for space (growth), and impeded water availability or scarcity (moisture). When it comes to functional communities, the

PFT concept is a useful way to look into various ecological processes, like how plant communities form, gather, and compete. Any cohabiting existence of plant life in any ecosystem is an outcome of various traits, for example, resource, including the growth form, pattern, and other traits that indicate competition and, hence, determine the extent of growth and survival of plant communities. A valuable role of PFT is to develop life strategies in which the species can be grouped into stress bearers and competitors since many species cohabitate in ecosystems. The survival of organisms is divided into three kinds of life strategies adapted to ecosystems, including growth types, resource acquisition mechanisms, and other critical competitive plus survival traits. With an emphasis on terrestrial ecology, PFTs are considered specific assets of species that respond similarly while producing similar effects, leading to functional harmony across autoecological and allometric diversities.

In recent years, the measurement and exploration of PFT have generated interest among scientific communities across the world. These traits are detrimental tore-expressing the ecological and physiological basis of research in plant science. Ecologists and climatologists use a PFT-based classification system to classify functional plant communities in any ecosystem. Here, the complex plant diversity and behavior are simplified by ecological modeling through categorization of plant communities under common functional groups. PFTs can also be viewed as a specific set of similar traits linked to specific variables (Kumar et al., 2022; Chauhan et al., 2023), with measurements assisting in determining their dominance. The dominant PFTs are mostly the indirect determinants of biogeochemical cycles. Examples of solid links between plant traits and ecosystem processes have accumulated in the past few decades (Kumar et al., 2015, 2016). From an ecological point of view, PFTs are adopted by various ecosystem modelers to represent species characterization based on common qualities such as plant morphology, growth, and functions such as photosynthetic pathway and evapotranspiration. The biogeochemical cycles of carbon, water, and nutrients are indirectly controlled by dominant PFTs because they are the main parts of the flow of matter and energy across ecosystems. That's why it is essential to select relevant traits and apply appropriate methods or techniques while classifying functional plant communities.

The concept of PFT is also applicable to exploring chronological distinctions such as early and late successional plant communities or plant communities evolved as an outcome of secondary succession and responses to the causative factors. Therefore, PFTs are inevitable trade-offs between possible trait combinations in forest successions, a better option to identify ecological response and the effect exerted over a longer time period. The idea of PFT application to gain valuable information of ecosystem functioning and services is familiar and well-established today. Using PFT as a special indicator for ecosystem functioning provides a deeper insight into the mechanism of ecological principles and services that allows management and forecasting of ecological service availability. Nevertheless, identifying the crucial and significant effects of any species is essential for accurately predicting its ecosystem services, thereby providing the most precise and holistic explanations. These traits have the potential to explore ecosystem services through scientific, defensible, and comparable estimations with interspecific/intraspecific traits being used as fixed determinants to any environmental change (Brown and Anand, 2022; Kumar et al., 2017, 2019).

1.2 CLASSIFICATION OF FUNCTIONAL TRAITS AND PFTs

PFTs are heritable characteristics that can be categorized into morphological, physiological, and biological characteristics, indicating how plants interact with the surrounding environment. It is asserted that PFT concept is embedded in a broader ecological framework that seeks to explain how species characteristics influence their ecological roles build on two key types of traits:

1.2.1 Response Traits (RTs)

The response traits determine how certain plants respond to their surrounding environmental conditions such as draught, nutrient availability and disturbance regime. RTs also include responses such as life form, seed morphology, mass, and tolerance level.

1.2.2 Effect traits (ETs)

Opposite to RTs, the ETs influence the way plants affect their local environmental properties and functions. For example, certain plants traits can influence nutrient cycling, productivity, and carbon sequestration abilities.

These distinctions help ecologists separate the mechanisms of plant adaptation (response traits) from their functional roles (effect traits) in ecosystems (Lavorel & Garnier, 2002). An accurate estimation and measurement of RT and ET (nutrient content and availability, biomass accumulation, carbon sequestration, mineralization, etc.) is essential in this manner, as these responses are effects correlated to any driver or disturbing agent (e.g., fire, stress, and excess), which influence the functioning of any ecosystem. Figure 1.1 provides a brief account of the classification of PFTs.

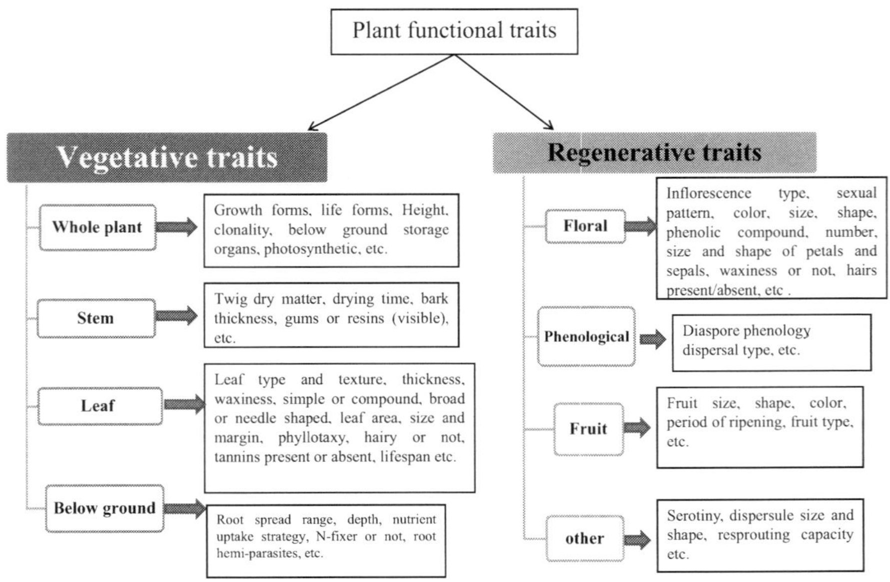

FIGURE 1.1 Plant functional traits (PFTs) classification system.

The above mentioned traits can simplify biological diversity and predict community and ecosystem processes. However, this system might be sufficient for placing species into distinct groups based on their dominant traits, such as life forms (Powers and Tiffin, 2010).

1.3 TECHNIQUES AND METHODS TO EXPLORE PFTS

1.3.1 Proteomics Techniques

Protein function depends not only on the molecular structure of the protein but also on its subcellular localization and post-translational modifications (PTMs) (Chauhan et al., 2023). Biotic stress constantly affects plants in both natural and agricultural environments, posing a threat to their survival and growth. In response to these changing circumstances, plants have evolved molecular programs to perceive and adapt to the environment quickly. The role of proteins in the plant biotic stress response is crucial to understand the physiological strategy for their survival. It is found that the plant's immune system reacts to biotic stress as a complicated system through interactions and crosstalk made up of many signals and a wide range of stress tolerance proteins that help the plant handle stress. Because of their unique ability to measure complex protein mixtures, mass spectrometry-based proteomics and PTMs have provided scientists with the tremendous capability to study plants more precisely.

Currently, proteomics has been transformed from an isolated field into a comprehensive tool for advanced research that can be used to explain more complicated biological mechanisms and functions. Several studies have successfully applied the power of proteomics as a discovery tool to uncover plant resistance mechanisms (Cherif et al., 2023; Kumar et al., 2021e). Growing evidence from scientific studies (Bravo et al., 2024) indicates that proteins' spatial proteome and PTMs directly affect the plant immune responses.

1.3.2 Image Spectroscopy and Remote Sensing

Remote sensing and GIS techniques are a new field of research for extracting PFTs. Nowadays, PFTs are used in remote sensing techniques as effective variables to explore the global carbon cycle, ecosystems, and climate change at a global level it mostly includes:

- Hyper spectral imaging – to estimate leaf traits (e.g. SLA, Chlorophyll, N content) at large scale
- LIDAR (Light Detection and Ranging) – to measure plant height, canopy structure and biomass
- Drones (UAVs) – to get fine scale trait mapping across landscapes
- Automated imaging platforms – to measure traits in large plant populations
- Non-distructive Spectroscopy

A Hyperrealistic spatial data can be acquired through the application of remote sensing techniques. With image spectroscopy's high spatial and temporal resolutions, a reliable monitoring system can be created to look at how the plants in a region are responding to changing weather conditions (Schweiger et al., 2016).

1.3.3 Machine Learning (ML) Approach

The ML approach is based on descriptive and predictive modeling. Statistical up-scaling and ML approaches based on relationships between FTs and environmental factors have recently been used to produce global maps for fewer FTs (mainly leaf traits). Recent decades have witnessed a noticeable lead in data stimulation and interpretation. This approach is applied to identify the most divergent traits among the species.

1.3.4 Earth-System Models (ESMs)

Within ESMs, PFTs are used to simplify terrestrial biosphere modeling in combination with soil information and other site characteristics (Nock et al., 2016, Cranko et al., 2024). In such modeling, the global land cover is a crucial variable to provide feedback on the status of natural resources, biodiversity, and climatic conditions. Although a big hindrance in the computation of such models is that the related dataset falls short due to multiple user needs while providing detailed thematic and spatial information to conceptualize and interpret ESMs (Poulter et al., 2015).

1.3.5 Data-Driven Methods (Single Trait/Multitrait Models)

Currently, advanced hyperspectral remote sensing is essential for future global monitoring applications. Available retrieval methods are of various types, ranging from data-driven statistical methods, such as hybrid methods, to radioactive transfer models (RTMs) that simulate light's interaction with vegetation properties. Such models can offer a reliable method to retrieve PFTs. Since data-driven models are independent of different traits, new approaches such as resolution-based and algo-modeling have emerged as the most powerful procedures to solve problems related to PFT retrieval in hyperspectral data. Exploitation of interrelationships between certain traits, such as overlapping and absorption in the same process, can be prevented through single trait/multitrait-based modeling (Cherif et al., 2023).

1.4 IMPORTANCE AND ROLE OF PFTs IN THE CONTEXT OF FOREST, ECOLOGY, AND CLIMATE CHANGE

PFTs serve as functional building blocks for understanding plant strategies, community dynamics and ecosystem functions. From leaf area to secondary metabolites, these traits provide a broadway mechanism to understand and explore how plant operates and influence their surrounding environment.

1.4.1 To Understand the Influence of Changing Climatic Conditions on Vegetation Cover

With rapidly changing climatic conditions, the world's natural resources and plant biodiversity are continuously diminishing with increasing and abrupt land use and resource exploitation as an outcome of the ever-increasing human population. PFTs are supposed to produce a significant impact on how any species responds to any change in the surrounding environment or how it functions as a guide in ecosystem restoration and land-use planning (Rao et al., 2024). PFTs also offer potential roles as the whole-plant syndrome indicator of vegetation responses to changing climatic conditions.

1.4.2 To Explore Distinct Plant Categories at Regional and Global Scales

PFTs are essential to explore as their traits play an important role in understanding ecological functions and ecosystem services. Based on various functional spectra, the possibility of distinct functional categories can be explored by applying the PFT concept in eco-modeling (Diaz and Cabido, 2001).

1.4.3 To Understand Trait-Based Functional Ecology

Violle et al. (2007) suggested the central role of FTs to those PFTs (morphological, physiological, chemical, and phenological) that indirectly contribute to the evolutionary fitness of a species. PFTs are keys to assess and monitor terrestrial ecosystem properties. Through an in-depth understanding of functional diversity, PFTs can enhance our perception of ecosystem functioning (Cherif et al., 2023). Rapidly developing trait-based ecology emphasizes the importance of PFTs in shaping community assembly, ecosystem properties, and services (Messier et al., 2010). Recording and quantitative estimation of PFTs in the context of the structural and functional characteristics of vegetation (Lausch et al., 2016) and modeling will further contribute to understand ecosystem functions such as nutrient absorption, gaseous exchange, stress, and tolerance, as well as the role of that particular species in regional climatic regulation.

1.4.4 To Describe Patterns and Processes in Plant Communities

PFTs have been explored multiple times to monitor the effect of environmental change on ecosystem management. PFTs are considered the key to explore community-oriented patterns and processes and functional classification of plant species (McMahan et al., 2011).

1.4.5 For Ecosystem and Biome Modeling

By representing gross areas, PFTs offer a fine level of modeling for various kinds of ecosystems such as deserts and deciduous or evergreen forests. To create a model with PFTs and satellite data, a minimum 1 km^2 area can be stipulated to define and elaborate existing plant communities there. In this case, for each PFT, key parameters such as competitiveness and re-absorption are defined, and their specific values are determined from observed plant characteristics such as height and leaf area.

1.4.6 To Understand Plant Interactions and Their Role in Ecological Succession

Differences in plant interactions and their role in certain natural processes, such as ecological succession, can be explored through the application of PFTs in modeling.

Similarly, any differentiation in early and late succession stages, leaf lifespan, and biological rhythms with distinct responses to anthropological, fire, or grazing pressure can be done by generating PFT-based datasets (Prentice and Cowling, 2013). From a handful of selected proxy traits, the ecologists have found success in summarizing interspecific variations in plant performance and functionality.

1.4.7 To Explore Vegetation Dynamics

It is important to note that the values of PFTs, which serve as functional indicators of vegetation responses, vary depending on the scale, as climate is considered a primary driver of vegetation dynamics (Gillison, 2019). In this regard, the number of selected traits can capture the leading dimensions of variations among multiple species in a zone. For example, a combination of three easily measured traits, specific leaf area (the ratio of leaf area to leaf dry mass), plant height, and seed mass, enables us to assess how different species use their resources, interact with neighbors, and disperse in time and space (Garnier and Navas, 2011).

1.4.8 To Estimate Community Role in Ecosystem Physiology

In order to explore ecosystem physiology, PFTs can also be applied to explore the role of plant community structure and species composition. Distinction among species with special emphasis on community categorization explores their contributions to physiological and architectural aspects of ecosystems. Such distinctions also affect the biochemical rates in plants through quantification and quality estimations. Annual *vs* perennial herbs, deciduous *vs* evergreen woody, etc. are certain examples of it (Kittel, 2013).

1.4.9 Community Response to Climate Change

Since PFTs have been found to vary along limited dimensions, they are supposed to share genetic controls or might be more dependent on alternative ecological strategies. Hence, exploring geographical patterns of PFT and their relationships to global climate change can provide a better-founded prediction on the effects of changing climatic conditions on community structure functioning and biodiversity patterns.

1.5 SUCCESSFUL CASE STUDIES

A study of the geographical patterns of six FTs across North and South America revealed that trait-climate relationships differ between woody and herbaceous plants. It also includes different climate predictors and different response shapes of FTs along climatic gradients. Li and Prentice (2024) also found that woody evergreen plants tend to have thicker and longer-lived leaves poleward, while woody deciduous plants show the opposite patterns. Majumder and Mason (2024) used random forest and gradient boosting machine classifiers, as well as recursive feature elimination, in a different study to look at trait data for the genus Helianthus. They used both descriptive and predictive ML methods. These approaches were applied at the genus level and within each of the three major clades within the genus to examine the variability in the principal axes of trait divergence in three independent species radiations. Earlier, Gillison (2019) used distribution, ordination, and regression modeling to predict values of plant climate indicators against ten bio-climatic variables. For this, the model PFTs and individual plant traits were found to be highly correlated and showed different patterns with bio-climate. Model PFT-based indicators were considered the best in each of the studied regions covered under their study. It was concluded that model traits such as photosynthesis in green stems, inclination angle of leaves, and stomatal distribution are the most important FTs. It proves that as an indicator of vegetation response, PFTs can be considered the best indicators to study the impact of climate change worldwide.

1.6 ISSUES AND CHALLENGES IN PFT-BASED MANAGEMENT TO MITIGATE CLIMATE CHANGE

From previous interpretations, it is clear that PFTs include all those biological, physio-morphological, and phenological characteristics that determine a plant's or a species' performance in terms of its fitness to attain survival and growth under specific climatic conditions. On the other hand, it is equally important to understand that apart from other alarming repercussions, constantly changing climatic conditions have also increased certain stressors in the environment that have been reported to reduce the resilient abilities of plants. As per Tyagi and Kumar (2024), it has further impacted the structure and the ecosystem services provided by the forests, causing the respective PFTs to change over time. Therefore, a periodic and progressive in-depth assessment is required to study how these modifications have altered the functioning of the ecosystem. Examining constant changes in PFTs can accomplish this. However, application of PFTs to study vegetation dynamics, community responses, and ecosystem functioning is not an easy task in the context of climate change. Following is a list of the issues and challenges associated with PFT-based management and conservation efforts:

1. Today, PFTs are widely used to understand plant responses and forecast shifts in plant distribution, phenology, and the possibilities of extinction due to changing environmental conditions, a critical step to detect intraspecific and interspecific sensitivity and their response to environmental change. In this context, a bottom-up approach of some vital FTs related to fitness (e.g., photosynthetic rate, hydraulic traits, plant height, and root depth) should be helpful. For example, suppose the plant height of each individual in a whole community has been surveyed. In that case, in decades we can construct a trait-biodiversity model based on species abundance, community biomass, ecosystem carbon storage, or regional biodiversity and then predict the dynamics in community structure, productivity or ecosystem resilience, and finally, the biodiversity when climate change (Liu et al., 2023). Due to different aspects of data in similar studies, the amalgamation of datasets and interpretation has become more complicated. Another major problem is that such dataset is naturally sparse and difficult to generate as all potential traits

have not been covered across different datasets using various techniques and modeling done so far.

2. To understand ecosystem functioning and functional diversity, it is very critical to access a wide range of information on several vegetation properties expressed by PFTs. It has become more complicated due to restricted access to globally representative datasets and their corresponding in situ measurements. Furthermore, it has been experienced by many scholars such as Cherif et al. (2023) that in the process of plant trait retrieval, most of the data-driven approaches cannot be extended easily with new training data. Such hindrances hamper continuous improvement and knowledge transferability through trait-based modeling.
3. Although advanced theoretical approaches with technological interventions have been developed to explore functional diversity in the context of spatial variations, more studies have to be conducted to examine temporal variations in the FTs. It is essential to explore their mechanisms and niche-complementary effects (Huxley et al., 2023).
4. Another major concern is the interpretation of vegetation dynamics in response to climatic shifts. Climatic shifts can be determinant on climatic responses of certain pathways such as energy and matter flows (Kittel, 2013). It would be immensely helpful for researchers to compare eco-physiological studies in different ecosystems with the meta-analysis approach to forecast future vegetation dynamics under changing climatic scenarios (Majumder and Mason, 2024).
5. Although the concept of PFT is frequently applied to explore a wide spectrum of plants and their ecological strategies, PFT-based methods follow certain limitations. For example, despite much advancement in eco-modeling, many climatologists and ecologists find it pretty difficult to establish which minimal characteristic can be utilized as the best precursor to model the accurate response of that plant groups as an adaptive mechanism to climate change. Also, researchers are still striving hard to capture diversity at each level (plant, species, and ecosystem by simplifying a few FTs). It is again very complicated to explain or represent scheduled (rare and endangered) species within a given environment. Nonetheless, researchers and scientists have attained much success in developing more sophisticated and pointed models such as trait-based models to address the above-mentioned challenges and concerns.

1.7 CONCLUSION

An in-depth understanding of FTs such as leaf characteristics, growth and life forms, and morphological variations that express the crucial characteristics and processes in higher plant forms is essential to estimate their role in predicting forest resilience against climate change. In the field of restoration and functional ecology, as an outcome of interaction with the external environment, the PFTs offer valuable insight into the importance of functional plant communities and their responsiveness to changing climatic conditions. PFT-based classification further helps in the identification, selection, and determination of morphological, heritable, and physiological traits, which determine plants' fitness. An in-depth study of region-wise variations in PFTs can further help explore trait-based resistance and adaptive system-mechanism among plant communities. It is valuable to develop and compare large-scale advanced trait-based ecological models in the future.

REFERENCES

Bravo, F., Werf, W., Tognetti, R., Mazía, N., Anten, N., Piazza, M.V., Tomelleri, E., and Pretzsch, H. 2024. Unleashing the power of plant structural and functional diversity: From common observations to theory and management models. *Food Energy Security*, 13 (5): 70005.

Brown, L.M., and Anand, M. 2022. Plant functional traits as measures of ecosystem service provision. *Ecosphere*, 13 (2): 1–13.

Bu, W., Zhang, C., Huang, J., Zang, R., Ding, Y., Xu, H., Lin, M., and Wang, Z. 2019. The influence of disturbance histories and soil properties on above ground and below ground biomass through plant functional traits in a tropical rainforest. *Forests*, 10 (9). https://www.mdpi.com/1999-4907/10/9/774

Butterfield, B,J. and Suding K.N. 2013. Single trait functional indices outperform multi-trait indices in linking environmental gradients and ecosystem, services in a complex landscape. *Journal of Ecology*, 101: 9–17.

Chauhan, C., Thakur, A., Sharma, V., Yadav, R.R., Sivakumar, K.B., Kumar, P., Verma, S.K., Panwar, R.K., Tewari, S.K., and Gautam, A. 2023. Proteomics as a tool for analyzing plant responses to abiotic and biotic stresses. In Husen, A., Ahmad, A. (Editors), *Plant Biology, Sustainability and Climate Change, Genomics, Transcriptomics, Proteomics and Metabolomics of Crop Plants*, Academic Press, New York, pp. 35–68.

Cherif, E., Feilhauer, H., Berger, K., Dao, P.D., Edwald, M., Hnak, T.B., He, Y, Kovach, K.R., Lu, B., Townsend, P.A., and Kattenborn, T. 2023. From spectra to plant functional traits: Transferable multi-trait models from heterogenous and sparse data. *Remote Sensing of Environment*, 292: 1–43.

Cranko, P.J., Abramowitz, G., De Kauwe, M.G., and Pitman, A.J. 2024. Are plant functional types fit for purpose? *Geophysical Research Letters*, 51, e2023GL104962.

Díaz, S. and Cabido, M. 2001. Vive la différence: plant functional diversity matters to ecosystem processes. *Trends in Ecology and Evolution* 16, 646–655.

Díaz, S., Hodgson, J.G., Thompson, K., et al. 2004. The plant traits that drive ecosystems: Evidence from three continents. *Journal of Vegetation Science*, 15(3), 295–304. https://doi.org/10.1111/j.1654-1103.2004.tb02266.x

Garnier, E., and Navas, M.L. 2011. A trait-based approach to comparative functional plant ecology: Concepts, methods and applications for agroecology: A review. *Agronomy for Sustainable Development*, 32: 365–399.

Gillison, A.N. 2019. Plant functional indicators of vegetation response to climate change, past present and future: II. Modal plant functional types as response indicators for present and future climates. *Flora*, 254: 474–478.

Huxley, J. D., White, C. T., Humphries, H.C., Weber, S.E., and Spasojevic, M.J. 2023. Plant functional traits are dynamic predictors of ecosystem functioning in variable environments. *Journal of Ecology*, 111 (12): 1597–2613.

Kittel, K.G.F. 2013. The voluntaribility of biochemistry to rapid climate change. In Roger, A.P. (Editor), *Climate Vulnerability*, Academic Press, New York, pp. 195-201.

Kumar, A., Dwivedi, G.K., Tewari, S., Paul, J., Anand, R., Kumar, N., Kumar, P., Singh, H., and Kaushal, R. 2020a. Carbon mineralization and inorganic nitrogen pools under *Terminalia chebula* Retz.-based agroforestry system in Himalayan foothills, India. *Forest Science*, 66 (5): 634–643.

Kumar, A., Dwivedi, G.K., Tewari, S., Paul, J., Sah, V.K., Singh, H., Kumar, P., Kumar, N., and Kaushal, R. 2020b. Soil organic carbon pools under *Terminalia chebula* Retz. based agroforestry system in Himalayan foothills, India", *Current Science*, 118 (7): 1098–1103.

Kumar, A., Kumar, P., Singh, H., Bisht, S., and Kumar, N. 2021a. Relationship of physiological plant functional traits with soil carbon stock in temperate forest of Garhwal Himalaya. *Current Science*, 120 (8): 1368–1373.

Kumar, A., Kumar, P., Singh, H., and Kumar, N. 2021b. Modulation of plant functional traits under essential plant nutrients during seasonal regime in natural forests of Garhwal Himalayas. *Plant and Soil*, 465: 197–212.

Kumar, A., Kumar, P., Singh, H., and Kumar, N. 2021d. Impact of plant functional traits on near saturated hydraulic conductivity of soil under different forests of Kempty watershed in Garhwal Himalaya. *Indian Journal of Soil Conservation*, 49 (1): 38–44.

Kumar, A., Singh, H., Kumari, G., Bisht, S., Malik, A., Kumar, N., Singh, M., Raturi, A., Barthwal, S., Thakur, A., and Kaushal, R. 2022. Adaptive resilience of roadside trees to vehicular emissions via leaf enzymatic, physiological, and anatomical trait modulations. *Environmental Pollution*, 313: 120191.

Kumar, A., Tewari, S., Singh, H., Kumar, P., Kumar, N., Bisht, S., Kushwaha, S., Tamta, N., and Kaushal, R. 2021c. Biomass accumulation and carbon stocks in different agroforestry system prevalent in Himalayan foothills, India. *Current Science*. 120 (6): 1083–1088.

Kumar, N., Jeena, N., and Singh, H. 2019. Elevated temperature modulates rice pollen structure: A study from foothill Himalayan agro-ecosystem in India. *3 Biotech*, 9: 175.

Kumar, N., Jeena, N., Kumar, A., Khairakpam, R., and Singh, H. 2021e. Comparative response of rice cultivars to elevated air temperature in Bhabar region of Indian Himalaya: Status on yield attributes. *Heliyon*, 7: e07474.

Kumar, N., Kumar, N., Shukla, A. Shankhdhar, S.C., and Shankhdhar, D. 2015. Impact of terminal heat stress on pollen viability and yield attributes of rice (*Oryza sativa* L.). *Cereal Research Communications*, 43(4): 616–626.

Kumar, N., Shankhdhar, S.C., and Shankhdhar, D. 2016. Impact of elevated temperature on antioxidant activity and membrane stability in different genotypes of rice (*Oryza sativa* L.). *Indian Journal of Plant Physiology (Now Plant Physiology Report)*, 21 (1): 37–43.

Kumar, N., Suyal, D.C., Sharma, I.P., Verma, A., and Singh, H. 2017. Elucidating stress proteins in rice (*Oryza sativa* L.) genotype under elevated temperature: A proteomic approach to understand heat stress response. *3 Biotech*, 7: 205.

Lausch, A., Bannehr, L., Beckmann, M., Boehm, C., Feilhauer, H., Hacker, J.M., Heurich, M., Jung, A., Klenke, R., Neumann, C., Pause, M., Rocchini, D., Schaepman, M.E., Schmidtlein, S., Schulz, K., Selsam, P., Settele, J., Skidmore, A.K., and Cord, A.F. 2016. Linking earth observation and taxonomic, structural and functional biodiversity: Local to ecosystem perspectives. *Ecological Indicators*, 70 (7): 317–339.

Lavorel, S., and Garnier, E. 2002. Predicting changes in community composition and ecosystem functioning from plant traits: Revisiting the Holy Grail. *Functional Ecology*, 16(5): 545–556. https://doi.org/10.1046/j.1365-2435.2002.00664.x

Li, J., Prentice, I.C. 2004. Global patterns of plant functional traits and their relationships to climate. *Community Biology*, 7, 1136. https://doi.org/10.1038/s42003-024-06777-3

Liu, H., Yin, D., He, P., Cadotte, W., and Ye, Q. 2023. Linking plant functional traits to biodiversity under environmental change. *Biological Diversity*, 1: 22–28.

Majumder, S., and Mason, C. 2024. A machine learning approach to study plant functional trait divergence. *Application in Plant Science*. https://doi.org/10.1101/2023.03.16.533012

McGill, B.J., Enquist, B.J., Weiher, E., and Westoby, M. 2006. Rebuilding community ecology from functional traits. *Trends in Ecology and Evolution*, 21 (4): 178–185.

McMahon, S.M., Harrison, S.P., Armbruster, W.S., Bartlein, P.J., Beale, C.M., Edwards, M.E., Kattge, J., Midgley, G., Morin, X., and Prentice, I.C. 2011. Improving assessment and modelling of climate change impacts on global terrestrial biodiversity. *Trends in Ecology & Evolution*, 26 (5): 249–259.

Messier, J., McGill, B.J., and Lechowicz, M.J. 2010. How do traits vary across ecological scales? A case for transit-based ecology. *Ecology Letters*, 13 (7): 838–848.

Nock, C.A., Richard, J.V., and Beisner, B. 2016. Functional traits. *eLS*, 1, 1–8.

Poulter, B, MacBean, N., Hartley, A., Khlystova, I., Arino, O., Betts, R., Bontemps, S., Boettcher, M., Brockmann, C., Defourny, P., and Hagemann, S. 2015. Plant functional type classification for earth system models: Results from the European space agency's land cover climate change initiative. *Geoscientific Model Development Discussions*, 8: 429–462.

Powers, J.S., and Tiffin, P. 2010. Plant functional type classifications in tropical dry forests in Costa Rica: Leaf habit versus taxonomic approaches. *Functional Ecology*, 24: 927–936.

Prentice, I.C., and Cowling, S.A. 2013. Dynamic global vegetation models. In Simon, A.L. (Editor), *Encyclopedia of Biodiversity* (2nd Edition), Academic Press, New York, pp. 670–689.

Rao, A.S., Chhawri, R., Chauhan, A., Yadav, S.S., Meena, K.C., and Bansal, P. 2024. Plant functional Traits: A key framework for understanding and managing ecosystem responses to global environmental change. In Kumar, N., Singh, H. (Editors), *Plant Functional Traits for Improving Productivity*. Springer, Singapore.

Schweiger, A.K., Schütz, M., Risch, A.C., Kneubühler, M., Haller, R., and Schaepman, M.E. 2016. How to predict plant functional types using imaging spectroscopy: Linking vegetation community traits, plant functional types and spectral response. *Methods in Ecology and Evolution*, 8 (1): 86–95.

Snapp, S. 2017. Agroecology: Principles and practice. In Snapp, S., Pound, B. (Editors), *Agricultural Systems* (2nd Edition), Academic Press, New York, pp. 33–72.

Tsakalos, J.L., Renton, M., Riveria, F., Veneklaas, J., Dobrowolski, M.P., and Mucina, L. 2019. Trait based formal definition of plant functional types and functional communities in the multispecies and multi trait context. *Ecological Complexity*, 40 (a): 100787.

Tyagi, J., and Kumar, M. 2024. Plant functional trait: Concept and significance. In Kumar, N., Singh, H. (Editors). *Plant Functional Traits for Improving Productivity*, Springer, Singapore. https://doi.org/10.1007/978-981-97-1510-7_1

Violle, C., Navas, M.L., Vile, D., Kazakou, E., Fortunel, C., Hummel, I., and Garnier, E. 2007. Let the concept of trait be functional. *Oikos*, 116 (5): 882–892.

2 Plant Functional Traits in Forest Ecosystems and Their Ecological Importance

A Review

Meenakshi Sati

2.1 INTRODUCTION

Plant functional traits (PFTs) are crucial for understanding plant's strategies and their impact on ecosystems, especially in complex and varied forests. Forest ecosystems are vital in maintaining biodiversity, storing carbon, and regulating some aspects of the global climate. Gitay and Noble (1997) define plant functional traits (FTs) as "plant properties that respond to dominant ecosystem processes." FTs include leaf size, toughness and longevity, seed size and distribution mechanism, canopy height and structure, resprout ability, and potential for symbiotic nitrogen fixation. Research on the functional characteristics of plants in forest ecosystems has taken off in the last few years, and it has helped us gain some significant new insights into ecosystem stability, resource use efficiency, and the mechanisms of species coexistence. Garnier et al. (2007) developed a methodology with FTs to analyse the impacts of land use change on vegetation and ecosystem function.

There is a trait-based approach to community assembly that makes it possible to combine functional ecology, gradient analysis, community ecology, and coexistence theory. Kraft et al. (2008) compared the co-occurrence patterns of over 1,100 tree species in a 25-hectare Amazonian forest plot to FTs they measured in the field. They showed that interspecific variation in trait-based ecological strategies maintains diversity in one of the world's most diverse tropical forests. Trait-based approaches to ecology, which look at traits that can be measured in a range of biotic and abiotic conditions, have shown how biodiversity affects the way ecosystems work. The complementarity of niches is a significant factor in controlling primary productivity in natural systems because communities with a wide range of plant characteristics have a large primary output. However, Ackerly and Cornwell (2007) found that dominant leaf traits affect nitrification rates more than trait diversity. So the sampling effect has a bigger role in nitrification rates than complementarity. Therefore, trait-based approaches can offer mechanisms that link biodiversity and ecosystem functioning.

 DOI: 10.1201/9781003545231-2

This article is a review of the classification of functional attributes in forest ecosystems, and it also explores the ecological consequences of these attributes in the regulation of species interactions, ecosystem processes, and reactions to environmental disturbances.

2.2 REVIEW OF LITERATURE

2.2.1 Classification of Plant Functional Traits

PFTs are morphological, physiological, or phenological features that influence plant function and ecological processes. They act as markers of how environmental factors affect a plant's ability to grow, reproduce, and survive. Because they offer a mechanistic knowledge of plant–environment interactions, community assembly, and ecosystem functioning, functional features have gained attention in ecological studies.

2.2.2 Categories of Plant Functional Traits

We can divide functional characteristics into many groups based on their importance to ecological processes and plant fitness. These consist of reproductive, root, leaf, and whole-plant characteristics. Each category reflects a distinct aspect of plant strategy and ecological niches.

2.2.3 Leaf Functional Traits

One must know the characteristics of a leaf to understand the photosynthetic efficiency and water-use strategy of a plant. These qualities of the leaves relate to how a plant fixes carbon, absorbs energy, and eventually cycles nutrients and carbon. The most often researched characteristics consist of

Specific Leaf Area (SLA): SLA is the leaf area to leaf dry mass ratio, and it should tell us something about the plant's growth strategy. Lower SLA correlates with slow-growing stress-tolerant species, while high SLA plants tend to have thinner leaves and grow faster (Kunstler et al., 2015).

Leaf Dry Matter Content (LDMC): This trait, often correlated with plant resource conservation, is expressed as the ratio of dry matter weight to fresh leaf weight.

Leaf Nitrogen Content (LNC): The LNC is important because nitrogen is a key part of making enzymes like Rubisco, and it is linked to photosynthetic capacity and growth rates (Cornelissen et al., 2003).

2.2.4 Root Functional Traits

Especially roots, they need them to get water and nutrients. Root characteristics, combined with leaf traits, give a good insight into how plants utilize resources under different conditions. Among the root qualities are:

Rooting Depth: Deep-rooted plants have the upper hand during drought times, they can access H_2O and nutrients from lower levels of the soil (Wang et al., 2020).

Root Diameter: Small roots are good at getting resources and fat ones are not. However, thicker roots are more resilient to environmental stress.

Root Tissue Density: This trait influences lifespan and the plant's allocation strategy in that it makes the root denser or looser (Weemstra et al., 2016).

2.2.5 Reproductive Traits

Different plant species have different reproductive methods, and functional characteristics related to reproduction have an impact on species ranges and population dynamics. Reproductive features are crucial for preserving species diversity and the stability of ecosystems because they have an impact on recruitment and establishment.

Seed Size: Although small seeds are produced in massive numbers, which encourages dispersal and the colonization of disturbed habitats, large seeds sometimes improve seedling survival in resource-poor environments (Li & Prentice, 2024).

Falling Time: According to Caradonna et al. (2014), this trait governs the phenological timing of reproduction and can serve as an indicator of how well a species has adapted to altered climate, changes in pollinator availability, and competitive relationships.

2.2.6 Whole-Plant Traits

Characteristics about the complete plant and its growth form are referred to as whole-plant attributes. The resistance to stress, competitive inter-relationships, and overall ecosystem productivity are all highly related to the whole-plant characteristics.

Plant Height: Taller plants have a competitive advantage over their shorter neighbours because they have better access to light, but at the same time, they must expend a great deal of energy on their structural tissues (Westoby et al., 2002)

Biomass Allocation: How much biomass a species allocates to leaves vs. stems vs. roots is a good predictor of how well it can capture resources and use them.

Life Form: Strategies that affect the structure and function of ecosystems are represented by functional kinds including trees, shrubs, herbs, and climbers.

2.3 TRAIT–ENVIRONMENT INTERACTIONS IN FOREST ECOSYSTEMS

Many of the FTs of plants are key to how animals overcome constraints in their environments, be they interactions with other organisms, acquiring resources, or coping with stress. Studying how these characteristics influence ecological processes is best done in forest environments, which are known for their high biodiversity and intricate interconnections.

2.3.1 Resource Acquisition and Utilization

The acquisition of resources in a forest ecosystem is dominated by traits that maximize light, water, and nutrient absorption. These are all highly influenced by environmental factors such as the availability of soil nutrients, moisture content, and canopy structure, and therefore, there is a plethora of strategies that enable plants to prosper in distinct ecological niches.

2.3.2 Leaf Traits and Light Utilization

Some leaf FTs, including SLA, LDMC, and chlorophyll concentration, are related to light acquisition. In the shaded understory, there are more SLA (light-coloured, thin leaves) species because they can exploit the low light intensities within the forest canopy. On the other hand, species with larger, denser leaves and low SLA are more efficient in high-light environments such as forest gaps. Certainly, species adapted to high-light environments, to minimize herbivory and water loss, must invest in thick, tough leaves, and species that can tolerate shade produce leaves that capture light as efficiently as possible (Poorter et al., 2018).

2.3.3 Root Architecture and Soil Resource Uptake

The function of the roots permits the acquisition of water and nutrients from the soil beneath the ground with characteristics such as root depth, specific root length (SRL), and root tissue density being the most relevant. Rapid nutrient uptake from the top layers of the soil would be of great advantage in moist and nutrient-rich conditions for shallow-rooted plants. On the contrary, during dry spells, deep-rooted plants would be able to extract water from lower soil levels, which makes them far more resilient to drought stress (Weemstra et al., 2016). Nutrient absorption is critically influenced by the characteristics of the roots. This is true, particularly in forest ecosystems, where soil heterogeneity highly exerts influence over nutrient availability.

Root FTs, such as rooting depth, SRL, and root tissue density, are necessary to acquire water and nutrients from soil layers below the ground. In moist, nutrient-rich conditions, shallow-rooted plants would do very well because they would be able to quickly absorb nutrients from the topmost soil. However, during dry spells, deep-rooted plants tap into the lower soil profile and are, therefore, more drought-tolerant (Weemstra et al., 2016). The root properties play a substantial role in nutrient acquisition. This is particularly true in forest ecosystems where soil heterogeneity plays a significant role in nutrient availability (Bardgett et al., 2014).

2.3.4 Mycorrhizal Associations and Nutrient Acquisition

The second most common method of acquiring nutrients in a forest ecosystem is via mutualistic associations with mycorrhizal fungi. They have a mutualistic relationship with the root systems of plants and allow for a much greater intake of nutrients (such as phosphate and nitrogen) from the soil. Arbuscular mycorrhizal (AM) fungi, common in tropical forests, assist in the uptake of phosphorus from inorganic forms, and ectomycorrhizal fungi, which dominate in temperate and boreal forests, are particularly efficient at acquiring nutrients from organic matter. Mycorrhizal associations are fundamental architects of tree communities due to their influence on resource partitioning and competitive interactions in tropical forests (Teste et al., 2020).

2.3.5 Adaptations to Environmental Stressors

There are various types of environmental stressors that can impact a forest, including fire, drought, herbivory, and climate variation. These stresses have led to the evolution of PFTs that not only allow species to survive but to flourish in a wide variety of habitats. After all, the existence of certain species and the stability of ecosystems in general are really due to adaptations to environmental stressors.

2.3.6 Drought Tolerance and Water-Use Efficiency

Some examples of these types of plants are seen in water-limited environments like temperate forests, which are drought-ridden due to global warming or seasonally dry tropical forests. Some characteristics that enable species to conserve water and access deep water reserves include low stomatal conductance, high wood density, and large rooting depth. The species with lower SLA and higher wood density are more drought-resistant due to higher hydraulic efficiency and lower risk of water stress (Fang et al., 2023). One of these is *Quercus ilex*, which has deep roots and efficient water transport, allowing it to withstand long droughts in Mediterranean forests (Sanchez-Gomez et al., 2006; Kumar et al., 2015, 2016, 2017).

2.3.7 Fire Resistance and Resilience

However, fire-adapted forests such as those in the savannas, Mediterranean zones, and some temperate ecosystems possess traits that not only let them endure a fire but thrive afterwards as well. This may include having a thick bark to protect the cambium from fire and the ability to reproduce from basal buds or underground structures if a fire occurs. Another adaptation frequently found in fire-prone ecosystems is serotiny, the characteristic of having seeds that are only released when a fire occurs (Pausas & Keeley, 2014). The fire-adapted species in the Australian forests are highly capable of resprouting and have serotinous seeds, which allow them to rapidly recolonize burned areas and thus maintain ecosystem function (Lamont & He, 2017).

2.3.8 Herbivory Defence Mechanisms

Herbivory is one of the major stresses in a lot of forest ecosystems, especially in the context of intense herbivore pressure as that found in tropical forests. Some defences that plants have developed against herbivores include physical defences (i.e., thorns and stiff leaves) and chemical defences (i.e. secondary metabolites such as alkaloids and tannins). Species with high leaf toughness and LDMC are more resistant to herbivory since their food is not very palatable and it takes herbivores more energy to eat (Carmona et al., 2011). These characteristics are essential in minimizing damage and promoting growth in tropical forests where there are abnormally high numbers of herbivores (Fine et al., 2004).

2.3.9 Interactions with Other Organisms

The FTs of plants serve as an intermediary between them and other organisms, from mutualistic associations with pollinators and mycorrhizal fungi to antagonistic interactions involving herbivores and pathogens. In forests, these interactions are vital in determining the members of communities and the functioning of ecosystems.

2.3.9.1 Mutualistic Interactions with Pollinators and Seed Dispersers

For successful reproduction and population dynamics, traits of seed dispersal and pollination are essential. Pollinator attractiveness is influenced by flower features like colour, shape, and nectar production, whereas seed coat properties and fruit size dictate the methods of dispersal. Plant species have evolved features that attract certain pollinators or seed dispersers in tropical forests, where animal-mediated pollination and seed dispersal are frequent. Examples of these qualities include vividly coloured flowers for avian pollinators and meaty fruits for mammal dispersers (McConkey & Drake, 2015). According to Brodie et al. (2021), maintaining seed distribution networks and fostering forest regeneration in tropical forests depend on mutualistic interactions between trees and large mammals.

Things like seed dispersal and pollination characteristics are significant when it comes to reproduction and population ecology. The colours, shapes, and nectar production of the flower all influence what pollinators it attracts, and the characteristics of the seed coat and the size of the fruit will both determine how the seeds are dispersed. Also common is the fact that many plants in a tropical rainforest have evolved to have certain characteristics that allow them to be pollinated or have their seeds dispersed only by certain animals. Examples include bright-coloured flowers for ornithophilous pollinators and fleshy fruits for mammalian dispersers (McConkey Drake, 2015). Seed dispersal mechanisms and forest rejuvenation in the tropics rely on these mutualistic associations between trees and large mammals (Brodie et al. 2021).

2.3.9.2 Competition and Niche Differentiation

In some forest habitats, there is a lot of competition for light, water, and nutrients. The FTs of plants are key indicators of their ability to compete. In resource-rich habitats, species that are conservative with their use of resources (e.g., low SLA and high wood density) are better at competing. However, species that grow quickly and have low wood density and high SLA often beat slower-growing species (Díaz et al., 2015). Niche differentiation based on functional features helps species live together by sharing resources, reducing direct competition, and increasing species diversity (Laughlin, 2014). According to Kraft et al. (2015), high levels of species richness are made possible by niche differentiation, which is largely driven by trait variation among coexisting tree species in tropical forests.

Light, water, and nutrients are all very competitive in various forest ecosystems, and PFTs are the best predictors of competitive ability. In productive environments, those species with conservative resource-use traits (e.g., low SLA and high wood density) tend to be the best competitors, but then there's also this other group of species that grow fast and have low wood density and high SLA and they often outcompete the slower-growing species (Díaz et al., 2015). Differentiation between species based on FTs in ecological niches allows species to live together by sharing resources and reducing direct competition, which increases species diversity (Laughlin, 2014). High species richness in tropical forests is mainly a result of niche differentiation, which in turn is primarily driven by interspecific trait variation among coexisting tree species (Kraft et al., 2015).

2.3.9.3 Pathogen Resistance

In wooded area ecosystems, pathogens often come into contact with plants, and these contacts can have a big effect on how populations change and how fit individuals are. Pathogen resistance is inspired by the aid of using traits such as the chemical makeup of the leaf, the thickness of the cuticle, and the presence of antimicrobial substances. Genetic range inside plant populations can enhance pathogen resistance with the aid of reducing the hazards of full-scale infection (Tack et al., 2012). Higher trait range species are more capable of facing up to pathogen outbreaks in tropical forests in which the disorder load is severe, including more suitable atmospheric resilience (Lamit et al., 2015).

2.4 ECOLOGICAL IMPORTANCE OF PLANT FUNCTIONAL TRAITS

The morphological, physiological, and phenological properties of plants greatly influence ecological activities. These characteristics impact the interaction of plants with their surroundings, mediating the flow of energy, matter, and resilience in ecosystems.

2.4.1 Influence on Primary Productivity

Primary productivity is the foundation of ecosystem function, which governs how plants use light, carbon dioxide, and water to convert into biomass through photosynthesis. Plant functional characteristics relating to photosynthesis, growth, and resource usage efficiency are significant determinants of productivity.

2.4.1.1 Photosynthetic Traits and Resource Efficiency

Primary productivity is closely associated with leaf-level characteristics such as SLA, photosynthetic rate, and LNC. Due to their rapid growth and higher resource acquisition efficiency, plants with high SLA can absorb less and generate more biomass in high-resource conditions. But the price for this is a shorter lifespan and less stress tolerance. Plants with high SLA predominate in resource-rich, early-successional habitats, which speed up the sequestration of carbon (Suter & Edwards, 2013). Conversely, species with low SLA, like many evergreen trees, grow more slowly but are better suited to environments with limited water or nutrients.

2.4.1.2 Root Traits and Water/Nutrient Uptake

The efficiency of root traits and water/nutrient uptake is the determining factor in plant growth. The determination of nutrients is based on the characteristics of the roots, including their length and depth. These two key factors contribute to primary productivity. Plants can be left in areas with low rainfall to maximize their productivity. In species with large roots, the ability to access nutrients and water results in larger or deeper root systems. In contrast, shallow roots are not viable (Bardgett et al., 2014). According to a meta-analysis, species with higher SRL are more likely to absorb nutrients efficiently, particularly in the gut and nutrient-poor soils contributing to increased productivity in limited environments (Weemstra et al., 2016; Kumar et al., 2019, 2021f, 2022; Giri et al., 2023).

2.4.1.3 Trait Diversity and Complementarity

The diversity of functional features within a group contributes to the development of ecosystem production. Species with different characteristics can use a greater range of resources and occupy multiple niches, making ecosystems with high functional diversity more productive. The combination of deeply rooted and diverse systems is unlike less complex or more diversified systems. In grassland ecosystems, shallow-rooted species can absorb more water and nutrients, resulting in increased plant growth, which in turn results in an increase in biomass production (Van den Pol-van Dasselaar et al., 2020).

2.4.2 Role in Nutrient Cycling

Plant functional characteristics have a significant impact on the flow and transformation of nutrients within the ecosystem, known as nutrient cycling. These characteristics control the intake of nutrients, the breakdown of litter, and the release of nutrients back into the soil. Their functional characteristics heavily influence the flow and transformation of nutrients within the ecosystem.

2.4.2.1 Leaf Litter Decomposition

Decomposition of leaf litter and its chemistry, including the quality or quantity of leaves, is very crucial. Lignin and nitrogen content are two crucial traits of plants that impact their rate of decomposition, which in turn affects the cycling of nutrients. Ecosystem fertility is increased by species that have high nitrogen content and are easily breakable (e.g., dry matter). A low level of lignin can lead to faster nutrient cycling. In faster-growing species, litter is commonly generated and disintegrates at a faster rate, leading to increased nutrient cycling rates.

2.4.2.2 Root Traits and Soil Nutrient Dynamics

The characteristics of roots, such as root architecture and mycorrhizal relationships, are important for the cycling of nutrients because they affect the intake of nutrients and the creation of soil organic matter in ecosystems where these species are prevalent. Plants with fine, spreading roots are better at cycling nutrients like nitrogen and phosphorus because their roots have more surface area for absorbing nutrients. In particular, in nutrient-deficient soils, through their symbiotic relationship with plant roots, mycorrhizal fungi enhance the uptake of nutrients by plants. The AM fungus supports the cycling of nutrients and enhances plant growth by aiding the absorption of phosphorus in forests located in tropical and temperate zones (Teste et al., 2020).

2.4.2.3 Nutrient Retention and Ecosystem Fertility

Slow-growing, resource-conserving plants with long lifespans and low SLA are often beneficial in retaining nutrients in ecosystems. These plants typically generate low-quality litter that undergoes gradual degradation and bit by bit, the soil absorbs nutrients. This encourages the building up of organic matter in soil, which leads to improved long-term soil fertility while slowing down the cycling of nutrients (Gross et al., 2021). Nutrient retention is essential for stabilizing ecosystem function in low-nutrient environments, such as the boreal woods, which are dominated by slow-growing species.

2.4.3 Contribution to Ecosystem Stability and Resilience

The ability of an ecosystem to withstand shocks is known as resilience, while ecological stability is defined as an ecosystem's capacity to uphold its structure and function despite disturbances. These factors significantly influence how plants respond to environmental stressors and disturbances like fire, drought, and others. Stability and resilience of ecosystems are closely linked to the FTs of plants.

2.4.3.1 Trait Responses to Disturbance

The functional response of a species in disturbed habitats is crucial for its survival. Factors that include drought tolerance, seed distribution strategies and resprouting capability are essential. Among drought-resistant plants, those with thick leaves and deep roots stand out. Roots and effective water usage strategies are more resilient against dry spells. Ecosystems with a greater number of drought-resistant species are more resilient to the impacts of climate change. Another example is the Mediterranean species *Quercus ilex*, which has a high water-retention capacity and maintains the stability of these ecosystems during prolonged dry spells (Valladares, 2008).

2.4.3.2 Functional Diversity and Ecosystem Resilience

Functional diversity increases the resilience of ecosystems by protecting against alterations and disturbances in the environment. Species that can resist or survive various shocks can be found in diverse populations with a range of functional features (Mori et al., 2013). For example, species with thick bark or serotinous cones that open when heated have evolved to withstand fire and can regrow rapidly after a fire, stabilizing the ecosystem and preserving production (Kumar et al., 2020a,b, 2021a–e; Pausas & Keeley, 2014). Similar to this, ecosystems that have an amalgamation of species that are complex to and tolerant of drought are better able to function during dry spells because the resilient species can keep growing and producing even when other species do not (Oliver et al., 2015).

2.4.3.3 Invasive Species and Trait–Environment Feedbacks

Since invasive species often differ functionally from native species, differences in stability characteristics between native-dominated and invasive-dominated communities could lead to long-term changes in ecosystem stability. These can also influence fire regimes, hydrological processes, and nutrient cycling. Invasive species that are adapted for fast growth, rapid N acquisition, and strong above- and below-ground competitive abilities can displace natives, thereby reducing biodiversity and altering ecosystem functions (Gaertner et al., 2009). Changes in fire regimes, for example, have been shown to create positive feedbacks that exacerbate the impact of some invasive grasses in the Mediterranean, increasing their spread and decreasing the resilience of the ecosystem to invasion.

2.5 FUNCTIONAL TRAIT DIVERSITY AND FOREST ECOSYSTEM FUNCTIONING

Extensive knowledge on such FT diversity, e.g., the identity composition, abundance, and type of functional features among plant species present in an ecosystem to understand their dynamics and functioning of forest ecosystems. It helps understand how species share a living space, how they utilize resources, and how ecosystems respond to alterations. This knowledge of trait diversity is more important than ever due to global environmental change and its effects on the stability and resilience of ecosystems.

2.5.1 Trait Diversity and Ecosystem Processes

For the ecological processes to work, such as nutrient cycling, primary productivity, and resilience, FT diversity is a must. It does this by regulating the interactions between different species, resources, and other factors in the environment.

A. **Primary Productivity and Resource Use Efficiency**: Different species work together to make the forest ecosystem more productive by using resources in ways that complement each other. This is possible because different species have a wide range of traits. Because of their functional characteristics, different species usually prefer using particular resources such as light, water, or nutrients, which reduces competition and thus makes the ecosystem function more efficiently. To make more out of resources and boost biomass output, for instance, species that have different root depths or leaf structures can utilize resources from different soil or canopy layers (Lohbeck et al., 2016). Ecosystems in tropical forests show wider root trait diversity with root length and diameter being the primary defector differences, which result in the main production processes of higher net primary productivity through excess nitrogen and phosphorus uptake (Freschet et al., 2021).
B. **Nutrient Cycling and Decomposition**: Functional diversity refers to different behaviours between plants that result in certain characteristics, such as varying rates of nutrient decomposition and litter quality, which, in turn,

have an impact on the entire nutrient cycle. The rapid degradation of the litter of those specific species with a short life span and nutrient-rich leaves influences the proper speed of nutrient turnover. Conversely, species with a lower growth rate and which are more resource-efficient help in the retention of the nutrients in the soil, and thus they contribute to a long-term stable system of soil nutrients (Cornwell & Ackerly, 2009). In addition, the rise of a larger functional diversity of the leaf, i.e., the toughness of the leaf and nutrient content, brings the opportunity to use to a greater extent the efficient decomposition process in tropical forests (Fortunel et al. 2014).

C. **Ecosystem Resilience and Stability**: Because they contain species with a wide range of adaptive strategies, ecosystems with high functionality are the most climate-resilient types of environments and are frequently more tolerant to disturbances like droughts, fires, or pest outbreaks. This idea, referred to as the "insurance hypothesis," is the view that organisms with different degrees of adaptation to various environmental perturbations will be more frequently found in the ecosystem with greater plant species richness (Mori et al., 2013). A study by Van den Pol-van Dasselaar et al. (2020) shows that temperate forests can frequently, safely, and consistently produce biomass when dealing with the problem of severe drought only if the spectrum of functionality connected to drought resistance, like wood density, and leaf water potential is the long one.

2.5.2 Implications for Biodiversity and Conservation

Especially considering habitat loss and climate change, the preservation of FT diversity has serious implications for biodiversity and conservation. Trait diversity allows for various ecological niches, and thus stability within ecosystems and ultimately the ability for different species to live together.

A. **Species Coexistence and Niche Partitioning**: Functional diversity boosts species cohabitation because the competition for resources is lower. When there are many distinct functional properties, the species in question can find the ecological niches that they can claim, thereby preventing supplementary exploitation. This niche differentiation stimulates higher species richness in forest settings. In hyper-diverse environments such as tropical rainforests, the differentiation in features like leaf size, root structure, and wood density allows the existence of various species together (Kraft et al., 2008). Therefore, functional differentiation sustains these forests by letting species divide the resources of time and space.

B. **Trait-Based Conservation Approaches**: Due to the increase in environmental problems, a significant portion of conservation strategies now emphasize different traits. To sustain ecosystem processes, conservation has increasingly focused on maintaining the diversity of FTs rather than species diversity. By preferring the species that contribute the most distinctively to the ecosystem, conservationists can enhance its resilience and functionality (Cadotte et al., 2011). For example, large-bodied, slow-growing trees are necessary for forest ecosystems primarily because they can store carbon much longer and occupy a smaller percentage of the forest area than fast growers by the nitrogen cycle (Poorter et al., 2018). Alongside the continuity of such diverse species, they are a barrier for nature against the background of biodiversity loss thus caused by environmental changes.

C. **Ecosystem Services and Human Well-Being**: The link between FT diversity and the proper functioning of the ecosystem in terms of carbon-storing, water filtering, and wood production, in turn, truly enriches human lives. Multiple ecosystem services are provided by different forest ecosystems, and these are the wellsprings of nature on earth that a human being can draw from to live joyfully along the life axis and in harmony with nature. According to Díaz et al. (2018), changing the important FTs in the forest fuel cage could lead to a loss of intensity and a lower emission of the benefits that people get. Conversely, we must acknowledge that maintaining functional diversity involves both biological and socioeconomic concerns.

2.6 TRAIT-BASED APPROACHES IN FOREST MANAGEMENT

Water conservation, protecting biodiversity, and carbon sequestration are some of the main ecosystem services forests offer. Understanding how plant functional characteristics affect ecosystem processes is essential for ensuring sustainable forest management. Plant biology and ecosystem function got a new dimension through trait-based methods, which makes decision-making in forest management easier (Díaz et al., 2004; Lavorel et al., 2013).

1. **Forest Resilience and Adaptive Capacity**: When forest management emphasizes functional attributes, resistance to environmental change is enhanced. For example, forests with species that exhibit high drought tolerance (phenotypic characteristics such as deep roots and thick leaves) are more resilient to prolonged dry periods. Leaf mass per area and wood density are characteristics that affect the flow of nutrients and the sequestration of carbon. These are important factors for figuring out how ecosystem services will change because of climate change (Poorter et al., 2018).
2. **Biodiversity and Ecosystem Function**: According to Cadotte et al. (2011), functional diversity, the diversity of traits in a community is a much better predictor of ecological function than species diversity alone. For example, under disturbance, a forest with a high diversity of typical types can still generate more (Loreau & de Mazancourt, 2013).
3. **Restoration Ecology and Reforestation Programs**: In restoration initiatives, trait-based methodologies are essential. Selecting species with certain attributes, such as drought tolerance and nitrogen fixation, can accelerate ecosystem recovery (Laughlin, 2014). Characteristics make it easier to determine which plants are best suited for soil regeneration or carbon sequestration by the new goals of international reforestation (Van den Pol-van Dasselaar et al., 2020).

2.7 FUTURE DIRECTIONS IN PLANT FUNCTIONAL TRAIT RESEARCH

2.7.1 Advancing Functional Trait Methodologies

A. **Technological Innovations**: Trait-based research is also being transformed with the new high-throughput trait assessment methods, such as drone technologies and remote sensing, allowing for temporal and spatial data to be more complete. Spectral imaging techniques allow for the high-throughput assessment of photosynthetic capacity and other leaf traits, thereby enhancing the efficiency of trait acquisition (Cavender-Bares et al., 2017).

B. **Trait Syndromes and Multivariate Approaches:** Future research should focus on understanding trait syndromes, or groups of correlated traits, and how these syndromes together affect ecological processes. For example, species that exhibit a syndrome of fast growth, low tissue density, and high leaf area will cause nutrients to cycle through the system in a different way than slow-growing conservative species (Wright & Sutton-Grier, 2012). However, multivariate models, which combine many traits and environmental factors, can improve predictions of a species' response to climate change and point the way for forest management.

C. **Trait–Environment Feedback Loops**: Research stresses the incorporation of feedback loops between traits and environments. Traits not only respond to environmental conditions but also have the potential to change them (e.g., through water uptake and litter quality) (Suding et al., 2015). This feedback approach can help model long-term ecosystem trajectories and forest dynamics under different management scenarios (Benson et al., 2022).

D. **Traits Under Global Change Scenarios**: As global change speeds up, one important area of study that is becoming more popular is how FTs affect how species react to things like warming, CO_2 levels, and changes in rainfall patterns (Fortini & Jacobi, 2018). To forecast future levels of biodiversity and ecosystem function, predictive models must incorporate characteristics related to plant physiology and resource-use strategies (Yousefpour & Hanewinkel, 2009).

2.7.2 Bridging the Trait Gap Between Species and Ecosystem Function

Even with advancements, there is still a gap in the understanding of how individual plant features relate to the functioning of large-scale ecosystems. The only way to overcome this is to incorporate trait data from the individual plant level, population level, and community level into ecosystem models (Funk et al., 2017). Trait-based forest management will continue to progress through the combined efforts of ecologists, foresters, and data scientists (Osborne et al., 2023).

2.7.3 Ecosystem Services and Human Well-Being

It will be essential to comprehend how functional features contribute to the preservation of ecosystem services, including carbon sequestration, water purification, and biodiversity protection to match research with societal objectives. This can serve as a guide for policymakers to find a balance between economic and conservation objectives (Petchey & Gaston, 2006).

2.8 CONCLUSION

The functional attributes of plants yield critical insights into the structure and functioning of forest ecosystems. By examining characteristics from multiple dimensions (leaf, root, and reproductive and whole plant), we can better learn about the ecological strategies that allow species to not only survive but also thrive in various forest habitats. Although drought, fire, and herbivory resistance characteristics may enhance a species' ability to survive in harsh conditions, leaf morphology, root architecture, and symbiotic interaction traits all maximize resource utilization in a wide range of environmental conditions. Furthermore, FTs affect both antagonistic and mutualistic interactions between plants and other species, which in turn affect the stability and structure of the ecosystem. To predict how forest ecosystems will react to long-term environmental stresses and to gather data, it is important to understand how the FTs of plants affect the environment. To predict how forest ecosystems will react to long-term environmental changes and come up with good conservation plans, we need to know what ecological effects PFTs have. Forest managers can develop techniques that enhance resistance, improve ecosystem performance, and promote diversity if they focus on plant functional attributes. While we have made significant progress in understanding trait–environment interactions, future research should concentrate on studying multivariate trait interactions, identifying trait syndromes, and incorporating feedback loops into predictive models. Also, emerging technologies such as remote sensing offer promising opportunities to extend trait-based research, which will in turn improve conservation and restoration efforts worldwide.

REFERENCES

Ackerly, D. D., & Cornwell, W. K. (2007). A trait-based approach to community assembly: partitioning of species trait values into within-and among-community components. *Ecology Letters*, 10(2): 135–145.

Bardgett, R. D., Mommer, L., & De Vries, F. T. (2014). Going underground: Root traits as drivers of ecosystem processes. *Trends in Ecology & Evolution*, 29(12): 692–699. https://doi.org/10.1016/j.tree.2014.10.006

Benson, D. L., King, E. G., & O'Brien, J. J. (2022). Forest dynamics models for conservation, restoration, and management of small forests. *Forests*, 13: 515. https://doi.org/10.3390/f13040515

Brodie, J. F., Williams, S., & Garner, B. (2021). The decline of mammal functional and evolutionary diversity worldwide. *Proceedings of the National Academy of Sciences of the United States of America*, 118(3): 1–6. https://doi.org/10.1073/pnas.1921849118

Cadotte, M. W., Carscadden, K., & Mirotchnick, N. (2011). Beyond species: Functional diversity and the maintenance of ecological processes and services. *Journal of Applied Ecology*, 48: 1079–1087. https://doi.org/10.1111/j.1365-2664.2011.02048.x

CaraDonna, P. J., Iler, A. M., & Inouye, D. W. (2014). Shifts in flowering phenology reshape a subalpine plant community. *Proceedings of the National Academy of Sciences*, 111(13): 4916–4921.

Carmona, D., Lajeunesse, M. J., & Johnson, M. T. J. (2011) Plant traits that predict resistance to herbivores. *Functional Ecology*, 25: 358–367.

Cavender-Bares, J., Gamon, J. A., Hobbie, S. E., Madritch, M. D., Meireles, J. E., Schweiger, A. K., & Townsend, P. A. (2017). Harnessing plant spectra to integrate the biodiversity sciences across biological and spatial scales. *American Journal of Botany*, 104: 966–969. https://doi.org/10.3732/ajb.1700061

Cornelissen, J. H. C., Lavorel, S., Garnier, E., Díaz, S., Buchmann, N., Gurvich, D. E., Reich, P. B., Ter Steege, H., Morgan, H. D., Van Der Heijden, M. G. A, & Poorter, H. (2003). A handbook of protocols for standardised and easy measurement of plant functional traits worldwide. *Australian Journal of Botany*, 51(4), 335. DOI: 10.1071/BT02124

Cornwell, W. K., & Ackerly, D. D. (2009). Community assembly and shifts in plant trait distributions across an environmental gradient in coastal California. *Ecological Monographs*, 79(1): 109–126. https://doi.org/10.1890/07-1134.1

Diaz, S., Hodgson, J. G., Thompson, K., Cabido, M., Cornelissen, J. H. C., Jalili, A., Montserrat-Martí, G., Grime, J. P., Zarrinkamar, F., Asri, Y., & Zak, M. R. (2004). The plant traits that drive ecosystems: Evidence from three continents. *Journal of Vegetation Science*, 15(3): 295–304. https://doi.org/10.1111/j.1654-1103.2004.tb02266.x

Díaz, S., Kattge, J., Cornelissen, J. H. C., Wright, I. J., Lavorel, S., Dray, S., Reu, B., Kleyer, M., Wirth, C., Colin Prentice, I., & Gorné, L. D. (2015). The global spectrum of plant form and function. *Nature*, 529(7585): 167–171. https://doi.org/10.1038/nature16489

Díaz, S., Pascual, U., Stenseke, M., Martín-López, B., Watson, R.T., Molnár, Z., Hill, R., Chan, K.M., Baste, I.A., Brauman, K.A., & Polasky, S. (2018). Assessing nature's contributions to people. *Science*, 359(6373): 270–272.

Fang, L., Marte, P., Jin, K., Du, X., Van der Putten, P. E. L., Yin. X., & Strik P. C. (2023). Neglecting acclimation of photosynthesis under drought can cause significant errors in predicting leaf photosynthesis in wheat. *Global Change Biology*, 29(2): 505–521. https://doi.org/10.1111/gcb.16488

Fine, P. V. A., Mesones, I., & Coley, P. D. (2004). Herbivores promote habitat specialization by trees in Amazonian forests. *Science*, 305: 663–665.

Fortini, L. B., & Jacobi, J. D. (2018). Identifying opportunities for long-lasting habitat conservation and restoration in Hawai'i's shifting climate. *Regional Environmental Change*, 18: 2391–2402.

Fortunel, C., Paine, C. E. T., Fine, P. V. A., Kraft, N. J. B., & Baraloto, C. (2014). Environmental factors predict community functional composition in Amazonian forests. *Journal of Ecology*, 102(1): 145–155. https://doi.org/10.1111/1365-2745.12160

Funk, J.L., Larson, J.E., Ames, G.M., Butterfield, B.J., Cavender-Bares, J., Firn, J., Laughlin, D.C., Sutton-Grier, A.E., Williams, L., & Wright, J. (2017). Revisiting the H oly G rail: using plant functional traits to understand ecological processes. *Biological Reviews*, 92(2): 1156–1173.

Gaertner, M., Breeyen A. D., Hui C., & Richardson D. M. (2009). Impacts of alien plant invasions on species richness in Mediterranean-type ecosystems: A meta-analysis. *Progress in Physical Geography*, 33: 319–338. https://doi.org/10.1177/0309133309341607

Garnier, E., Lavorel, S., Ansquer, P., Castro, H., Cruz, P., Dolezal, J., Eriksson, O., Fortunel, C., Freitas, H., Golodets, C., & Grigulis, K. (2007). Assessing the effects of land-use change on plant traits, communities and ecosystem functioning in grasslands: a standardized methodology and lessons from an application to 11 European sites. *Annals of Botany*, 99(5): 967–985.

Giri, K., Mishra, G., Suyal, D. C., Kumar, N., Doley, B., Das, N., Baruah, R. C., Bhattacharyya, R., & Bora, N. (2023). Performance evaluation of native plant growth-promoting rhizobacteria for paddy yield enhancement in the Jhum fields of Mokokchung, Nagaland, North East India. *Heliyon*, 9: e14588.

Gitay, H., & Noble, I. R. (1997). What are functional types and how should we seek them? In: Smith, T. M., Shugart, H. H., and Woodward, F. I. (Eds.) *Plant Functional Types: Their Relevance to Ecosystem Properties and Global Change*. Cambridge University Press, Cambridge, UK, pp. 3–19.

Gross, A., Bromm, T., & Glaser, B. (2021). Soil organic carbon sequestration after biochar application: A global meta-analysis. *Agronomy*, 11: 2474. https://doi.org/10.3390/agronomy11122474

Kraft, N.J., Adler, P.B., Godoy, O., James, E.C., Fuller, S., & Levine, J.M. (2015). Community assembly, coexistence and the environmental filtering metaphor. *Functional Ecology*, 29(5): 592–599.

Kraft, N.J., Valencia, R., & Ackerly, D.D. (2008). Functional traits and niche-based tree community assembly in an Amazonian forest. *Science*, 322 (5901): 580–582.

Kumar, A., Dwivedi, G. K., Tewari, S., Paul, J., Anand, R., Kumar, N., Kumar, P., Singh, H., & Kaushal, R. (2020a). Carbon mineralization and inorganic nitrogen pools under *Terminalia chebula* Retz.-Based agroforestry system in Himalayan foothills, India. *Forest Science*, 66 (5): 634–643.

Kumar, A., Dwivedi, G. K., Tewari, S., Paul, J., Sah, V. K., Singh, H., Kumar, P., Kumar, N., & Kaushal, R. (2020b) Soil organic carbon pools under *Terminalia chebula* Retz. based agroforestry system in Himalayan foothills, India. *Current Science*, 118(7), 1098–1103.

Kumar, A., Kumar, P., Singh, H., Bisht, S., & Kumar, N. (2021a). Relationship of physiological plant functional traits with soil carbon stock in temperate forest of Garhwal Himalaya. *Current Science*, 120(8), 1368–1373.

Kumar, A., Kumar, P., Singh, H., & Kumar, N. (2021b). Modulation of plant functional traits under essential plant nutrients during seasonal regime in natural forests of Garhwal Himalayas. *Plant and Soil*, 465: 197–212.

Kumar, A., Kumar, P., Singh, H., & Kumar, N. (2021d). Impact of plant functional traits on near saturated hydraulic conductivity of soil under different forests of Kempty watershed in Garhwal Himalaya. *Indian Journal of Soil Conservation*. 49(1): 38–44.

Kumar, A., Kumar, P., Singh, H., & Kumar, N. (2021e). Adaptation and mitigation potential of roadside trees with bio-extraction of heavy metals under vehicular emissions and their impact on physiological traits during seasonal regimes. *Urban Forestry & Urban Greening*, 58: 126900.

Kumar, A., Singh, H., Kumari, G., Bisht, S., Malik, A., Kumar, N., Singh, M., Raturi, A., Barthwal, S., Thakur, A., & Kaushal, R. (2022). Adaptive resilience of roadside trees to vehicular emissions via leaf enzymatic, physiological, and anatomical trait modulations. *Environmental Pollution*, 313: 120191.

Kumar, A., Tewari, S., Singh, H., Kumar, P., Kumar, N., Bisht, S., Kushwaha, S., Tamta, N., & Kaushal, R. (2021c). Biomass accumulation and carbon stocks in different agroforestry system prevalent in Himalayan foothills, India. *Current Science*, 120(6): 1083–1088.

Kumar, N., Jeena, N., & Singh, H. (2019). Elevated temperature modulates rice pollen structure: A study from foothill Himalayan agro-ecosystem in India. *3 Biotech*, 9: 175.

Kumar, N., Jeena, N., Kumar, A., Khairakpam, R., & Singh, H. (2021f). Comparative response of rice cultivars to elevated air temperature in Bhabar region of Indian Himalaya: Status on yield attributes. *Heliyon*, 7: e07474.

Kumar, N., Kumar, N, Shukla, A., Shankhdhar, S. C., & Shankhdhar, D. (2015). Impact of terminal heat stress on pollen viability and yield attributes of rice (*Oryza sativa* L.). *Cereal Research Communications*, 43(4): 616–626.

Kumar, N., Shankhdhar, S. C., & Shankhdhar, D. (2016). Impact of elevated temperature on antioxidant activity and membrane stability in different genotypes of rice (*Oryza sativa* L.). *Indian Journal of Plant Physiology (Now Plant Physiology Report)*, 21(1): 37–43.

Kumar, N., Suyal, D. C., Sharma, I. P., Verma, A., & Singh, H. (2017). Elucidating stress proteins in rice (*Oryza sativa* L.) genotype under elevated temperature: A proteomic approach to understand heat stress response. *3 Biotech*, 7: 205.

Kunstler, G., Falster, D., Coomes, D. A., Hui, F., Kooyman, R. M., Laughlin, D. C., Poorter, L., Vanderwel, M., Vieilledent, G., Wright, S. J., & Westoby, M. (2015). Plant functional traits have globally consistent effects on competition. *Nature*, 529(7585): 204–207. https://doi.org/10.1038/nature16476

Lamit, L. J., Busby P. E., Lau M. K., Compson Z. G., Wojtowicz T., Keith A. R., Whitham, T. G. (2015). Tree genotype mediates covariance among communities from microbes to lichens and arthropods. *Journal of Ecology*, 103: 840–850.

Lamont, B. B., & He, T. (2017). Fire-proneness as a prerequisite for the evolution of fire-adapted traits. *Trends in Plant Science*, 22(4): 278–288. https://doi.org/10.1016/j.tplants.2016.11.004

Laughlin, D. C. (2014). Applying trait-based models to achieve functional targets for theory-driven ecological restoration. *Ecology Letters*, 17(7): 771–784. https://doi.org/10.1111/ele.12288

Lavorel, S., Storkey, J., Bardgett, R. D., de Bello, F., Berg, M. P., Le Roux, X., Moretti, M., Mulder, C., Pakeman, R. J., Díaz, S, & Harrington, R. (2013). A novel framework for linking functional diversity of plants with other trophic levels for the quantification of ecosystem services. *Journal of Vegetation Science*, 24(5): 942–948. https://doi.org/10.1111/jvs.12083

Li, J., & Prentice, I. C. (2024). Global patterns of plant functional traits and their relationships to climate. *Communications Biology*, 7: 1136. https://doi.org/10.1038/s42003-024-06777-3

Lohbeck, M., Bongers, F., Martinez-Ramos, M., & Poorter, L. (2016). The importance of biodiversity and dominance for multiple ecosystem functions in a human-modified tropical landscape. *Ecology*, 97: 2772–2779.

Loreau, M., & de Mazancourt, C. (2013). Biodiversity and ecosystem stability: A synthesis of underlying mechanisms. *Ecology Letters*, 16: 106–115. https://doi.org/10.1111/ele.12073

McConkey, K. R., & Drake, D. R. (2015). Low redundancy in seed dispersal within an island frugivore community. *AoB Plants*, 7: plv088. https://doi.org/10.1093/aobpla/plv088

Mori, A. S., Furukawa, T., & Sasaki, T. (2013). Response diversity determines the resilience of ecosystems to environmental change. *Biological Reviews*, 88(2): 349–364. https://doi.org/10.1111/brv.12004

Oliver, T. H., Heard, M. S., Isaac, N. J. B., Roy, D. B., Procter, D., Eigenbrod, F., Freckleton, R., Hector, A., Orme, C. D. L., Petchey, O. L., & Bullock, J. M. (2015). Biodiversity and resilience of ecosystem functions. *Trends in Ecology and Evolution*, 30: 673–684. https://doi.org/10.1016/j.tree.2015.08.009

Osborne, P., Aquilué, N., Mina, M., Moe, K., Jemtrud, M., & Messier, C. (2023). A trait-based approach to both forestry and timber building can synchronize forest harvest and resilience. *PNAS Nexus*, 2(8). https://doi.org/10.1093/pnasnexus/pgad254

Pausas, J. G., & Keeley, J. E. (2014). Evolutionary ecology of resprouting and seeding in fire-prone ecosystems. *New Phytologist*, 204(1): 55–65. https://doi.org/10.1111/nph.12921

Petchey O. L., & Gaston K. J. (2006). Functional diversity: Back to basics and looking forward. *Ecology Letters*, 9: 741–758.

Poorter, L., Castilho, C. V., Schietti, J., Oliveira, R. S., & Costa, F. R. C. (2018). Can traits predict individual growth performance? A test in a hyperdiverse tropical forest. *New Phytologist*, 219(1): 109–121. https://doi.org/10.1111/nph.15206

Sanchez-Gomez, D., Valladares, F., & Zavala, M. A. (2006). Functional traits and plasticity in response to light in seedlings of four Iberian forest tree species. *Tree Physiology*, 26(11): 1425–1433. https://doi.org/10.1093/treephys/26.11.1425

Suding, K., Higgs, E., Palmer, M., Callicott, J. B., Anderson, C. B., Baker, M., Gutrich, J. J., Hondula, K. L., LaFevor, M. C., Larson, B. M., & Schwartz, K. Z. S. (2015). Committing to ecological restoration. *Science*, 348: 638–640. https://doi.org/10.1126/science.aaa4216

Suter, M., & Edwards, P. J. (2013). Convergent succession of plant communities is linked to species' functional traits. *Perspectives in Plant Ecology, Evolution and Systematics*, 15(4): 217–225. https://doi.org/10.1016/j.ppees.2013.05.001

Tack, A. J. M., Thrall, P. H., Barrett, L. G., Burdon, J. J., & Laine, A. L. (2012). Variation in infectivity and aggressiveness in space and time in wild host–pathogen systems: Causes and consequences. *Journal of Evolutionary Biology*, 25(10), 1918–1936. https://doi.org/10.1111/j.1420-9101.2012.02588.x

Teste, F. P., Jones, M. D., &Dickie, I. A. (2020). Dual-mycorrhizal plants: Their ecology and relevance. *New Phytologist*, 225(25): 1835–1851. https://doi.org/10.1111/nph.16190

Valladares, F. (2008). A mechanistic view of the capacity of forests to cope with climate change. In: Bravo, F. (Ed.), *Managing Forest Ecosystems: The Challenge of Climate Change*. Kluwer Academic Publishers, The Netherlands, pp. 11–35.

Van den Pol-van Dasselaar, A., Hennessy, D., & Isselstein, J. (2020). Grazing of dairy cows in Europe—An in-depth analysis based on the perception of grassland experts. *Sustainability*, 12(3): 1098. https://doi.org/10.3390/su12031098

Wang, J., Wang, Y., He, N., Ye, Z., Chen, C., Zang, R., Feng, Y., Lu, Q., & Li, J. (2020). Plant functional traits regulate soil bacterial diversity across temperate deserts. *Science of The Total Environment*, 715: 136976. https://doi.org/10.1016/j.scitotenv.2020.136976

Weemstra, M., Mommer, L., Visser, E. J. W., van Ruijven, J., Kuyper, T. W., Mohren, G. M. J., & Sterck, F. J. (2016). Towards a multidimensional root trait framework: A tree root review. *New Phytologist*, 211(4): 1159–1169. https://doi.org/10.1111/nph.14003

Westoby, M., Falster, D. S., Moles, A. T., Vesk, P. A., & Wright, I. J. (2002). Plant ecological strategies: Some leading dimensions of variation between species. *Annual Review of Ecology and Systematics*, 33(1): 125–159. https://doi.org/10.1146/annurev.ecolsys.33.010802.150452

Wright, J. P., & Sutton-Grier, A. (2012). Does the leaf economic spectrum hold within local species pools across varying environmental conditions? *Functional Ecology*, 26(6): 1390–1398. https://doi.org/10.1111/1365-2435.12001

Yousefpour, R., & Hanewinkel, M. (2009). Modelling of forest conversion planning with an adaptive simulation-optimization approach and simultaneous consideration of the values of timber, carbon and biodiversity. *Ecological Economics*, 68(6): 1711–1722. https://doi.org/10.1016/j.ecolecon.2008.12.009

3 Anatomical Traits
Adaptations for Resource Acquisition and Allocation

Deepa Bisht, Upasna Sharma, and Narendra Kumar

3.1 INTRODUCTION

Plants rely on resource acquisition (water, light, nutrients, and carbon) and allocation (growth, storage, and reproduction) for survival. Despite the vast diversity among plant species, all seed plants share a fundamental body plan. This plan consists of three main body parts: stem, root, and leaves. The stem grows upward and provides a structural framework, the root grows downward to provide anchorage and absorb water and nutrients, and the leaves grow laterally from the stems and serve as the primary photosynthetic organ. Each body part contains specialized cells and tissues that facilitate essential physiological functions, including water, minerals, and photosynthate transport; food assimilation and storage; mechanical support; and adaptation to environmental stress. The internal or anatomical structure of these parts is organized to perform these functions and highlights the relationship between plant structure and function. Their various adaptations determine the efficiency of these processes and help the plants to thrive in diverse ecosystems and under various kinds of stresses.

Plants acquire four primary resources: Water, light, nutrients, and carbon dioxide. CO_2 is absorbed from the atmosphere through stomata and used in photosynthetic leaves (sources) to produce sucrose. This sucrose is then transported to non-photosynthetic tissues (sinks) via phloem to carry out various functions such as growth and reproduction. In contrast, water and minerals are absorbed from the soil with the help of roots and transported to the aerial parts of the plant through the xylem (Ruan et al., 2013). The anatomical structures of leaves and roots are specialized to efficiently acquire these resources, while the stem and vascular system (xylem and phloem) ensures their transport and distribution based on the plant's temporal demands.

Plants are broadly classified into three habitat forms, hydrophytes, mesophytes, and xerophytes, based on water availability in their habitat (Sharma & Kumar, 2023). Their anatomical and morphological adaptation enables them to survive and thrive in their respective habitats. Hydrophytes, adapted to aquatic habitats, possess large air spaces (aerenchyma) in their tissues for buoyancy and efficient gas exchange, along

DOI: 10.1201/9781003545231-3

with reduced or absent cuticles and stomata to facilitate water absorption (Sharma & Kumar, 2023). Mesophytes, which grow in moderate conditions, exhibit well-developed vascular tissues, balanced stomatal distribution, and moderately thick cuticles to optimize water and nutrient uptake while preventing excessive loss (Sharma & Kumar, 2023). In contrast, xerophytes, adapted to dry/arid environments, feature thick cuticles, sunken stomata, and reduced leaf surface areas to minimize water loss. These structural modifications ensure efficient utilization of available resources, enhancing survival and productivity in their respective ecosystems. Additionally, Table 3.1 provides a comprehensive overview of morpho-anatomical adaptations in hydrophytes, mesophytes, and xerophytes for resource acquisition and allocation (Saini, 2015).

Plants face significant trade-offs in their growth strategies, and the balance between above-ground resources (sunlight, CO_2) and below-ground resources (water, minerals) influences how plants allocate resources for growth. A critical factor in these trade-offs is identifying which nutrient is most limiting for plant growth in an environment (Hu et al., 2021). As per Liebig's Law of the Minimum, the resource in the shortest supply is most crucial, as it determines plant growth in an ecosystem. Thus, a plant that can efficiently acquire that limiting resource with minimal metabolic cost can allocate more resources and energy towards growth, defence, and reproduction. Previous research suggests that when light or CO_2 is limited, plants tend to distribute more resources and biomass to aerial organs. But, when water or nutrients are limiting below ground, plants allocate more biomass to root development. Resource allocation is a key strategy reflecting how plants grow and adapt to their environment (Lynch, 2015).

The chapter reviews the major anatomical structures and traits in plants that facilitate resource acquisition and allocation across different stages of growth and development. Identifying these traits can aid crop breeders in making informed selections, enhancing plant adaptability for efficient resource capture and improved yield (Van der Bom et al., 2020).

3.2 MAJOR ANATOMICAL TRAITS AND THEIR FUNCTIONAL ROLE

3.2.1 Root System Adaptations

The root system is the key trait for plant survival as it serves as the primary interface for soil resource acquisition. It plays a vital role in capturing water and minerals and providing stability through anchorage. In species such as legumes and some higher plants, the efficiency of water and mineral absorption is enhanced by soil microorganisms, such as mycorrhiza and root nodules (Miyasaka & Habte, 2001). The primary function of water and mineral absorption is performed by root hairs, which are thin, hair-like protrusions of root epidermal cells in the distal regions. These structures enhance the surface area, facilitating efficient uptake of nutrients and water. Anatomically, root hairs show adaptations that are well suited to their function. They have a freely permeable thin cell wall made of pectin and cellulose allowing substances to enter and exit the cell, a selectively permeable plasma membrane, and a

TABLE 3.1
Anatomical Adaptations in Hydrophytes, Mesophytes, and Xerophytes for Resource Acquisition and Allocation (Saini, 2015)

Traits	Hydrophytes (Aquatic Plants)	Mesophytes (Moderate Water Plants)	Xerophytes (Drought-Adapted Plants)
Root system (water and nutrient acquisition)	i. Roots may be **reduced or absent** in floating hydrophytes (*Eichhornia*). ii. **Aerenchyma (air spaces)** facilitates oxygen transport in submerged roots (*Nelumbo*). iii. Root hairs are absent or poorly developed since water and nutrients are directly absorbed from the surroundings. iv. In anchored hydrophytes, **shallow fibrous roots** absorb nutrients from the sediment.	i. Well-developed root system with **moderate branching**. ii. Root hairs enhance **surface area** for efficient absorption of water and minerals. iii. Vascular tissues (xylem and phloem) are well-balanced for nutrient and water transport.	i. **Extensive root systems** for maximum water absorption (deep taproots in *Prosopis*, wide-spreading fibrous roots in *Cactus*). ii. Some plants have **succulent roots** for water storage (*Asparagus*). iii. **Mycorrhizal associations** help increase nutrient and water uptake. iv. **Epidermal modifications** (e.g., velamen in epiphytes) assist in moisture absorption.
Shoot system (structural support and resource transport)	i. Weak, **flexible, non-lignified stems** for easy movement in water (*Hydrilla*). ii. **Aerenchyma** in stems aids in gaseous exchange and buoyancy (*Nymphaea*). iii. **Reduced vascular tissues**, as water transport is less demanding.	i. Well-developed, **lignified stems** for mechanical support. ii. Vascular tissues are balanced, ensuring efficient transport of water, nutrients, and food. iii. **Cuticle present**, but not excessively thick.	i. **Succulent stems** for water storage (*Euphorbia, Cactus*). ii. **Well-developed xylem** for efficient water conduction (*Acacia*). iii. **Reduced phloem** in some cases to conserve energy. iv. **CAM photosynthesis in stems** (e.g., *Opuntia*) for water-use efficiency. v. Thick, waxy **cuticle** to reduce transpiration.
Leaves (light capture and gas exchange)	i. Thin, **broad floating leaves** with large intercellular spaces (*Nymphaea*). ii. Stomata present only on the **upper surface** of floating leaves for gas exchange. iii. **Highly dissected submerged leaves** increase surface area for underwater gas exchange (*Ceratophyllum*). iv. **Chloroplasts present in epidermal cells**, enhancing light absorption.	i. **Broad leaves** with a well-developed mesophyll for efficient photosynthesis. ii. Stomata present **on lower epidermis** to balance water loss and gas exchange. iii. Moderate **cuticle thickness** prevents excessive water loss.	i. **Small, thick, or needle-like leaves** to reduce transpiration (*Pinus, Nerium*). ii. **Sunken stomata** in deep pits minimize water loss (*Nerium*). iii. **Thick, waxy cuticle** to conserve moisture (*Aloe vera*). iv. **Leaves modified into spines** (*Opuntia*) to prevent water loss. v. Some xerophytes shed leaves during dry seasons to conserve resources (*Ocotillo*).

(*Continued*)

TABLE 3.1 (*Continued*)
Anatomical Adaptations in Hydrophytes, Mesophytes, and Xerophytes for Resource Acquisition and Allocation (Saini, 2015)

Traits	Hydrophytes (Aquatic Plants)	Mesophytes (Moderate Water Plants)	Xerophytes (Drought-Adapted Plants)
Stomatal adaptations (water regulation and CO_2 uptake)	i. Stomata are **absent or reduced** in submerged species (*Ceratophyllum*). ii. Floating hydrophytes have **stomata only on the upper leaf surface** (*Victoria*). iii. Stomatal opening is regulated by surrounding water conditions.	i. Stomata present on **lower epidermis**, ensuring balanced gas exchange and water conservation. ii. Regular stomatal activity during daytime photosynthesis.	i. **Sunken stomata** create humid microenvironments to prevent excess water loss (*Nerium*). ii. **Fewer stomata**, reducing transpiration rates. iii. **CAM photosynthesis** (e.g., *Agave, Opuntia*) allows stomata to open at night, minimizing water loss.
Cuticle and epidermal adaptations (water retention and protection)	i. **Very thin or absent cuticle**, as water loss is not a concern. ii. Epidermis may contain chloroplasts for photosynthesis in some submerged species.	i. **Moderate cuticle thickness** to prevent excess transpiration. ii. Single-layered epidermis with normal wax deposition.	i. **Thick, waxy cuticle** to minimize water loss (*Aloe vera*). ii. **Multiple-layered epidermis** in some species (*Nerium*). iii. Epidermal hairs (**trichomes**) trap moisture and reduce transpiration (xerophytic grasses).
Photosynthetic adaptations (carbon fixation and efficiency)	i. Photosynthesis occurs in **submerged parts** in some cases (*Hydrilla*). ii. Floating leaves have **large palisade mesophyll** for efficient light absorption. iii. Submerged species often rely on **CO_2 diffusion from water** rather than atmospheric sources.	i. Typical **C_3 photosynthesis pathway**, optimized for moderate conditions. ii. Well-developed mesophyll with palisade and spongy layers for effective photosynthesis.	i. Many xerophytes use **CAM photosynthesis**, storing CO_2 at night as malic acid for daytime use (*Opuntia, Agave*). ii. Some desert plants exhibit **C_4 photosynthesis**, which is more water-efficient (*Amaranthus, Sorghum*).
Resource allocation and storage	i. Energy is primarily allocated to **buoyancy structures (aerenchyma)** rather than root development. ii. Minimal investment in mechanical support due to water buoyancy.	i. **Balanced allocation** of resources between roots, stems, and leaves for optimal growth. ii. Resources are efficiently used in response to seasonal changes.	i. **High resource allocation to root growth** for water absorption. i. ii.**Water storage tissues** in stems, roots, and leaves (succulents like *Aloe*). iii. Leaves may be reduced or lost to prioritize **stem-based photosynthesis**.

large vacuole that makes it more efficient to take in water and nutrients. Root hairs also facilitate the efficient uptake of immobile nutrients such as phosphorus (Peterson & Farquhar, 1996; Zhu et al., 2010). During low phosphorus and iron conditions, plants produce long and dense root hair for their efficient uptake (Cruz-Ramírez et al., 2009; Ramachandran et al., 2024). The long-haired genotypes of maize under low P availability were reported to have significantly greater plant growth, P uptake, and specific P absorption rates than short-haired genotypes. Mycorrhizal associations of species ensure extended root development and improved uptake of immobile nutrients like P, Cu, and Zn (Miyasaka & Habte, 2001; Zhu et al., 2010).

The root system, technically called root architecture, is an essential trait that directly influences the soil resource acquisition. It refers to the spatial organization and structure of plant roots in the soil (Ho et al., 2005). It is a key determinant of how effectively plants can attain water and nutrients from the surrounding environment. The architecture of roots includes various factors such as root depth, branching patterns, root hair development, root system plasticity, and the distribution of roots in soil layers. Understanding root architecture is crucial for improving plant productivity, especially in environments where water is scarce or irregularly distributed, making it a significant focus in agricultural research, particularly in the context of water mining (efficient extraction of water from the soil). The distribution of immobile resources like phosphorus in shallow soil layers and mobile resources such as water and nitrate in deeper soil layers (Galindo-Castañeda et al., 2022) highlights the adaptations of root architecture. The architectural root ideotype with shallow root growth angle and more lateral root branching named as "topsoil foraging" would promote P acquisition (Holz et al., 2024). In contrast, phenotypes with steep rooting angles and low lateral branching named as "steep, cheap and deep" ideotype exhibit improved water and nitrate uptake and demonstrate significant drought tolerance (Saengwilai et al., 2014; Zhan & Lynch, 2015; Lynch, 2018; Holz et al., 2024). Ho et al. (2005) noted that root localization, root size, and root carbon allocation are effective means of altering root architecture.

Root anatomy plays a crucial role in understanding the root architecture and function. The anatomical structure of roots regulates selective ion uptake. Endodermis is the innermost layer of the cortex, distinguished by the Casparian bands, which consist of suberin deposits in the anticlinal walls of endodermal cells. During the mineral and water uptake, endodermis prevents free apoplastic movement of various ions, heavy metals, and fluorescent dyes. The movement of water and selected ions is then directed to the symplast pathway. Also, endodermis prevents the flow of ions backwards from the stele. Like endodermis, when the hypodermal cells form Casparian bands, this layer is known as exodermis, and it performs the same function as endodermis. During the drought and salinity stress, root growth rate decreases and wall deposition of endodermis and exodermis starts relatively close to the root apex, resulting in a greater proportion of exodermis and endodermis in the mature root (Enstone et al., 2002).

Hydrophytes and plants in waterlogged conditions often develop root aerenchyma interconnected air spaces in root cortex (Shiono et al., 2008). Depending on the species, aerenchyma may develop throughout the entire plant or in some regions, creating an internal pathway for O_2, CO_2, and ethylene (Videmšek et al., 2006). In such conditions, aerenchyma maintains buoyancy in plants. Hydrophytes also develop specialized root structures, such as hypodermis or exodermis, which act as a barrier for solute uptake, while aerenchyma facilitates gas exchange. Species like *Nymphaea odorata* form suberized exodermis with astrosclereids for support and protection of aerenchyma. Flood-tolerant crops like rice and *Zea nicaraguensis* also form suberized exodermis and aerenchyma that prevent toxic ion uptake and oxygen loss (Herzog et al., 2016).

Plants living in arid and desert conditions are called xerophytes. They have various root adaptations that enable plants to thrive in desert environments (Ramachandran et al., 2024). Species like *Opuntia* and *Agave* develop specialized rain roots, some of which form aerenchyma to prevent water loss (Dubrovsky et al., 1998). Many desert plants establish a broad lateral root network early on, aided by determinate primary roots and meristems, enhancing seedling survival. Anatomically, desert roots exhibit extensive cell wall modifications, including suberin and lignin deposition in ground tissue layers (epidermis, exodermis, and endodermis), which help prevent water loss and regulate ion transport (AbdElhalim et al., 2016; Xiao et al., 2019). The occurrence of thick cuticle, sunken stomata, and sclerenchymatus hypodermis are a few anatomical xerophytic adaptations that are common to all plants. A unique desert adaptation, root contraction, occurs in *Cactaceae*, *Agavaceae*, and *Asteraceae* families, where root tissues undergo coordinated shortening, pulling the shoot deeper into the soil to mitigate heat stress. These anatomical modifications collectively enhance drought resilience and resource acquisition in extreme desert environments (Ramachandran et al., 2024).

Additionally, plants have anatomically developed traits that are adapted to physiological conditions. For example, in *Zygophyllum album*, fibre cells near xylem vessels protect vascular tissues from embolisms. Lignified cell walls in *Phoenix dactylifera* (date palm) contribute to structural support and water retention (AbdElhalim et al., 2016).

Mangroves exhibit specialized anatomical root adaptations to survive in tidal environments. Pioneer species like *Avicennia marina* develop a thin exodermis and aerenchyma to supply oxygen to submerged roots (Colmer, 2002). In contrast, *Rhizophora* species develop thicker, lignified, and suberized exodermis to withstand salinity and heavy metal stress (Cheng et al., 2014). Additionally, species like *R. mangle* produce mucilage in xylem sap to regulate water transport and prevent salt toxicity (Zimmermann et al., 2002).

Based on these adaptations, key root anatomical traits such as cortex thickness and stele diameter determine root functions like absorption, transport, and mycorrhizal colonization (Wang et al., 2024). Roots adapt to environmental conditions by modifying their anatomy; for instance, thinner roots are favoured in nutrient-poor or compact soils to enhance exploratory efficiency. Similarly, cortex thickness tends to increase in fertile soils, improving resource storage, while stele diameter decreases in cold conditions to minimize metabolic costs. These structural adaptations highlight the role of root anatomy in optimizing resource acquisition under varying environmental constraints (Wang et al., 2024).

3.2.2 Stem and Vascular Tissue Adaptations

The stem serves as the structural framework of the plant, facilitating water transport, providing mechanical support, promoting growth through meristems, and ensuring efficient nutrient distribution. Plants have evolved specialized vascular tissues to adapt to different environmental conditions to ensure both efficiency and tolerance against stress. These adaptations primarily involve xylem and phloem:

Xylem consists of various cell types: vessel elements and tracheids for water conduction, fibres for mechanical support, and parenchyma for storage. For water conduction, Gymnosperms predominantly rely on tracheids, which also provide mechanical support (Sperry et al., 2006). In contrast, angiosperms contain vessels that are wider and allow rapid water movement. Hence, vessels show higher efficiency in water transport but are more prone to embolism, while tracheids, being elongated and narrow, reduce the risk of cavitation, a condition where air bubbles block water flow making them well-suited for cold or arid environments. Xylem parenchyma cells contribute not only to storage but also to embolism recovery when placed near them. They create an energy gradient that facilitates water movement into empty vessels or tracheids, aiding in their refilling (Secchi et al., 2017).

The transport and allocation of sugars and nutrients are carried out by phloem tissue. Sieve tube elements are the primary conducting cell that functions efficiently in coordination with companion cells. They exhibit several structural and functional adaptations to ensure efficient translocation. The symplastic and apoplastic loading mechanisms in different plant species further enhance the efficiency of sugar transport, ensuring a continuous supply of energy and nutrients to growing tissues, storage organs, and developing seeds. In temperate environments, seasonal changes significantly influence phloem structure and resource allocation from the new growth (leaf, buds, fruits, etc.) in spring to senescence in autumn and winter, allowing the plant to adapt to temperature variations throughout the year (Ray & Savage, 2021). During the active growth period in spring and summer, active phloem cells transport high volumes of sugars to support rapid growth and the development of leaves, flowers, and fruits. In response to colder temperatures in autumn and winter, the activity of phloem cells and transport declines due to slow growth and metabolism (Savage, 2020). It leads to anatomical changes, such as the deposition of callose on sieve plates. In certain plant species, callose deposition is irreversible, leading to the permanent sealing of sieve elements, which subsequently lose their functionality. This form of callose is commonly known as definitive callose. In contrast, other species exhibit a reversible process where callose, known as dormancy callose, is enzymatically degraded in the spring, temporarily obstructing phloem flow during winter dormancy. Regardless of the species, new sieve elements are generated each growing season by the vascular cambium, ensuring the restoration of phloem function (Savage, 2020).

3.2.3 Leaf Adaptations

Plants exhibit various leaf anatomical modifications for the uptake of resources (CO_2 and sunlight) and play an important role in regulating photosynthesis, transpiration, and protection against environmental stress. One such adaptation is the deposition of cuticle, a hydrophobic lipid layer deposited on the outer cell walls of the aerial parts, especially leaf epidermis (Taiz et al., 2015). This structure prevents water loss and serves as a barrier against pathogens and insects. The cuticle also reflects light, reducing heat load. Plants in water deficit conditions develop thicker cuticle, which decreases water loss through transpiration, as seen in Agave (Fernández et al., 2016; Taiz et al., 2015). Similarly, many leaves and stems possess trichomes, hair-like structures from epidermal cells that help reduce water loss by reflecting light and wind exposure near the surface (Taiz et al., 2015).

Leaf is the primary site of photosynthesis. It is a vital biochemical process on Earth, wherein plants convert the solar energy into chemical energy. Photosynthesis consists of two major reactions: light reactions, which occur in grana, and carbon reactions, which take place in the stroma of chloroplast. To optimize carbon fixation, plants have evolved three distinct pathways C_3, C_4, and Crassulacean acid metabolism (CAM), each adapted to different environmental conditions (Taiz et al., 2015). Among these, C_4 plants such as maize, sugarcane, and sorghum possess a specialized leaf structure where mesophyll and bundle sheath cells are arranged concentrically around the vascular bundles. This anatomical adaptation is known as Kranz anatomy and enhances the efficiency of C_4 plants in CO_2 fixation. Unlike C_3 plants, where the Calvin cycle occurs only in mesophyll cells, C_4 plants exhibit a spatial separation for carbon fixation. The mesophyll cells initially capture atmospheric CO_2 and convert it into a four-carbon compound (such as malate or aspartate) using phosphoenol pyruvate carboxylase enzyme. This enzyme has a high affinity for CO_2 with minimal oxygen interference. The four-carbon compound is then transported to the bundle sheath cells, where it releases CO_2 creating a high concentration around RuBisCO the enzyme responsible for catalyzing the carbon fixation reaction in the C_3 cycle. This adaptation significantly reduces photorespiration, thereby enhancing the photosynthetic efficiency, particularly under conditions of high temperature, intense light, and low CO_2 availability. By optimizing carbon fixation, Kranz anatomy enables C_4 plants to thrive in arid and tropical environments, making them more productive than C_3 plants in such conditions (Taiz et al., 2015).

In another adaptation, CAM plants show the remarkable water-use efficiency, resulting in widespread diversification and adaptation to arid environments. Xerophytic CAM plants, such as cacti, have evolved a unique strategy to minimize water loss by closing their stomata during the daytime and opening them at night when temperatures are lower. Thus, these plants absorb CO_2 at night time through stomatal opening and store it in vacuoles as malic acid. When daylight becomes available, stored malic acid is broken down by decarboxylating enzymes, releasing CO_2 for use in the C_3 cycle. The internally released CO_2 is effectively retained within the leaf since the stomata remain closed during the day, preventing its escape. This process not only conserves water but also maintains a high internal CO_2 concentration, optimizing its fixation by RuBisCO and thereby preventing photorespiration. This adaptation enables CAM plants to thrive in extreme conditions where water is scarce (Taiz et al., 2015).

Other key traits, such as the thickness of palisade and spongy tissue, influence light absorption and gas exchange, respectively (He et al., 2018). A thicker palisade tissue, rich in chloroplasts, enhances photosynthetic efficiency, while a well-developed spongy tissue with larger intercellular spaces facilitates CO_2 diffusion and water movement (He et al., 2018). Additionally, stomatal traits, stomatal density, and size regulate transpiration and carbon assimilation, directly impacting WUE and gross primary productivity (He et al., 2018). Stomatal size decreases under drought conditions, while stomatal density shows more plasticity in varying environmental conditions (De Micco & Aronne 2012). Variations in these anatomical traits across different environmental conditions enable plants to adapt to resource limitations, improving survival and ecosystem functioning (He et al., 2018). To understand large-scale variations and their relationships with ecosystem functions, He et al. (2018) examined eight essential leaf anatomical traits: leaf thickness, adaxial and abaxial epidermis thickness, palisade and spongy tissue thickness, palisade tissue leaf thickness ratio, palisade–spongy tissue ratio, and spongy tissue–leaf thickness ratio. Their findings contributed to bridge the data gap in plant anatomical traits at regional scales, enhanced the understanding of plant adaptation strategies, and provided new insights into the connection between plant traits and ecosystem functioning in natural communities.

3.3 TRADE-OFFS IN ANATOMICAL ADAPTATIONS

Plants exhibit a trade-off between hydraulic efficiency and safety in the xylem, particularly within the same plant (Pockman & Sperry, 2000; Martinez-Vilalta et al., 2002; Loepfe et al., 2007). Wider vessels enhance hydraulic conductivity as per the Hagen–Poiseuille law, allowing plants to transport water more efficiently over long distances (Awad et al., 2010; Wang et al., 2025). However, this advantage comes at a cost; wider vessels are more vulnerable to embolism, a condition where air bubbles enter the xylem, disrupting water flow and leading to hydraulic failure (Figure 3.1b). Embolism risk increases under drought stress, freezing conditions, or mechanical damage, as air-seeding through pit membranes can cause rapid spread of cavitation to the neighbouring xylem conduits. The balance between efficiency and safety

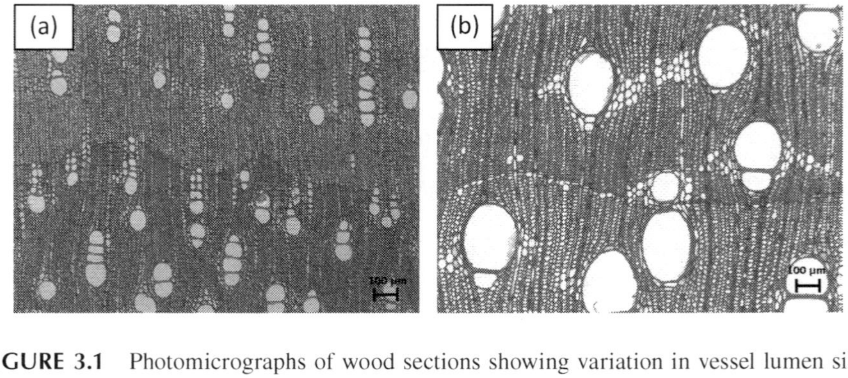

FIGURE 3.1 Photomicrographs of wood sections showing variation in vessel lumen size. (a) *Anogeissus sericea*: T.S. showing radial multiples of vessels at 5x. The wood sample shows narrow vessels with vessel grouping, an anatomical adaptation that ensures hydraulic safety. (b) *Terminalia bialata:* T.S showing solitary vessels at 5x. The wood sample shows wider solitary vessels, which are more hydraulically efficient but prone to embolism, an anatomical adaptation that thrives in environments with optimal water availability.

varies among species based on their ecological strategies; for example, fast-growing species in wet environments prioritize hydraulic efficiency, while drought-tolerant species in arid regions emphasize safety through narrower, more resistant conduits (Figure 3.1a). Gleason et al. (2016) tested the safety–efficiency trade-off across 335 angiosperm and 89 gymnosperm species. They observed a weak correlation between safety and efficiency; however, no species showed both high efficiency and high safety, reinforcing the concept of safety–efficiency trade-off.

In recent times, plant physiologists have shown growing interest in the study of quantitative wood anatomical traits and their plasticity under varying environmental conditions. Unlike qualitative traits, which tend to be conservative within a species, quantitative wood anatomical traits, in various combinations, contribute to species adaptation both inter- and intraspecifically across different environments (De Micco & Aronne, 2012). Over the years, wood anatomists have extensively studied shrubs and trees to identify anatomical features that support hydraulic efficiency or safety. Hydraulic efficiency, determined by hydraulic conductivity, and hydraulic safety, assessed through vulnerability to embolism (De Micco & Aronne, 2012), can be evaluated both theoretically and through laboratory-based studies. Increased efficiency is typically associated with features such as wider vessel lumen size and simple perforation plates, whereas safety is enhanced by traits like narrow vessel lumens, helical thickenings, scalariform perforation plates, reduced intervessel pitting area, increased vessel grouping or frequency, and the presence of vasicentric tracheid (Lens et al., 2011; De Micco & Aronne, 2012).

Another trade-off that plants exhibit under stress is root investment vs. shoot growth, where plants maintain a high root–shoot ratio through allocating more resources to the roots and a low root–shoot ratio through allocating resources to their shoots (Kalra et al., 2024). Plants under stress conditions like drought and nutrient deficiency maintain a high root-to-shoot ratio, which allows for a greater allocation of resources to the root system at the expense of above-ground productivity to enhance water and nutrient absorption (Kalra et al., 2024; Holz et al., 2024). Under these conditions, roots often become metabolically more active and undergo structural modifications to enhance water uptake. They send biochemical signals to other parts of the plant, triggering an appropriate physiological response. This intricate root-to-shoot communication system ensures that resources are distributed efficiently, enabling the plant to adapt and survive in a stressed environment (Kalra et al., 2024; Holz et al., 2024). However, during water and nutrient availability, plants maintain a low root–shoot ratio and allocate more biomass to their shoots (leaves and stems) and promote aerial growth.

3.4 IMPLICATIONS FOR CROP BREEDING

Anatomical traits can aid crop breeders in making informed species selections, enhancing plant adaptability for efficient resource capture and improved yield (Lynch, 2019; Van der Bom et al., 2020). Optimizing root traits such as enhanced root hair development, deeper root systems with lateral root growth and fostering mycorrhizal associations can improve nutrient use efficiency, reduce fertilizer dependence, and promote sustainable agriculture, especially in nutrient-deficient areas (Lynch, 2005). Similarly, selecting species with an optimal combination of root, leaf, and stem anatomical traits would help in adaptation to diverse environmental conditions.

Crop breeding strategies for resource utilization encompass both root system modifications and aerial adaptations. These adaptive strategies highlight the challenge of breeding crops for improved drought and nutrient efficiency, as optimizing root and rhizosphere traits may inadvertently impact carbon allocation to above-ground structures, potentially influencing yield (Holz et al., 2024). Developing plastic traits in response to drought and nutrient limitations requires significant carbon investment, which becomes challenging when carbon availability is constrained. Therefore, effective trait optimization must balance carbon efficiency, prioritization of carbon assimilation, and the equilibrium between photosynthesis and respiration (Amthor et al., 2019; Holz et al., 2024). Further research is needed to identify and apply anatomical traits that enhance yield, especially in nutrient–deficient regions and under rising global temperatures. With increasing global food demand, a more comprehensive insight into these traits and their physiological functions will be crucial for advancing both agriculture and forestry.

3.5 CONCLUSION

Plant anatomical adaptations across roots, stems, vascular tissues, and leaves demonstrate their ability to optimize resource utilization and allocation to withstand various environmental stresses. Root systems adjust their anatomical structure to enhance water and nutrient uptake, often prioritizing below-ground investment under stress. Xylem adaptations, such as vessel and tracheid modifications, contribute to species' survival strategies, with angiosperms prioritizing rapid water transport and gymnosperms emphasizing embolism resistance. Similarly, phloem adaptations, including sieve tube elements and companion cells, regulate seasonal shifts in resource allocation, ensuring sustained growth and survival under fluctuating conditions. These vascular adaptations highlight the dynamic balance between physiological efficiency and stress tolerance, crucial for plant resilience in diverse habitats.

Leaf anatomical adaptations, such as cuticle thickening, trichomes, and specialized photosynthetic pathways (C_4 and CAM), enhance water-use efficiency and carbon assimilation, allowing plants to thrive in diverse climatic conditions. The evolution of Kranz anatomy in C_4 plants and nocturnal CO_2 uptake in CAM species demonstrates nature's ability to optimize photosynthetic efficiency in response to environmental constraints. Additionally, variations in leaf structure, such as palisade and spongy tissue thickness, significantly impact light absorption, gas exchange, and overall productivity. Stomatal structure and density also show plasticity to adapt to different environmental conditions. These adaptations play an important role in plant performance, ecosystem dynamics, and agricultural productivity. Trade-offs in plant anatomical traits further illustrate the complex interplay between resource efficiency and survival. The hydraulic efficiency and safety trade-off and root-to-shoot investment strategy determine how plants cope with drought, nutrient scarcity, and other abiotic stresses.

The balance between hydraulic efficiency, photosynthetic optimization, and stress tolerance is critical for plant success. Understanding these anatomical adaptations and their functions can aid in phenotypic selection and predicting species resilience in specific environments. Traits can be analysed in relation to efficient resource uptake and allocation, providing a basis for species selection through field/laboratory experiments. These studies are invaluable for ecological research, crop improvement, and reforestation, particularly in the face of climate change. Future research should integrate anatomical, physiological, and genetic insights to enhance plant resilience for crop breeding and reforestation efforts.

REFERENCES

AbdElhalim, M. E., Abo-Alatta, O. K., Habib, S. A., & Abd Elbar, O. H. (2016). The anatomical features of the desert halophytes *Zygophyllum album* LF and *Nitraria retusa* (Forssk.) Asch. *Annals of Agricultural Sciences*, *61*(1), 97–104. https://doi.org/10.1016/j.aoas.2015.12.001

Amthor, J. S., Bar-Even, A., Hanson, A. D., Millar, A. H., Stitt, M., Sweetlove, L. J., & Tyerman, S. D. (2019). Engineering strategies to boost crop productivity by cutting respiratory carbon loss. *The Plant Cell*, *31*(2), 297–314. https://doi.org/10.1105/tpc.18.00743

Awad, H., Barigah, T., Badel, E., Cochard, H., & Herbette, S. (2010). Poplar vulnerability to xylem cavitation acclimates to drier soil conditions. *Physiologia Plantarum*, *139*(3), 280–288. https://doi.org/10.1111/j.1399-3054.2010.01367.x

Cheng, H., Jiang, Z. Y., Liu, Y., Ye, Z. H., Wu, M. L., Sun, C. C., … & Wang, Y. S. (2014). Metal (Pb, Zn and Cu) uptake and tolerance by mangroves in relation to root anatomy and lignification/suberization. *Tree physiology*, *34*(6), 646–656. https://doi.org/10.1093/treephys/tpu042

Colmer, T. (2002). Aerenchyma and an inducible barrier to radial oxygen loss facilitate root aeration in upland, paddy and deep-water rice (*Oryza sativa* L.). *Annals of Botany*, *91*(2), 301–309. https://doi.org/10.1093/aob/mcf114

Cruz-Ramírez, A., Calderón-Vázquez, C., & Herrera-Estrella, L. (2009). Effect of nutrient availability on root system development. *Annual Plant Reviews Volume 37: Root Development*, *37*, 288–324. https://doi.org/10.1002/9781444310023

De Micco, V., & Aronne, G. (2012). Morpho-anatomical traits for plant adaptation to drought. In: Aroca, R. (ed.) *Plant Responses to Drought Stress*. Springer, Berlin, Heidelberg. https://doi.org/10.1007/978-3-642-32653-0_2

Dubrovsky, J. G., North, G. B., & Nobel, P. S. (1998). Root growth, developmental changes in the apex, and hydraulic conductivity for *Opuntia ficus-indica* during drought. *The New Phytologist*, *138*(1), 75–82. https://doi.org/10.1046/j.1469-8137.1998.00884.x

Enstone, D. E., Peterson, C. A., & Ma, F. (2002). Root endodermis and exodermis: Structure, function, and responses to the environment. *Journal of Plant Growth Regulation*, *21*, 335–351. https://doi.org/10.1007/s00344-003-0002-2

Fernández, V., Guzmán-Delgado, P., Graça, J., Santos, S., & Gil, L. (2016). Cuticle structure in relation to chemical composition: Re-assessing the prevailing model. *Frontiers in Plant Science*, *7*, 427. https://doi.org/10.3389/fpls.2016.00427

Galindo-Castañeda, T., Lynch, J. P., Six, J., & Hartmann, M. (2022). Improving soil resource uptake by plants through capitalizing on synergies between root architecture and anatomy and root-associated microorganisms. *Frontiers in Plant Science*, *13*, 827369. https://doi.org/10.3389/fpls.2022.827369

Gleason, S. M., Westoby, M., Jansen, S., Choat, B., Hacke, U. G., Pratt, R. B., … & Zanne, A. E. (2016). Weak tradeoff between xylem safety and xylem-specific hydraulic efficiency across the world's woody plant species. *New Phytologist*, *209*(1), 123–136. https://doi.org/10.1111/nph.13646

He, N., Liu, C., Tian, M., Li, M., Yang, H., Yu, G., … & Hou, J. (2018). Variation in leaf anatomical traits from tropical to cold-temperate forests and linkage to ecosystem functions. *Functional Ecology*, *32*(1), 10–19. https://doi.org/10.1111/1365-2435.12934

Herzog, M., Striker, G. G., Colmer, T. D., & Pedersen, O. (2016). Mechanisms of waterlogging tolerance in wheat–a review of root and shoot physiology. *Plant, Cell & Environment*, *39*(5), 1068–1086. https://doi.org/10.1111/pce.12676

Ho, M. D., Rosas, J. C., Brown, K. M., & Lynch, J. P. (2005). Root architectural tradeoffs for water and phosphorus acquisition. *Functional Plant Biology*, *32*(8), 737–748. https://doi.org/10.1071/FP05043

Holz, M., Zarebanadkouki, M., Benard, P., Hoffmann, M., & Dubbert, M. (2024). Root and rhizosphere traits for enhanced water and nutrients uptake efficiency in dynamic environments. *Frontiers in Plant Science*, *15*, 1383373. https://doi.org/10.3389/fpls.2024.1383373

Hu, J., Yu, H., Li, Y., Wang, J., Lv, T., Liu, C., & Yu, D. (2021). Variation in resource allocation strategies and environmental driving factors for different life-forms of aquatic plants in cold temperate zones. *Journal of Ecology*, *109*(8), 3046–3059. https://doi.org/10.1111/1365-2745.13719

Kalra, A., Goel, S., & Elias, A. A. (2024). Understanding role of roots in plant response to drought: Way forward to climate-resilient crops. *The Plant Genome*, *17*(1), e20395. https://doi.org/10.1002/tpg2.20395

Lens, F., Sperry, J. S., Christman, M. A., Choat, B., Rabaey, D., & Jansen, S. (2011). Testing hypotheses that link wood anatomy to cavitation resistance and hydraulic conductivity in the genus *Acer*. *New Phytologist*, *190*(3), 709–723. https://doi.org/10.1111/j.1469-8137.2010.03518.x

Loepfe, L., Martinez-Vilalta, J., Piñol, J., & Mencuccini, M. (2007). The relevance of xylem network structure for plant hydraulic efficiency and safety. *Journal of Theoretical Biology*, *247*(4), 788–803. https://doi.org/10.1016/j.jtbi.2007.03.036

Lynch, J. P. (2015). Root phenes that reduce the metabolic costs of soil exploration: Opportunities for 21st century agriculture. *Plant, Cell & Environment*, *38*(9), 1775–1784. https://doi.org/10.1111/pce.12451

Lynch, J. P. (2018). Rightsizing root phenotypes for drought resistance. *Journal of Experimental Botany*, *69*(13), 3279–3292. https://doi.org/10.1093/jxb/ery048

Lynch, J. P.(2019). Root phenotypes for improved nutrient capture: an underexploited opportunity for global agriculture. *New Phytologist*, *223*(2): 548–564.

Lynch, C. A., & Lippincott, J. K. (2005). Institutional repository deployment in the United States as of early 2005. *D-lib Magazine*, *11*(9): 1–11.

Martinez-Vilalta, J., Prat, E., Oliveras, I., Pinol, J., (2002). Xylem hydraulic properties of roots and stems of nine Mediterranean woody species. *Oecologia 133*, 19–29. https://doi.org/10.1007/s00442-002-1009-2

Miyasaka, S. C., & Habte, M. (2001). Plant mechanisms and mycorrhizal symbioses to increase phosphorus uptake efficiency. *Communications in Soil Science and Plant Analysis*, *32*(7–8), 1101–1147. https://doi.org/10.1081/CSS-100104105

Peterson, R. L., & Farquhar, M. L. (1996). Root hairs: Specialized tubular cells extending root surfaces. *The Botanical Review*, *62*, 1–40. https://doi.org/10.1007/BF02868919

Pockman, W. T., & Sperry, J. S., (2000). Vulnerability to xylem cavitation and the distribution of Sonoran Desert vegetation. *American Journal of Botany*. *87*, 1287–1299. https://doi.org/10.2307/2656722

Ramachandran, P., Ramirez, A., & Dinneny, J. R. (2024). Rooting for survival: How plants tackle a challenging environment through a diversity of root forms and functions. *Plant Physiology*, kiae586. https://doi.org/10.1093/plphys/kiae586

Ray, D. M., & Savage, J. A. (2021). Seasonal changes in temperate woody plant phloem anatomy and physiology: Implications for long-distance transport. *AoB Plants*, *13*(4), plab028. https://doi.org/10.1093/aobpla/plab028

Ruan, Y. L., Patrick, J. W., Shabala, S., & Slewinski, T. L. (2013). Uptake and regulation of resource allocation for optimal plant performance and adaptation to stress. *Frontiers in Plant Science*, *4*, 455. https://doi.org/10.3389/fpls.2013.00455

Saengwilai, P., Nord, E. A., Chimungu, J. G., Brown, K. M., & Lynch, J. P. (2014). Root cortical aerenchyma enhances nitrogen acquisition from low-nitrogen soils in maize. *Plant physiology*, *166*(2), 726–735. https://doi.org/10.1104/pp.114.241711

Saini, A. (2015). *Plant Ecology* (New ed.). Truman Publication, New Delhi.

Savage, J. A. (2020). It's all about timing—or is it? Exploring the potential connection between phloem physiology and whole plant phenology. *American Journal of Botany*, *107*(6), 848–851. https://doi.org/10.1002/ajb2.1480

Secchi, F., Pagliarani, C., & Zwieniecki, M. A. (2017). The functional role of xylem parenchyma cells and aquaporins during recovery from severe water stress. *Plant, Cell & Environment*, *40*(6), 858–871. https://doi.org/10.1111/pce.12831

Sharma, P. K., & Kumar, S. (2023). *Soil Physical Environment and Plant Growth.* Springer, Berlin/Heidelberg, Germany.

Shiono, K., Takahashi, H., Colmer, T. D., & Nakazono, M. (2008). Role of ethylene in acclimations to promote oxygen transport in roots of plants in waterlogged soils. *Plant Science*, *175*(1–2), 52–58. https://doi.org/10.1016/j.plantsci.2008.03.002

Sperry, J. S., Hacke, U. G., & Pittermann, J. (2006). Size and function in conifer tracheids and angiosperm vessels. *American Journal of Botany*, *93*(10): 1490–1500.

Taiz, L., Zeiger, E., Møller, I. M., & Murphy, A. (2015). Plant physiology and Development, 6th Ed.; Sinauer Associates, Inc.: Sunderland, MA, USA, ISBN 978-1-60535-255-8.

Van der Bom, F. J., Williams, A., & Bell, M. J. (2020). Root architecture for improved resource capture: Trade-offs in complex environments. *Journal of Experimental Botany*, *71*(19), 5752–5763. https://doi.org/10.1093/jxb/eraa324

Videmšek, U., Boris, T. U. R. K., & Vodnik, D. (2006). Root aerenchyma–formation and function. *Acta agriculturae slovenica*, *87*(2), 445–453. https://doi.org/10.14720/aas.2006.87.2.15124

Wang, H., Ma, J., Wang, Y., Wang, J., Xie, T., Niu, F., … & Shan, L. (2025). Xylem anatomical structure as a determinant of hydraulic trait variation in C3 plant *Reaumuria soongorica* and C4 plant *Salsola passerina*. *Environmental and Experimental Botany*, *229*, 106055. https://doi.org/10.1016/j.envexpbot.2024.106055

Wang, X., Liu, X., Chen, S., Zhu, J., Yuan, Y., Zhu, R., … & Zhang, S. (2024). Elevational variation in anatomical traits of the first-order roots and their adaptation mechanisms. *Plant Diversity*. https://doi.org/10.1016/j.pld.2024.09.008

Xiao, T. T., Raygoza, A. A., Pérez, J. C., Kirschner, G., Deng, Y., Atkinson, B., … & Blilou, I. (2019). Emergent protective organogenesis in date palms: A morpho-devo-dynamic adaptive strategy during early development. *The Plant Cell*, *31*(8), 1751–1766. https://doi.org/10.1105/tpc.19.00008

Zhan, A., & Lynch, J. P. (2015). Reduced frequency of lateral root branching improves N capture from low-N soils in maize. *Journal of Experimental Botany*, *66*(7), 2055–2065. https://doi.org/10.1093/jxb/erv007

Zhu, J., Zhang, C., & Lynch, J. P. (2010). The utility of phenotypic plasticity of root hair length for phosphorus acquisition. *Functional Plant Biology*, *37*(4), 313–322. https://doi.org/10.1071/FP09197

Zimmermann, U., Wagner, H. J., Heidecker, M., Mimietz, S., Schneider, H., Szimtenings, M., … & König, W. (2002). Implications of mucilage on pressure bomb measurements and water lifting in trees rooting in high-salinity water. *Trees*, *16*, 100–111. https://doi.org/10.1007/s00468-001-0135-5

4 Physiological Traits
Response to Environmental Stress and Resource Utilization

Pragya

4.1 INTRODUCTION

Plant physiological traits constitute a very vast group of characteristics essential for the development and evolution as well as survival of plants under variable environmental conditions. Plants in the environment under extreme stressors such as drought, nutrient deficiency, and extreme temperatures develop certain morphological, anatomical, and physiological characters, which cannot be pushed aside. They are somehow relevant in improving the performance of plants in the field, encountering varied and changing climatic conditions. These include photosynthesis, respiration, nutrient transpiration, and water metabolism processes. All these processes are interconnected with those mechanisms that ensure the maintenance of homeostasis and optimum resource-use efficiency in a plant. For instance, root architecture largely determines how effectively a plant absorbs moisture from the soil. This is especially important in times of drought so that deeper roots can absorb moisture from the lower soil layers. Stomatal regulation, in which the amount of water lost through vaporization and gas exchange at the leaf surface, is controlled. This, in turn, affects the health and development of the plant. Four morphological ways in which plants adapt to environmental stresses are identified as drought escape, dehydration avoidance and desiccation tolerance, or temporal postponement of the events of damage to the plant's life organs. Each type of strategy employs a particular physiological mechanism that enables plants to complete their life cycles before severe stress begins or that allows them to maintain metabolic functions without life activities even during stark stress.

The physiological characteristics of plants represent a wide group of traits, including one of the important traits, i.e., growth traits. Such traits include the acquisition and allocation of resources in planning the health and productivity of the plants, for example, carbon assimilation, water-use efficiency, and nutrient uptake (Ackerly et al., 2000). Some of them are photoperiod and vernalization, which are typical for many plants in latitudes that vary in seasons. Traits such as stomatal conductance and carbohydrate metabolism are also critical in understanding plant response under

 DOI: 10.1201/9781003545231-4

environmental stress, such as during a drought (Mencuccini et al., 2015). The physiological traits may interact and restrain or enhance the ability of a plant to stretch and recover from environmental stress conditions, growth, and yield potential in non-stressed and stressed conditions, respectively. In general, these traits are needed in any plant breeding program directed toward the enhancement of plant performance under a range of environments. Furthermore, since plants interact with their physiological traits and these external conditions, such knowledge is fundamental in elucidating or forecasting plant behaviors under climate change situations. For example, increased concentrations of CO_2 in the atmosphere may lead to changes in water consumption by woods due to the hierarchy of the stomatal conductance, which is a significant determinant of the rate of the transpiration process. In addition to that, the adjustment of the parameters within the models that encompass these particular physiological or growth responses, etc., may make it possible to simulate potential evaporation more effectively in different ecosystems, providing a better understanding of the behavior of water under different climatic conditions within those ecologies. Such complex adaptive mechanisms highlight the necessity of combining physiological information with the surrounding environment in forest ecology predictive models. This is important in creating better conservation management policies directed toward biodiversity protection. Integrating the response of the plant to different environmental factors will facilitate building better structures to predict and address the effects of changing climatic conditions on forests.

Environmental stress can be defined as any situation that interferes with the equilibrium that an organism strives to sustain life itself. It can involve such factors as very high or very low temperatures, inadequate water supply, lack of nutrients, contamination, and pollution. Numerous physiological adaptations have been developed by organisms to deal with various environmental pressures. A plant can return to its initial state once the stress has passed, restoring homeostasis (Fraire-Velázquez et al., 2011). Due to their sessile nature, plants had to contend with various biotic and abiotic environmental variables. Throughout the course of development, plants have evolved extremely complex and effective coping mechanisms to deal with natural environmental stressors. Physiological adaptations, such as Crassulacean acid metabolism, to arid environments are adapted by plants to enhance water conservation. Whereas under abiotic stress, stomatal closure minimizes loss through the reduced opening of stomata, decreases photosynthesis, and generates oxidative stress responses and early vigor and leaf senescence are highly correlated with grain yield during drought stress in agricultural species; therefore, they play a significant role in stress resistance and are critical for adaptation, the rapid pace of environmental change poses challenges, necessitating ongoing research to enhance plant resilience and resource efficiency.

4.2 KEY PHYSIOLOGICAL TRAITS

4.2.1 Photosynthesis

Two essential physiological functions that are present in all plants are photosynthesis and respiration. During photosynthesis, atmospheric CO_2 is converted into carbohydrates and other carbon-bearing compounds through the use of light energy and water. Respiration is the process in which these compounds are broken down to releasable energy, producing carbon precursors for biosynthetic processes in use (Barber, 2009). These processes are connected and are important for natural development and growing conditions for plants.

On the one hand, plants need to conserve water, and on the other hand, they should be able to acquire CO_2. In extreme arid conditions, this balance can be lethal for the plants as the growth and moistening of their tissues is a very important factor (Cowan, 1982). Among these environmental stressors, one directly impacts the photosynthetic system, for instance, disrupting all photosynthesis-related processes, such as thylakoid electron transport, the carbon fixation cycle, and the stomatal restriction of CO_2, as well as causing damage to lipid oxidation, excess carbohydrate storage, and disruption of water homeostasis (Allen & Ort, 2001). Natural selection has over time led to the emergence of a class of plants that are even more resistant to harsh environments because they do not strive for quick growth. Other plants have emerged in less humid conditions with relatively higher rates of photosynthesis; these trade-offs aid in the explanation of the various plant species and their adaptations to the various environmental conditions.

Drought stress has been shown to significantly reduce plant's photosynthetic activity. This results in the reduction of leaf gas exchange and is usually explained in terms of the stomatal and non-stomatal factors (Samarah, 2016). Due to moisture stress, stomata that are extremely vital for the uptake of CO_2 for the process of photosynthesis and transpiration tend to close due to drought stress, thus resulting in less photosynthesis. Such a closure reduces the amount of CO_2, which leads to enhanced photorespiration rather than carbon fixation. Previous studies have established the fact that the stomata tend to close progressively as the intensity of drought stress increases. A notable relationship between the stomatal conductance and leaf water potential is also widely accepted even in drought stresses. Non-stomatal mechanisms that can explain the decline in net photosynthesis in drought stress situations include factor changes in chlorophyll; structural and functional changes in chloroplasts; and obstructions in the processes of accumulation, transport, and distribution of assimilates. Stomatal closure mechanisms also work hand-in-hand with these mechanisms to reduce the efficiency of photosynthesis. Chlorophyll is contained in the membranes of the chloroplasts of plants and is essential for photosynthesis. Its levels determine the photosynthetic activities of plants. However, drought stress increases the chances of low chlorophyll levels due to oxidative stress, photo-oxidative bleaching of pigments, and chlorophyll breakdown. All these pigments contribute significantly to the absorption of light and in the production of reducing power. This decrease in chlorophyll content increases the aggravating effects of drought on photosynthetic capability (Ahmadi & Joudi, 2007).

4.2.2 Plant Water Relations

Plant water relations refer to the mechanisms of water uptake, movement, and loss in plants, which significantly contribute to plant growth and health throughout the entire life cycle. These relations greatly help counter water availability problems in different types of ecosystems. Osmotic gradients and root architecture are the two important processes by which plant roots take up water from the soil (Kirkham, 2023).

Turgor pressure acts as an upward force exercised by the cell contents against the cell wall, which is the most crucial factor controlling the growth and morphogenesis of plant cells. It impacts the rates of cell expansion, water uptake, and growth of the plant cells. The knowledge of plant development and its response to some environmental factors requires an understanding of how turgor pressure relates to growth, and also, turgor and growth are related, such that the root and shoot structures indicate responses to water stress, as Kramer (1988) states. Other critical factors, that define the growth of the plants, are turgor pressure. Mechanisms of opening flowers, dehiscence, and tubes growing pollen require tightly controlled turgor pressure (Beauzamy et al., 2014).

The interaction between vegetation and access to water forms an essential element of plant physiology, affecting growth, distribution patterns, and overall vitality. This review addresses the mechanisms by which plants interact with water within their ecosystem. The availability of water is decided by the atmospheric requirements and the moisture content in soil, and different crop species (Teare & Peet, 1983). Water stress affects growth and photosynthesis but triggers reactive oxygen species (ROS) and antioxidant responses, thereby showing a type of acclimation that may occur over time. ROS are produced mainly by two processes, including respiration and photosynthesis, and are common by-products of regular cellular metabolism. Although ROS is involved in various essential cellular functions, excessive levels of ROS cause oxidative stress that can damage cells, proteins, lipids, and DNA. Another way is that fungal interactions can also influence plant water relations, modify host water dynamics, and lead to disease induction (Ayres & Boddy, 2011).

This is very relevant to plant water status. It is regarded as one of the reliable parameters for plant's response to water stress. The leaf water potential has a close correlation with the growth of the leaves; that is, there exists a specific range of water potentials that guarantee ideal growth. For instance, the sunflower leaves require a minimum turgor equivalent to a water potential of about –3.5 bars for growth. The leaf water potential can be influenced by a number of environmental factors, such as soil moisture and ambient temperature. As an instance, high soil moisture tensions or reduced soil temperatures can cause alterations in leaf water potential (LWP) readings integrating deviations from their expected nominal values. Sinclair and Ludlow (1985) suggested that leaf relative water content (RWC) is a more representative measure of water status than water potential. RWC is the amount of water necessary to saturate a plant to artificial full saturation and its relationship with water potential is dependent on the type and age of the plant (González & González-Vilar, 2001). Canopy temperature can also be linked to water deficiency. According to Ehrler et al. (1978), measuring plants with various degrees of water stress, the canopy temperature of wheat matched its water potential. This research sought to determine the water relations of wheat in the presence of varied drought stress levels. The drought stress negatively impacted the water potential and RWC of the wheat leaves, and these resulted in variations in the photosynthetic rate. Drought stress can also affect leaf and canopy temperature, which may have happened because of the increase in respiration and drop in transpiration because of stomatal closure (Siddique et al., 2000). Water deficit is sufficient to close stomata and reduce photosynthesis and also the reduction of dark respiration. The reduction in dark respiration in leaves is less than the reduction in photosynthesis, but the reduction is similar throughout the plant. Photorespiration is unaffected by short-term water deprivation, but is eventually reduced by water deprivation resulting from consumption of photorespiratory substrates (Troughton & Slatyer, 1969).

The total area of stomatal pores only represents a small portion of the leaf surface, typically less than 3%, from where 98% of all CO_2 is taken up and water lost passes through these pores. They can mediate an evaporation rate that is half that of a wet surface of the same area when they are fully open. Early studies illustrated that photosynthetic rates were correlated with stomatal conductance (gs) when other factors were not limiting (Wong et al., 1979). Stomatal conductance is a critical physiological process in plants that regulates the gas exchange, including H_2O and CO_2, between the plant and its surrounding environment. This process is essential for photosynthesis and transpiration, influencing plant water-use efficiency and overall growth. Low gs slows down the absorption rate by limiting the diffusion of CO_2 into the leaf. During the growing season, this will affect the leaf's carbohydrate status which will have an impact on crop yield. Well-hydrated plant stomata have lower photosynthetic rates in the field by roughly 20% for the majority of C_3 species and less for C_4 plants (Jones, 1987).

Nevertheless, even this limitation has been shown to significantly affect yield. For instance, Fischer et al. (1998) established a strong connection between gs and yield across eight different cultivars of wheat (*Triticum aestivum*). These studies underscored the influence gs can have on crop yield, not only by limiting CO_2 diffusion but also through its effects on water loss and the leaf's evaporative cooling. Improving photosynthesis by just 2%–3% can greatly enhance plant growth and biomass throughout the growing season (Zhu et al., 2007).

Many studies show that smaller stomata respond more rapidly compared to the larger ones. This finding has been explained by surface-to-volume ratios and the need for transportation of solute for proper movement (Drake et al., 2013). Whereas fern's stomata are larger than those of many angiosperm species, they react quickly to variations in the vapor pressure difference, or the difference in the vapor pressure of water inside and outside the leaf. In contrast, angiosperms react to light considerably more slowly than ferns do (McAdam & Brodribb, 2012).

4.2.3 Nutrient Uptake and Assimilation

The processes of nutrient uptake and assimilation are critical for the development and growth of plants. These mechanisms are based on the uptake of the nutrients available in the soil and their utilization by plant tissues. A better comprehension of the mechanisms of nutrient uptake and factors that stimulate or limit nutrient assimilation can be of help in improving practices in the farms as well as the health of the plants. Nutrient application in plants is more active through strategies of root agriculture whereby roots employ the mechanisms of ion transporters to incorporate nutrients with N, P, and K being among the many minerals sourced from the soil in the surroundings. Assortative root absorption mechanisms consist of various physiological processes that enable the plant to colonize soils and obtain ions and organic materials residing in them. Several transport systems are available for various nutrients; for example, for the absorption of nitrogen by roots, several active transporter systems are used, such as transporters for the nitrates, ammonium transporters, and amino acid transporters that are included in the nitrogen cycle in plants (Glass et al., 2002; Kumar et al., 2020a,b, 2021a–e).

Potassium uptake is via high and low-affinity uptake mechanisms, but the high-affinity one is proton transport-based, which makes it possible to have a wide range of K ion accumulation (Maathuis & Sanders, 1996). Root exudates like oxalate have also been implicated in affecting the uptake of heavy metals, including copper, as the latter has been shown to either enhance or reduce metal bioavailability for pH levels in the soil (Seuntjens et al., 2004). Further, the assimilation of radionuclides such as technetium and rhenium indicates that soluble forms of these are being assimilated through mechanisms associated with water absorption (Tagami & Uchida, 2010). The development of roots and the plant's ability to respond to nutrient gradients existing in the soil will determine the efficacy of nutrient uptake; thus, root growth should be coordinated with regions that are nutrient-rich. The absorption of water and nutrients is largely determined by the morphological, physiological, and genetic traits of roots. A nutrient must be absorbed by the root even if it is present in the soil solution. The primary factor influencing this is the water entering the roots. When there are drought conditions, water flows passively to and across the roots quickly, as demonstrated by high transpiration and open stomata (Lyzenga et al., 2023).

The primary part for plants to adjust to drought is thought to be the roots themselves. According to Turner et al. (2001), a root's capacity to absorb water from deeper soils and/or to do so rapidly increases with its size and number of roots. High rates of water uptake improve a plant's capacity to absorb nutrients from the soil. Accordingly, the response of plants to drought stress in the acquisition of water and nutrients depends critically on the size, development, proliferation, density, and distribution of their roots (Kavar et al., 2008). On a sandy loam, Greenwood et al. (1982) developed equations that describe the root distribution of seven vegetable crops. Root density frequently decreases exponentially with soil depth, they confirmed. They also demonstrated that there is a considerable correlation between root length and plant weight.

Once absorbed, nutrients are transported through the vascular system, with water acting as the medium for long-distance movement to shoots. Nutrient transport inside the plant involves the acquisition of important elements such as nitrogen and phosphorus through the roots, by transporters and symbionts. Nitrogen enters the plant mainly in two forms: nitrate and ammonium; assimilation requires nitrate reductase and glutamine synthetase, creating amino acids—hence, essential elements that guarantee growth in a plant. The uptake of phosphorus is significantly enhanced by arbuscular mycorrhizal fungi, allowing the exchange of nutrients between soil and roots of the plant, hence underlining the transport of nitrogen and phosphorus interplay (Rui et al., 2022; Kumar et al., 2015, 2016, 2017).

Nitrogen, phosphorus, and sulfur in the soil primarily exist in organic forms. While it is widely recognized that organic nitrogen compounds in the soil and rhizosphere contribute to the nitrogen supply for plants (Kreuzwieser & Gessler, 2010), the availability of inorganic nitrogen forms, such as nitrate and ammonium, is still considered essential (Lucash et al., 2007). The incorporation of essential nutrients implies complex biochemical responses, specifically protein phosphorylation, which alters the activity of nutrient transporters to become more functional in situations of nutrient scarcity (Hao et al., 2023). In addition, gene expression disparities in nutrient acquisition and assimilation-related genes play an important role in maximizing nutrient utilization efficiency, such as in wheat varieties that have resilience to such stresses (Kiri & Hausa, 2023).

Several factors affect nutrient uptake, including the exposure of roots to ions and the ability of the plant to assimilate the nutrients along with its specific nutritional requirements. The movement of ions within the plant cells takes place through active transport as well as passive transport mechanisms. These processes of transport and, therefore, the amount of nutrients available are dependent on environmental factors such as the soil or water content. These mechanisms are crucial for improving nutrient-use efficiency to achieve maximum yield because they are genetically controlled and vary among species of the plant (Kiri & Hausa, 2023).

To sum up, understanding how nutrients are absorbed and utilized is very significant for the intensification and sustenance of agricultural practices. These systems interact with internal signaling in the plant and varied external environmental factors, which make it complex yet fundamental for the investigation on the advancement of farming techniques and plants (Prado, 2021).

4.3 COPING WITH ENVIRONMENTAL STRESS: MECHANISMS AND ADAPTATIONS

Plants react much more widely to different environmental conditions such as water deficiency, cold or hot temperatures, salinity, and pollution, which strongly affect their growth and productivity. In the face of desiccation and salinity, osmotic changes as well as transcription factors that control the expression of different genes are activated. High temperatures, on the other hand, induce the formation of heat-shock proteins that help maintain the structure of the cells during stressful conditions. Moreover, salinity affects the absorption of water and nutrients and can alter the redox level of the cells, thus growth inhibition occurs (Nawaz et al., 2023). Growth inhibition is also possible due to heavy metal toxicity although plants have certain adaptations such as vacuolar compartmentation and phytochelatin synthesis to minimize such damage. As a concluding note, it is clear how the presence of these stressors together makes it imperative to understand plant responses, and how there is a need for sophisticated omics tools to understand the complex signaling pathways involved in stress tolerance (Morsy & El-Sherif, 2024) (Figure 4.1).

4.3.1 DROUGHT

To survive and even thrive in drought-prone regions, plants have to undergo a series of complex physiological and biochemical changes. It is well known that prolonged water deficit can adversely affect the vegetative growth and production of crops; however, plants have developed complex signaling and regulatory mechanisms enabling them to sense environmental changes and respond appropriately (Bhatt et al., 2020).

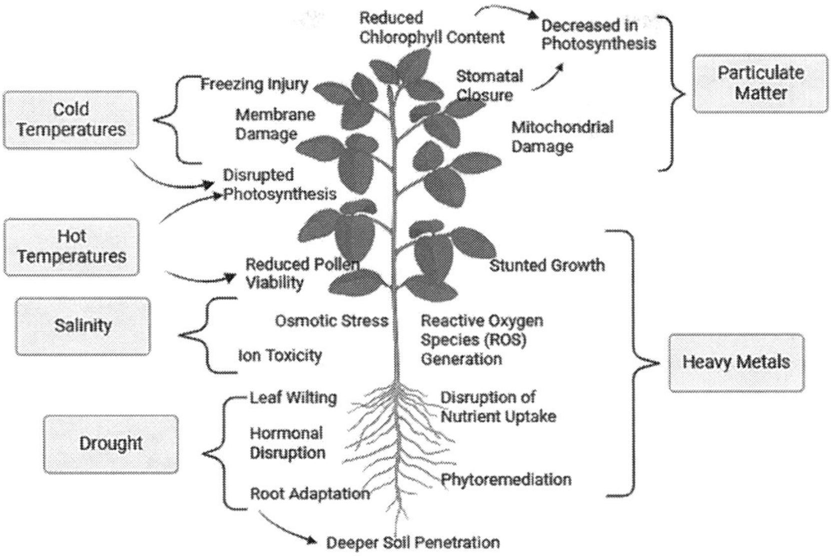

FIGURE 4.1 Key physiological responses of plants to environmental stresses.

Drought resistance refers to the overall ability of plants to respond to and cope with the challenging conditions induced by drought stress. This trait is complex, involving adaptations across various levels, from plant morphology and anatomical features to physiological and biochemical processes (Blum, 2009). Over time, plants have developed a range of drought resistance strategies, enabling them to adjust to specific environments in ways that promote their growth and development.

Water scarcity in soil during drought conditions affects both evaporation and evapotranspiration, which in turn influences the precipitation patterns. This occurs when high temperatures, low relative humidity, and reduced soil moisture prevail. Drought stress arises when water demand exceeds supply, which is further influenced by factors such as soil depth and water availability. Evapotranspiration and evaporation both play a crucial role in determining the water requirements of plants. While evapotranspiration is essential for photosynthesis, much of the water is lost through transpiration without being used for growth (Blum, 2011). Throughout their life cycle, all terrestrial plants experience varying degrees of water stress, prompting the development of adaptive strategies to cope with drought (Hussain et al., 2018). The primary impact of drought on plants is slowed growth, primarily due to the inhibition of shoot development, which limits the plant's metabolic demands. Similarly, drought stress alters root development, encouraging plants to extract water from deeper soil layers. The suppression of lateral root growth is an adaptive response, as it promotes the growth of the primary root, allowing the plant to access more water from deeper soil reserves (Farooq et al., 2009).

Plants continually absorb water (and nutrients) from the soil through their root systems, making roots crucial in responding to water deficit stress. Some plants exhibit a strong ability to enhance root growth early in drought conditions, enabling them to access water from deeper soil layers (Hu & Xiong, 2014). A positive correlation has been observed between root penetration ability and drought resistance in *Phaseolus acutifolius* (Mohamed et al., 2002). Additionally, root length, weight, volume, and density have been linked to drought resistance in various crops (Forster et al., 2005).

The morphological and physiological responses of leaves to drought stress are vital for minimizing water loss and improving water-use efficiency. When plants experience significant water deficiency, their leaves may droop or curl due to a loss of turgor pressure, a phenomenon known as wilting (Kumar et al., 2019, 2021f, 2022; Giri et al., 2023). High transpiration rates can temporarily lead to insufficient water supply, causing some plants to wilt during the day. However, as transpiration decreases, the water deficit is alleviated at night, and the leaves gradually re-expand as they recover.

These adaptations include adjustments to hormone homeostasis, activating mechanisms to prevent photodamage, and increasing the expression of stress-responsive genes. Such changes may include the induction of antioxidant enzymes, alteration of their stomatal aperture to reduce water loss, and accumulation of compatible solutes like proline and betaines to maintain cellular balance. The pattern of development is also changed and redirected toward survival rather than productivity by diverting resources from growth to defense mechanisms in the case of drought stress.

Abscisic acid, an important plant hormone, is involved in many physiological activities, including responses to stress, stomatal closure, and seed dormancy. As such, the role that ABA signaling plays in inducing systemic resistance to pathogens and in the process of drought stress acclimatization can be explained by the concept of system priming. The role of ABA in improving plant defenses against drought is through physiological behavioral modifications such as stomatal closure and root adjustment (Aslam et al., 2022).

Prior evidence suggests that external application of ABA enhances the defense mechanism of *Arabidopsis thaliana* with respect to root-knot nematode *Meloidogyne paranaensis*. Also, induced systemic resistance, which is far broader in the scope of protection against potential pathogens, is based on the integration of ABA with other plant hormones such as jasmonate and ethylene (Panpatte et al., 2020). Hence, ABA signaling is essential in plants not only in respect of defending from pathogens but also in respect of tolerating drought. The mechanisms through which plants tolerate drought stress are multifaceted in nature. For any given species, plants typically employ a combination of mechanisms to achieve drought resistance at various developmental stages. At each stage, plant drought resistance is linked to a series of processes, including stomatal regulation, photosynthesis, osmotic adjustment at the cellular level, and the synthesis of protective macromolecules and antioxidants, all of which occur across morphological, physiological, and molecular dimensions. Furthermore, natural drought stress is both dynamic and unpredictable, which makes it challenging to comprehensively and accurately assess the overall drought resistance of a plant species.

4.3.2 Temperature Extremes

Extreme temperatures have a highly significant impact on the physiology of the plant and bring forth separate heat and cold-stress responses. It is critically important to study these mechanisms to develop more resilient crop varieties when facing the challenge of climate change. Under heat stress, plants are known to produce heat-shock proteins that help in the preservation of cellular functions and the integrity of proteins at high temperatures. Some species will even display adaptive behaviors such as changing the direction of their leaves to not be exposed to direct sunlight and reducing absorption of heat (Teskey et al., 2015).

The pollination phase is particularly sensitive to high temperatures. Maize pollen viability declines when exposed to temperatures exceeding 35°C (Dupuis & Dumas, 1990). This temperature effect is exacerbated by high vapor pressure deficits, as pollen viability before silk reception depends on its moisture content, which is closely influenced by the vapor pressure deficit (Fonseca & Westgate, 2005). When temperatures exceed 30°C, maize kernels suffer damage through disruptions in cell division and amyloplast replication, which reduces the grain sink size and overall yield (Commuri & Jones, 2001). These studies highlight the negative impact of heat stress on crop productivity, especially in critical crops like maize, which are vital for food security.

On the other hand, cold stress conditions trigger different physiological changes in plants such as the production of antifreeze proteins, which stop the formation of ice crystals within plant cells and thus allow plants to survive frozen conditions. The second very important response is dormancy, where plants enter a state of reduced metabolic activities that ensure survival in low-temperature conditions. However, the interaction of these responses is important because high-temperature anomalies lead to physiological stress and thus influence growth and survival (Buckley & Huey, 2016). Besides, both types of heat and cold stresses can cause abnormal functions in plants and the ability to adapt to such conditions plays a crucial role in the resilience of plants in shifting climates. Understanding the plant responses to temperature extremes is important for agricultural science and ecology. Such understanding will feed into improvement strategies in crops and conservation measures concerning global climate change to ensure food security and ecosystem stability against rising temperature variability.

4.3.3 Salinity

Soil salinization is a global issue, affecting approximately 1,100 million hectares (Mha) of land, or about 7% of the Earth's total land area. This phenomenon results from both natural geochemical processes and secondary human activities (Balasubramaniam et al., 2023). Typically, soil salinization occurs due to the accumulation of water-soluble salts, such as sodium (Na^+), potassium (K^+), chloride (Cl^-), and sulfate ($SO4^{2-}$), in the root zone. This accumulation leads to osmotic changes that impair the ability of plant roots to absorb water from the soil (Kamran et al., 2019). The presence of these salt ions induces hyperionic salt stress, which damages plant cells. Among the salts contributing to soil salinity, Na^+ and Cl^- are the primary ions responsible, with excess Na^+ in exchangeable cations also leading to sodicity.

It strongly impacts the productivity of plants and trees by responding and adapting in different ways to salinity stress. The mechanisms are osmoregulation, ionic homeostasis, and antioxidant defense, which must address high salinity levels. For example, osmolytes prevent the destabilization of cellular functions and osmotic stress, and phytohormones play a great role in the signaling and regulation of these adaptive responses in plants. Plants and trees have salinity stress coping mechanisms through modulation of ion homeostasis, ion compartmentalization, export, and osmoprotectant biosynthesis, which increase their tolerance and survival in saline environments (Balasubramaniam et al., 2023).

Salt stress, like other abiotic stresses, also promotes plant growth reduction, and the rate of growth reduction is controlled by several factors. Some factors include species, developmental stage, and salt concentration (Yadav et al., 2019). Stunted growth is an adaptive mechanism for survival, which allows plants to combat salt stress (Munns, 2002). Certain plants "panic" immediately upon salt-stress exposure, react quickly, and stop growing. In contrast, others do not respond adequately and face the danger of dying by continuing to grow under severe salt-stress conditions (Chinnusamy & Zhu, 2003).

When salt-stress conditions are there in plants, their cells immediately shrink and dehydrate, though they may later recover. However, even with recovery, there is a persistent reduction in root and leaf growth, primarily due to disruptions in cell elongation and, to a lesser extent, cell division. After conditions of salinity stress, there are noticeable differences in overall growth and damage between salt-stressed plants and their non-stressed counterparts, particularly in the enlargement of lateral shoots. This response is linked to changes in cell–water relations resulting from osmotic changes outside the root, referred to as the osmotic effect, which decreases the plant's ability to absorb water. Salt stress also interferes with the uptake of key mineral nutrients, especially nitrogen, phosphorus, and potassium, in a complex manner. Nitrogen, a crucial nutrient and constituent of plant cellular components, is particularly affected under saline conditions. Increased chloride (Cl^-) uptake can hinder nitrogen absorption due to the antagonistic interaction between chloride (Cl^-) and nitrate (NO_3^-) ions (Munns, 2002).

In addition to affecting plant growth, salinity stress impairs processes like protein synthesis, energy metabolism, and cell signaling. This increased metabolic cost, required for adaptation to the stress, ultimately limits plant growth and productivity, resulting in a decline in agricultural yield (Munns & Gilliham, 2015). Many species, like *Launaeas armentosa*, display enhanced growth at lower salinities but less biomass at higher salinities exposing a threshold-like interaction in tolerance. More importantly, *Brassica* species have made several different morphological and biochemical adaptations that activate salt-tolerant genes to enhance tolerance to saline conditions (Roy, 2023). Salt tolerance can only be achieved with an understanding of the interaction between these various factors and mechanisms.

4.4 CONCLUSION

Plants are sessile organisms that have an excellent adaptation range to survive in different environmental conditions. Their ability to respond and adapt to environmental stresses such as drought, temperature, salinity, and pollution is essential for their survival and contribution to an ecosystem. A better understanding of such adaptations may lead to various applications in agriculture, horticulture, and other fields of environmental conservation. The study of plant physiology leads to designing strategies for the enhancement of crop production, the creation of resilient landscapes, and the mitigation of climate change impacts. Plants have developed various defense strategies against environmental stresses. These include morphological, physiological, and biochemical changes that allow plants to withstand drought, temperature extremes, salinity, and pollution. In fact, from the production of heat-shock proteins and antifreeze proteins to combat temperature extremes, and complex root systems for drought and salinity tolerance, all these offer a long range of adaptive strategies in plants. Photosynthesis, plant-water relations, nutrient absorption, and stress-coping mechanisms are closely interlinked. The plant has to balance between photosynthesis and efficient intake and transportation of water, nutrient absorption, and cope with various environmental stresses.

For instance, photosynthesis and water loss are interlinked via leaf stomatal aperture regulation, while the uptake of nutrients may be impeded by salinization of soils and dry periods. However, it should be noted that studying plant modifications is important for sustainable agriculture and environmental management. Such knowledge will further empower people to engineer more resilient crop varieties, improve agricultural practices, and restore degraded ecosystems. To sum up, plant adaptation studies represent a highly relevant area of research with important perspectives regarding solutions to the current global problems. Moreover, considering how plants grow under essential factors such as water scarcity, heat, salinity, and high pollution, it shows how remarkably resilient plants are.

REFERENCES

Ackerly, D. D., Dudley, S. A., Sultan, S. E., Schmitt, J., Coleman, J. S., Linder, C. R., Sandquist, D. R., Geber, M. A., Evans, A. S., Dawson, T. E., & Lechowicz, M. J. (2000). The evolution of plant ecophysiological traits: Recent advances and future directions: New research addresses natural selection, genetic constraints, and the adaptive evolution of plant ecophysiological traits. *Bioscience*, *50*(11), 979–995.

Ahmadi, A., & Joudi, M. (2007). Effects of timing and defoliation intensity on growth, yield and gas exchange rate of wheat grown under well-watered and drought conditions. *Pakistan Journal of Biological Sciences (PJBS)*, *10*(21), 3794–3800.

Allen, D. J., & Ort, D. R. (2001). Impacts of chilling temperatures on photosynthesis in warm-climate plants. *Trends in Plant Science*, *6*(1), 36–42.

Aslam, M. M., Rashid, M. A. R., Siddiqui, M. A., Khan, M. T., Farhat, F., Yasmeen, S., Khan, I. A., Raja, S., Rasool, F., Sial, M. A., & Yan, Z. (2022). Recent insights into signaling responses to cope drought stress in rice. *Rice Science*, *29*(2), 105–117.

Ayres, P. G., & Boddy, L. (Eds.). (2011). *Water, Fungi and Plants: Symposium of the British Mycological Society Held at the University of Lancaster, April 1985*. Cambridge University Press.

Balasubramaniam, T., Shen, G., Esmaeili, N., & Zhang, H. (2023). Plants' response mechanisms to salinity stress. *Plants*, *12*(12), 2253.

Barber, J. (2009). Photosynthetic energy conversion: Natural and artificial. *Chemical Society Reviews*, *38*(1), 185–196.

Beauzamy, L., Nakayama, N., & Boudaoud, A. (2014). Flowers under pressure: Ins and outs of turgor regulation in development. *Annals of Botany*, *114*(7), 1517–1533.

Bhatt, D., Nath, M., Sharma, M., Bhatt, M. D., Bisht, D. S., & Butani, N. V. (2020). Role of growth regulators and phytohormones in overcoming environmental stress. In:Roychoudhury, A., & Tripathi D.K.(eds.) *Protective Chemical Agents in the Amelioration of Plant Abiotic Stress: Biochemical and Molecular Perspectives* (pp. 254–279), Chichester: John Wiley & Sons Ltd.

Blum, A. (2009). Effective use of water (EUW) and not water-use efficiency (WUE) is the target of crop yield improvement under drought stress. *Field Crops Research*, *112*(2–3), 119–123.

Blum, A. (2011). Drought resistance–is it really a complex trait? *Functional Plant Biology*, *38*(10), 753–757.

Buckley, L. B., & Huey, R. B. (2016). Temperature extremes: Geographic patterns, recent changes, and implications for organismal vulnerabilities. *Global Change Biology*, *22*(12), 3829–3842.

Chinnusamy, V., & Zhu, J. K. (2003). Plant salt tolerance. In Hirt, H., &Shinozaki, K. (eds.) *Plant Responses to Abiotic Stress* (pp. 241–270). Berlin, Heidelberg: Springer Berlin Heidelberg.

Commuri, P. D., & Jones, R. J. (2001). High temperatures during endosperm cell division in maize: A genotypic comparison under in vitro and field conditions. *Crop Science*, *41*(4), 1122–1130.

Cowan, I. R. (1982). Regulation of water use in relation to carbon gain in higher plants. In Lange, O.L., Nobel, P.S., Osmond, C.B., & Ziegler, H. (eds.) *Physiological Plant Ecology II: Water Relations and Carbon Assimilation* (pp. 589–613). Berlin, Heidelberg: Springer Berlin Heidelberg.

de Mello Prado, R. (2021). *Mineral Nutrition of Tropical Plants* (Vol. 1). Cham, Switzerland: Springer.

Drake, P. L., Froend, R. H., & Franks, P. J. (2013). Smaller, faster stomata: Scaling of stomatal size, rate of response, and stomatal conductance. *Journal of Experimental Botany*, *64*(2), 495–505.

Dupuis, I., & Dumas, C. (1990). Influence of temperature stress on in vitro fertilization and heat shock protein synthesis in maize (*Zea mays* L.) reproductive tissues. *Plant Physiology*, *94*(2), 665–670.

Ehrler, W. L., Idso, S. B., Jackson, R. D., & Reginato, R. J. (1978). Diurnal changes in plant water potential and canopy temperature of wheat as affected by drought 1. *Agronomy Journal*, *70*(6), 999–1004.

Farooq, M., Aziz, T., Wahid, A., Lee, D. J., & Siddique, K. H. (2009). Chilling tolerance in maize: Agronomic and physiological approaches. *Crop and Pasture Science*, *60*(6), 501–516.

Fischer, R. A., Rees, D., Sayre, K. D., Lu, Z. M., Condon, A. G., & Saavedra, A. L. (1998). Wheat yield progress associated with higher stomatal conductance and photosynthetic rate, and cooler canopies. *Crop Science*, *38*(6), 1467–1475.

Fonseca, A. E., & Westgate, M. E. (2005). Relationship between desiccation and viability of maize pollen. *Field Crops Research*, *94*(2–3), 114–125.

Forster, B. P., Thomas, W. T. B., & Chloupek, O. (2005). Genetic controls of barley root systems and their associations with plant performance. In: Foulkes, J., Russell, G., Hawkesford, M., Gooding, M., Sparkes, D., & Stockdale, E. (eds.) *Journal article Conference paper, UK, 0265-1491, (No.73), Aspects of Applied Biology*. Scottish Crop Research Institute, Invergowrie, Dundee DD2 5DA, UK.

Fraire-Velázquez, S., Rodríguez-Guerra, R., & Sánchez-Calderón, L. (2011). Abiotic and biotic stress response crosstalk in plants. In Shanker, A.K., & Venkateswarlu, B. (eds.) *Abiotic Stress Response in Plants—Physiological, Biochemical and Genetic Perspectives* (pp. 3–26), Croatia: InTech.

Giri, K., Mishra, G., Suyal, D. C., Kumar, N., Doley, B., Das, N., Baruah, R. C., Bhattacharyya, R., Bora, N. (2023). Performance evaluation of native plant growth-promoting rhizobacteria for paddy yield enhancement in the jhum fields of Mokokchung, Nagaland, North East India. *Heliyon*. 9, e14588.

Glass, A. D., Britto, D. T., Kaiser, B. N., Kinghorn, J. R., Kronzucker, H. J., Kumar, A., Okamoto, M., Rawat, S., Siddiqi, M. Y., Unkles, S. E., & Vidmar, J. J. (2002). The regulation of nitrate and ammonium transport systems in plants. *Journal of Experimental Botany*, *53*(370), 855–864.

González, L., & González-Vilar, M. (2001). Determination of relative water content. In Reigosa Roger, M.J. (ed.) *Handbook of Plant Ecophysiology Techniques* (pp. 207–212). Dordrecht: Springer Netherlands.

Greenwood, D. J., Gerwitz, A., Stone, D. A., & Barnes, A. (1982). Root development of vegetable crops. *Plant and Soil*, *68*, 75–96.

Hao, X., Wang, X., Chen, C., Liu, R., Yin, Y., Yao, J., Xiao, Z., Liu, X., Shen, X., & Liu, X. (2023). Synthetic bacterial communities reshape microbial communities and enhance nutrient supply in desertified land of Northwest China. *Applied Soil Ecology*, *189*, 104972.

Hu, H., & Xiong, L. (2014). Genetic engineering and breeding of drought-resistant crops. *Annual Review of Plant Biology*, *65*(1), 715–741.

Hussain, M., Ahmad, S., Hussain, S., Lal, R., Ul-Allah, S., & Nawaz, A. (2018). Rice in saline soils: Physiology, biochemistry, genetics, and management. *Advances in Agronomy*, *148*, 231–287.

Jones, H. G. (1987). Breeding for stomatal characters. In Zeiger, E., Farquhar, G.D., &Cowan, I.R. (eds.) *Stomatal Function* (pp. 431–443), Kent: CABI Databases.

Kamran, M., Parveen, A., Ahmar, S., Malik, Z., Hussain, S., Chattha, M. S., Saleem, M. H., Adil, M., Heidari, P., & Chen, J. T. (2019). An overview of hazardous impacts of soil salinity in crops, tolerance mechanisms, and amelioration through selenium supplementation. *International Journal of Molecular Sciences*, *21*(1), 148.

Kavar, T., Maras, M., Kidrič, M., Šuštar-Vozlič, J., & Meglič, V. (2008). Identification of genes involved in the response of leaves of *Phaseolus vulgaris* to drought stress. *Molecular Breeding*, *21*, 159–172.

Kiri, I., & Hausa, K. (2023). Mechanisms of nutrient uptake and assimilation processes in some plants: A review. *Dutse Journal of Pure and Applied Sciences*, *9*, 223–237.

Kirkham, M. B. (2023). *Principles of Soil and Plant Water Relations*. London: Elsevier.

Kramer, P. J. (1988). Changing concepts regarding plant water relations. *Plant,* Cell & Environment, *11*(7), 565.

Kreuzwieser, J., & Gessler, A. (2010). Global climate change and tree nutrition: Influence of water availability. *Tree Physiology*, *30*(9), 1221–1234.

Kumar, A., Dwivedi, G. K., Tewari, S., Paul, J., Anand, R., Kumar, N., Kumar, P., Singh, H., & Kaushal, R. (2020a). Carbon mineralization and inorganic nitrogen pools under *Terminalia chebula* Retz.-Based agroforestry system in Himalayan foothills, India. *Forest Science*, *66*(5), 634–643.

Kumar, A., Dwivedi, G. K., Tewari, S., Paul, J., Sah, V. K., Singh, H., Kumar, P., Kumar, N., & Kaushal, R. (2020b) Soil organic carbon pools under *Terminalia chebula* Retz. based agroforestry system in Himalayan foothills, India", *Current Science*, *118*(7), 1098–1103.

Kumar, A., Kumar, P., Singh, H., Bisht, S., & Kumar, N. (2021a). Relationship of physiological plant functional traits with soil carbon stock in temperate forest of Garhwal Himalaya. *Current Science*, *120*(8), 1368–1373.

Kumar, A., Kumar, P., Singh, H., & Kumar, N. (2021b). Modulation of plant functional traits under essential plant nutrients during seasonal regime in natural forests of Garhwal Himalayas. *Plant and Soil*, *465*, 197–212.

Kumar, A., Kumar, P., Singh, H., & Kumar, N. (2021d). Impact of plant functional traits on near saturated hydraulic conductivity of soil under different forests of Kempty watershed in Garhwal Himalaya. *Indian Journal of Soil Conservation*. *49*(1), 38–44.

Kumar, A., Kumar, P., Singh, H., & Kumar, N. (2021e). Adaptation and mitigation potential of roadside trees with bio-extraction of heavy metals under vehicular emissions and their impact on physiological traits during seasonal regimes. *Urban Forestry & Urban Greening*, *58*, 126900.

Kumar, A., Singh, H., Kumari, G., Bisht, S., Malik, A., Kumar, N., Singh, M., Raturi, A., Barthwal, S., Thakur, A., & Kaushal, R. (2022). Adaptive resilience of roadside trees to vehicular emissions via leaf enzymatic, physiological, and anatomical trait modulations. *Environmental Pollution*, *313*, 120191.

Kumar, A., Tewari, S., Singh, H., Kumar, P., Kumar, N., Bisht, S., Kushwaha, S., Tamta, N., & Kaushal, R. (2021c). Biomass accumulation and carbon stocks in different agroforestry system prevalent in Himalayan foothills, India. *Current Science*, *120*(6), 1083–1088.

Kumar, N., Jeena, N., & Singh, H. (2019). Elevated temperature modulates rice pollen structure: A study from foothill Himalayan agro-ecosystem in India. *3Biotech*, *9*, 175.

Kumar, N., Jeena, N., Kumar, A., Khairakpam, R., & Singh, H. (2021f). Comparative response of rice cultivars to elevated air temperature in Bhabar region of Indian Himalaya: Status on yield attributes. *Heliyon*, *7*, e07474.

Kumar, N., Kumar, N., Shukla, A., Shankhdhar, S. C., & Shankhdhar, D. (2015). Impact of terminal heat stress on pollen viability and yield attributes of rice (*Oryza sativa* L.). *Cereal Research Communications*, *43*(4), 616–626.

Kumar, N., Shankhdhar, S. C., & Shankhdhar, D. (2016). Impact of elevated temperature on antioxidant activity and membrane stability in different genotypes of rice (*Oryza sativa* L.). *Indian Journal of Plant Physiology (Now Plant Physiology Report)*. *21*(1), 37–43.

Kumar, N., Suyal, D. C., Sharma, I. P., Verma, A., & Singh, H. (2017). Elucidating stress proteins in rice (*Oryza sativa* L.) genotype under elevated temperature: A proteomic approach to understand heat stress response. *3 Biotech*, *7*, 205.

Lucash, M. S., Eissenstat, D. M., Joslin, J. D., McFarlane, K. J., & Yanai, R. D.(2007). Estimating nutrient uptake by mature tree roots under field conditions: challenges and opportunities. *Trees*, *21*, 593–603.

Lyzenga, W. J., Liu, Z., Olukayode, T., Zhao, Y., Kochian, L. V., & Ham, B. K. (2023). Getting to the roots of N, P, and K uptake. *Journal of Experimental Botany*, *74*(6), 1784–1805.

Maathuis, F. J., & Sanders, D. (1996). Mechanisms of potassium absorption by higher plant roots. *PhysiologiaPlantarum*, *96*(1), 158–168.

McAdam, S. A., & Brodribb, T. J. (2012). Stomatal innovation and the rise of seed plants. *Ecology Letters*, *15*(1), 1–8.

Mencuccini, M., Minunno, F., Salmon, Y., Martínez-Vilalta, J., & Hölttä, T. (2015). Coordination of physiological traits involved in drought-induced mortality of woody plants. *New Phytologist*, *208*(2), 396–409.

Mohamed, M. F., Keutgen, N., Tawfika, A. A., & Noga, G. (2002). Dehydration-avoidance responses of tepary bean lines differing in drought resistance. *Journal of Plant Physiology*, *159*(1), 31–38.

Morsy, A., & El-Sherif, N. (2024). Integrating omics approaches for abiotic stress tolerance in plants. In Hasanuzzaman, M., & Nahar, K. (eds.) *Abiotic Stress in Crop Plants-Ecophysiological Responses and Molecular Approaches*. London: Intech Open.

Munns, R. (2002). Comparative physiology of salt and water stress. *Plant, Cell & Environment*, *25*(2), 239–250.

Munns, R., & Gilliham, M. (2015). Salinity tolerance of crops–what is the cost? *New Phytologist*, *208*(3), 668–673.

Nawaz, M., Sun, J., Shabbir, S., Khattak, W. A., Ren, G., Nie, X., Bo, Y., Javed, Q., Du, D., & Sonne, C. (2023). A review of plants strategies to resist biotic and abiotic environmental stressors. *Science of the Total Environment*, *900*, 165832.

Panpatte, D. G., Jhala, Y. K., & Vyas, R. V. (2020). Signaling pathway of induced systemic resistance. In Sharma, V., Salwan, R., & Tawfeeq Al-Ani, L.K. (eds.) *Molecular Aspects of Plant Beneficial Microbes in Agriculture* (pp. 133–141). Cambridge, MA: Academic Press.

Rui, M., Jing, Y., Jiang, H., & Wang, Y. (2022). Quantitative system modeling bridges the gap between macro-and microscopic stomatal model. *Advanced Biology*, *6*(10), 2200131.

Roy, S. (2023).Fundamental response and salinity tolerance in Brassicaceae Plants.*ChemRxiv*, doi:10.26434/chemrxiv-2023-0xhnx

Samarah, N. H. (2016). Understanding how plants respond to drought stress at the molecular and whole plant levels. In Hossain, M., Wani, S., Bhattacharjee, S., Burritt, D., & Tran, L.S. (eds.) *Drought Stress Tolerance in Plants, Vol 2: Molecular and Genetic Perspectives* (pp. 1–37), Cham: Springer.

Seuntjens, P., Nowack, B., & Schulin, R. (2004). Root-zone modeling of heavy metal uptake and leaching in the presence of organic ligands. *Plant and Soil*, *265*, 61–73.

Siddique, M. R. B., Hamid, A. I. M. S., & Islam, M. S. (2000). Drought stress effects on water relations of wheat. *Botanical Bulletin of Academia Sinica*, *41*, 35–39.

Sinclair, T. R., & Ludlow, M. M. (1985). Who taught plants thermodynamics? The unfulfilled potential of plant water potential. *Functional Plant Biology*, *12*(3), 213–217.

Tagami, K., & Uchida, S. (2010). Can elemental composition data of crop leaves be used to estimate radionuclide transfer to tree leaves? *Radiation and Environmental Biophysics*, *49*, 583–590.

Teare, I. D., & Peet, M. M. (Eds.). (1983). *Crop-Water Relations* (547 pp).

Teskey, R., Wertin, T., Bauweraerts, I., Ameye, M., McGuire, M. A., & Steppe, K. (2015). Responses of tree species to heat waves and extreme heat events. *Plant, cell & environment*, *38*(9), 1699–1712.

Troughton, J. H., & Slatyer, R. O. (1969). Plant water status, leaf temperature, and the calculated mesophyll resistance to carbon dioxide of cotton leaves. *Australian Journal of Biological Sciences*, *22*(4), 815–828.

Turner, N. C., Wright, G. C., & Siddique, K. H. M. (2001). Adaptation of grain legumes (pulses) to water-limited environments, *Advances in Agronomy*, *71*, 193–231.

Wong, S. C., Cowan, I. R., & Farquhar, G. D. (1979). Stomatal conductance correlates with photosynthetic capacity. *Nature*, *282*(5737), 424–426.

Yadav, S. P., Bharadwaj, R., Nayak, H., Mahto, R., Singh, R. K., & Prasad, S. K. (2019). Impact of salt stress on growth, productivity and physicochemical properties of plants: A Review. *International Journal of Chemical Studies*, *7*(2), 1793–1798.

Zhu, X. G., De Sturler, E., & Long, S. P. (2007). Optimizing the distribution of resources between enzymes of carbon metabolism can dramatically increase photosynthetic rate: A numerical simulation using an evolutionary algorithm. *Plant Physiology*, *145*(2), 513–526.

5 Genetic Basis of Functional Traits Variation in Forest Tree Species

Jharna Maiti and Amit Joshi

5.1 INTRODUCTION

Forest ecosystems are vital for the storage of carbon, preservation of biodiversity worldwide, and regulation of the hydrological and climatic cycles (Figure 5.1a). A forest tree species' functional characteristic, such as growth rates, wood density, drought tolerance, disease timing of reproduction resistance, and timing of reproduction, have a significant role in determining its ecological fitness, productivity, and resistance to environmental changes (Figure 5.1b) (Zhang et al., 2024b). These traits influence not just the structure of forests but also the relationships between species, ecological processes, and the performance of individual trees. These functional characteristics vary as a result of complex genetic systems interacting with environmental

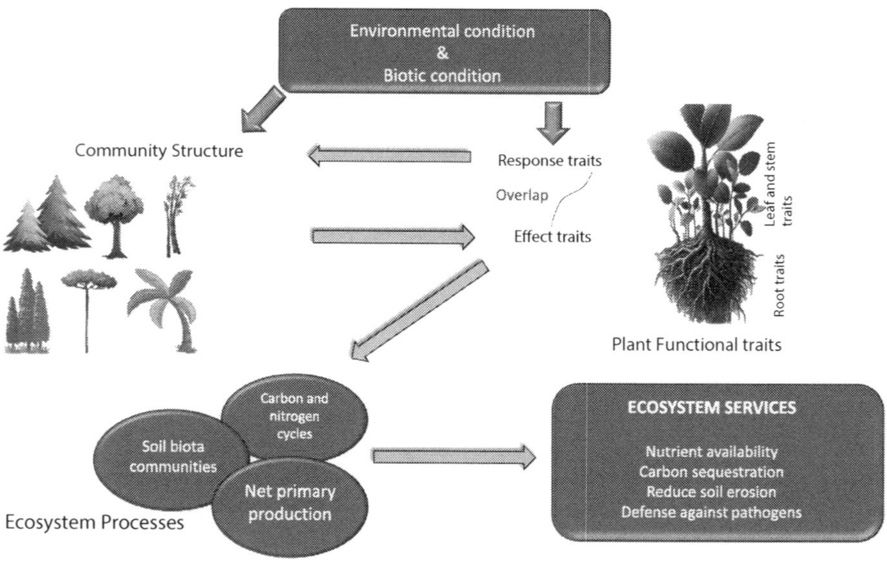

FIGURE 5.1 Plant traits: (a) Ecosystem processes, (b) Functional traits and their ecosystem properties.

(Continued)

DOI: 10.1201/9781003545231-5

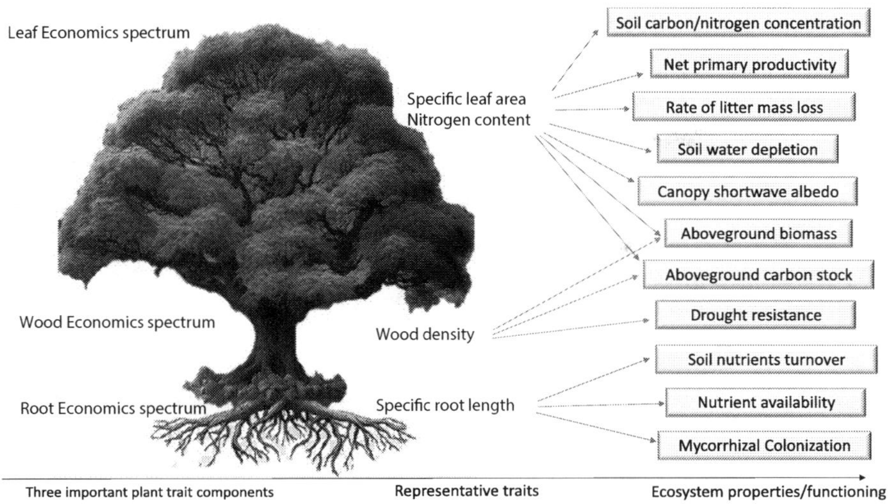

FIGURE 5.1 (*Continued*) Plant traits: (a) Ecosystem processes, (b) Functional traits and their ecosystem properties.

factors. Genetic diversity is the cornerstone of adaptability, enabling trees to cope with changing environmental conditions such as pests, changing weather patterns, and habitat loss (Bouchard et al., 2024). Understanding the genetic basis of this trait variation is essential to predicting how forest species will react to challenges in the future and to developing effective management and conservation strategies. The fields of genomics and molecular biology have advanced significantly, changing our knowledge of the genetic composition of functional traits in trees (Climent et al., 2024). Using methods such as transcriptomics, genome-wide association studies (GWAS), and quantitative trait loci (QTLs) mapping, researchers can identify the genes and regulatory networks responsible for trait variation. These findings are crucial for generating more resilient tree populations, providing direction for breeding programs, and overseeing conservation initiatives meant to maintain genetic variety and forest resilience (Sharma et al., 2024).

This chapter examines the genetic foundations of functional feature variation in forest tree species, the evolutionary mechanisms that determine it, and the consequences of this variation for forest management in a rapidly changing global environment. By combining genetics with ecological and evolutionary perspectives, we may be able to better understand how trees adapt to their environments and ensure the sustainability of forest ecosystems for future generations (Aitken et al., 2024).

5.2 FUNCTIONAL TRAITS IN FOREST TREES

Functional traits are fundamental attributes of organisms that influence their growth, survival, reproduction, and role in ecosystems. Functional traits in forest tree species are important for understanding how trees interact with their environment, adapt to stressors, and contribute to ecosystem services including carbon sequestration and biodiversity support (Lindh et al., 2024). These traits can be broadly classified into four categories: morphological, physiological, reproductive, and phenological traits (Table 5.1). Each category of traits provides insight into the ways trees adapt to their surroundings, ensuring survival and reproductive success under varying environmental conditions (Ye et al., 2024).

5.3 GENETIC VARIATION IN FOREST TREE SPECIES

Due to their long lifespans and high population densities, forest trees can preserve a significant amount of genetic diversity. Among the major causes of genetic variation in trees are the following: The majority of tree species are outcrossers, which means they have systems in place to promote cross-pollination, increasing genetic diversity. Natural selection means that environmental pressures like drought, pests, and competition lead to the selective survival of trees with beneficial traits (Scotti-Saintagne et al., 2024). Geographic distribution means that trees distributed over large geographic areas exhibit local adaptation to environmental gradients, resulting in diverse gene pools within species. For tree species to be able to adapt to various environmental circumstances and stressors, there must be genetic variety.

5.3.1 Quantitative Genetics of Functional Traits

In forest trees, both genetic and environmental influences can affect functional qualities, which are usually governed by many genes (polygenic inheritance). Because these characteristics can exhibit wide individual variation in expression, they are regarded as quantitative features. Key ideas in comprehending how these qualities are regulated by genetics include: Heritability is a measure of how much of a trait's variance may be attributed to individual genetic variations. In forest tree species, for instance, growth rate, wood density, and drought tolerance frequently show different degrees of heritability. These features are known as QTLs, which are particular areas of the genome linked to the expression of quantitative traits. Researchers have discovered genetic areas connected to important functional characteristics in trees, like disease resistance and growth performance, through QTL mapping (Lagattuta et al., 2024). It is possible to forecast how populations will react to selecting pressures or breeding efforts by having an understanding of the heritability and QTL associated with attributes.

5.3.2 Adaptive Traits and Local Adaptation

Forest trees frequently exhibit local adaptation, in which populations acquire particular features appropriate for their particular surroundings. These adaptations are frequently seen in characteristics like: drought tolerance is the result of the identification of genes in drought-tolerant species that control stomatal conductance, root development, and water-use efficiency. Tolerance for cold means genes regulating bud dormancy, leaf senescence, and cold acclimation have been discovered to affect cold tolerance in frost-exposed species (Love & Ferris, 2024). To forecast how trees will react to climate change, it is essential to investigate genetic variation in adaptive

TABLE 5.1
Classification of Functional Traits

Category	Trait	Description	Examples & References
Morphological	Leaf size and shape	Influences photosynthesis, water regulation, and temperature control. Trees in dry environments have smaller, thicker leaves, while trees in shaded/moist areas have larger, broader leaves.	*Quercus* (oak) species differ in leaf size based on habitat (Akram et al., 2023).
	Wood density	Determines structural support, resilience to environmental stress, and carbon storage. Higher density increases resistance to wind, pests, and diseases.	*Eucalyptus* species have high wood density (Hart et al., 2023).
	Root architecture	Influences water and nutrient acquisition and anchorage. Deep roots help in arid environments; shallow roots thrive where nutrients are concentrated near the surface.	*Pinus* (pines) species have deep roots, allowing access to water in dry conditions (Lu et al., 2024).
Physiological	Tree height and crown shape	Affects light competition and interactions with other species. Tall trees with broad crowns dominate canopies, while short trees with compact crowns may thrive in the understory.	*Fagus sylvatica* (European beech) has a tall height and broad crown, dominating canopy layers (Larocque et al., 2024).
	Photosynthetic efficiency	Refers to the rate at which trees convert light into chemical energy. Species with higher rates grow faster in environments with abundant sunlight and water.	*Populus* (poplars) exhibit high photosynthetic efficiency in favorable environments (Kumar et al., 2024).
	Water-use efficiency (WUE)	The amount of carbon gained per unit of water lost. Trees with high WUE are better suited for arid conditions.	*Pinus ponderosa* (ponderosa pine) shows high WUE and thrives in dry, mountainous regions (Altieri et al., 2024).
	Stomatal conductance	Regulates gas exchange and water loss. Trees that adjust stomatal aperture can optimize gas exchange under varying environmental conditions.	*Eucalyptus* species can rapidly adjust stomatal conductance (Rahman et al., 2024).
	Nutrient uptake and use	Efficient nutrient acquisition is essential in nutrient-poor soils. Some species form associations with fungi to enhance nutrient uptake.	*Pinus sylvestris* (Scots pine) forms mycorrhizal associations for improved phosphorus uptake (Pena & Tibbett, 2024).

(Continued)

TABLE 5.1 (*Continued*)
Classification of Functional Traits

Category	Trait	Description	Examples & References
Reproductive	Flowering phenology	Timing of flowering affects pollinator attraction and resource competition. Early flowering can provide a competitive advantage.	*Acer rubrum* (red maple) flowers early, taking advantage of early-season resources (Maple et al., 2024).
	Seed size and number	Trade-off between seed size and number. Large seeds enhance seedling survival but are produced in smaller numbers, while small seeds are more numerous and can disperse widely.	*Quercus robur* (pedunculate oak) produces fewer large seeds, while *Betula pendula* (silver birch) produces many small seeds (Wang et al., 2024).
	Dispersal mechanisms	Seeds can be dispersed by wind (anemochory), animals (zoochory), or other means. Dispersal mechanisms influence a tree's ability to colonize new areas.	*Pinus sylvestris* uses wind dispersal, while *Quercus* species rely on animals for seed dispersal (Navarro-Ramos et al., 2024).
Phenological	Bud burst and leaf senescence	Timing of bud burst and leaf fall is important for optimizing growth in response to seasonal changes. Early bud burst can extend the growing season but risks frost damage.	*Fagus sylvatica* (European beech) exhibits early bud burst to maximize growing season (Meger et al., 2024).
	Dormancy and cold hardiness	Trees enter dormancy to survive winter. Cold hardiness allows trees to withstand freezing temperatures, especially in high latitudes or altitudes.	*Picea abies* (Norway spruce) enters dormancy during winter for protection from freezing (Stuke et al., 2024).
	Response to climate variability	Trees with phenotypic plasticity can adjust their phenology in response to climate changes, increasing their likelihood of survival.	*Betula pendula* (silver birch) adjusts bud burst timing in response to warming temperatures (Sullivan et al., 2024).

features. While local adaptation guarantees tree populations' survival and growth in particular environmental niches, it also raises questions about how quickly these populations will be able to adjust to rapidly changing conditions.

5.4 GENETIC BASIS OF ENVIRONMENTAL STRESS TOLERANCE

The ecosystems of forests are seriously threatened by environmental pressures such as frost, drought, and diseases. For the purpose of managing forests, it is essential to determine the genetic foundation of tree stress tolerance. Important discoveries in this field include: genes that control root development and water-use efficiency, which are essential for a tree's survival in dry environments, have been found through genetic studies associated with drought tolerance. Another one is frost tolerance, which refers to a tree's ability to withstand winter frost in temperate and boreal climates. In species like *Pinus* and *Populus*, genes linked to early bud burst, dormancy control, and cold acclimation have been found (Shah et al., 2024). Finally, resistance to disease means another important characteristic is pathogen resistance, which has been linked to genetic loci in a number of tree species that are resistant to diseases like canker and rust.

5.5 ENVIRONMENTAL INFLUENCE ON GENETIC EXPRESSION

Functional characteristics in forest trees, such as growth rate, stress tolerance, and timing of reproduction, are influenced by both genetic and environmental factors. The way these features appear is controlled by interactions between environmental influences and the underlying genetic structure (Li et al., 2024; Kumar et al., 2019, 2021f; Giri et al., 2023). Understanding how trees adapt to different conditions, especially in the face of climate change, requires an understanding of how the environment influences DNA expression.

5.5.1 Gene–Environment Interactions (GxE)

The phenomenon known as gene–environment interactions, or GxE, is the result of both environmental and genetic factors influencing a trait's expression. Because forest trees have a long lifespan and are subject to changing environmental factors like temperature, GxE is especially important in these plants. Tree growth may be accelerated in warmer climates, although this effect may differ based on the genetic composition of individual trees or populations (Blumstein, 2024). The growth rate and phenology the timing of life cycle events like leaf flush and flowering may differ depending on whether the genotype is more sensitive to temperature rises than another (Table 5.2). Additionally, water accessibility: Even while trees in drought-prone areas may have genetically adapted to use water more efficiently, drought-related environmental stress can still have a significant impact on how these qualities manifest (Mattioli et al., 2024).

Even within the same species, these genetic–environment interactions can lead to notable phenotypic diversity among individuals in various settings.

TABLE 5.2
Example of Some Gene Response to Temperature

Gene	Plant Example	Function	Response to Temperature	Impact on Growth/Phenology
FT (flowering locus T)	*Arabidopsis thaliana*	Regulates flowering time based on environmental cues	Warmer temperatures induce FT expression, promoting earlier flowering	Controls reproductive timing, adjusts flowering under warmer conditions (Susila et al., 2024)
CO (constans)	Rice (*Oryza sativa*)	Regulates photoperiodic flowering, influenced by temperature	Certain alleles promote earlier flowering in response to higher temperatures	Alters flowering time to optimize reproduction in warmer climates (Qin et al., 2024)
PIF4 (phytochrome-interacting factor 4)	*Arabidopsis thaliana*	Regulates growth (cell elongation) in response to temperature	Activated by increased temperatures, leading to thermomorphogenesis (elongation, leaf flush)	Promotes growth and earlier leaf flush under warmer temperatures (Ebrahimi et al., 2024)
SVP (short vegetative phase)	Apple (*Malus domestica*)	Regulates bud break and flowering timing	Temperature-sensitive, warmer conditions cause early bud flush and flowering	Earlier flowering can lead to frost damage, affecting yield (Kumar et al., 2024)
TPS1 (trehalose-6-phosphate synthase 1)	Maize (*Zea mays*)	Regulates sugar metabolism and flowering time	Higher temperatures accelerate sugar signaling, leading to earlier flowering	Adjusts flowering time under heat stress, impacts reproductive success (Rojas et al., 2023)

5.5.2 Phenotypic Plasticity in Forest Trees

The ability of an organism to modify its phenotype, or observable qualities, in response to environmental changes without causing a change in the underlying genetic code is known as phenotypic plasticity. Forest trees have to adapt to a wide range of environmental situations, which means they frequently show high degrees of plasticity. For instance, the shape of leaves and the architecture of roots can vary significantly in response to environmental conditions. Trees may grow larger, thinner leaves in leaf morphology situations with limited light availability in order to maximize light capture, while smaller, thicker leaves may be developed in brighter locations in order to decrease water loss and trees can modify their root systems to grow deeper in times of drought, allowing them to collect water that is not readily available at the surface (Kibria et al., 2024). Although these root adjustments are a direct reaction to environmental stimuli, their adaptability is made possible by underlying genetic potential. Forest trees' phenotypic plasticity enhances their adaptability and buffers the effects of environmental stressors, allowing them to live and prosper in various settings (Kalra et al., 2024).

5.5.3 Epigenetics and Environmental Stress

Drought, high temperatures, and nutrient shortages are examples of environmental stresses that can cause epigenetic modifications, which change gene expression without changing the DNA sequence (Abdulraheem et al., 2024). The synthesis and activity of small RNAs (sRNAs), such as microRNAs (miRNAs) and small-interfering RNAs (siRNAs), which are essential for controlling gene expression at the post-transcriptional level, can be influenced by epigenetic processes including DNA methylation and histone modification (Figure 5.2). These epigenetic changes give forest trees a malleable response to environmental stressors, enabling the trees to gradually modify their growth, development, and stress tolerance. The process of adding methyl groups to DNA, usually at cytosine bases, is known as DNA methylation. This process has the

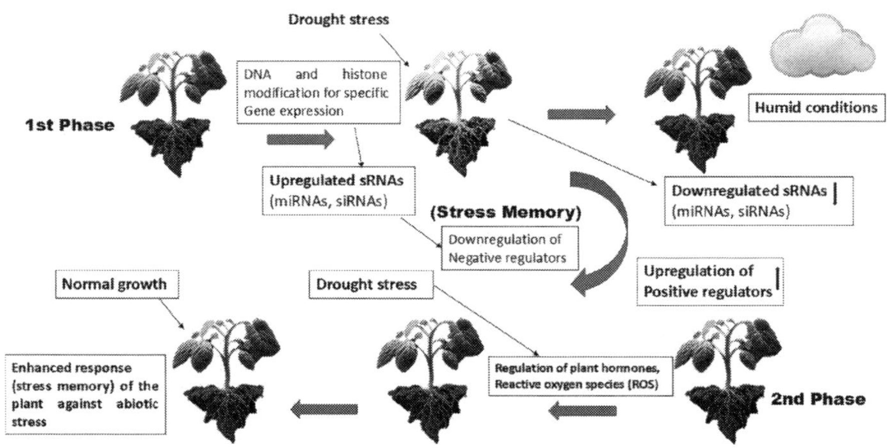

FIGURE 5.2 Epigenetic changes.

ability to either activate or suppress the expression of particular genes. The amounts of sRNAs can be increased or decreased by this mechanism, which affects how important they are for regulating genes. Methylation in the promoter regions of genes expressing sRNAs, like miRNAs, can lower their production, therefore degrading sRNAs (Peck & Sork, 2024). Transcript factors cannot access the DNA to start sRNA creation when DNA is hypermethylated in these areas, which lowers the amounts of sRNA. Consequently, these sRNAs have an impaired ability to modulate their target mRNAs, thereby affecting post-transcriptional gene silence. Additionally, in the process of upgrading sRNAs, DNA methylation occasionally causes siRNA synthesis. For instance, methylation of transposable elements or repetitive regions in the genome might result in the production of siRNAs, which, by silencing these potentially dangerous elements, aid in the maintenance of genome integrity. In a feedback loop, these siRNAs have the ability to strengthen DNA methylation, which amplifies the silencing effect. Histone modifications, which entail adding or removing chemical groups (such as acetyl, methyl, or phosphate groups) from the histone proteins that surround DNA, are another type of epigenetic mechanism (Miryeganeh & Armitage, 2024). These modifications may affect DNA accessibility for transcription, which may control sRNA expression.

Histone modifications, such as histone deacetylation or methylation (H3K9me3, H3K27me3), can downgrade sRNAs by forming heterochromatin, a densely packed chromatin structure that blocks the transcription machinery from accessing sRNA-coding genes translational suppression. As a result, sRNA production is reduced, their regulatory function on target genes is weakened, and sRNAs are upgraded. However, certain chromatin alterations, such as histone acetylation (H3K9ac) and histone methylation (H3K4me3), encourage euchromatin, a more open chromatin structure that makes sRNA-coding genes simpler to access. This may result in more sRNAs being produced, which would improve their capacity to control gene expression by directing mRNAs toward translational repression or decay. Trees can quickly adjust to changes in their environment because of epigenetic modifications, which can either be transient or last for extended periods of time (Zhang et al., 2024a). Transgenerational adaptation may be possible in certain situations when these alterations are even inherited by subsequent generations.

5.5.4 Stress-Induced Genetic Expression

In forest trees, the expression of certain genes linked to growth, survival, and reproduction is highly influenced by environmental stress. Important environmental factors influencing genetic expression are as follows:

- **Drought**: Genes that regulate stomatal closure, root development, and water uptake are activated during drought circumstances (Blanca-Reyes et al., 2024). These genes are more robustly expressed in drought-tolerant species, such as *Quercus* and *Pinus*, when water is scarce, which enhances the trees' capacity to withstand protracted dry spells.
- **Temperature Extremes**: The expression of genes involved in temperature regulation, such as those governing frost resistance and bud dormancy, can

be impacted by cold or heat stress. Genes that control antifreeze proteins and the timing of bud burst may be increased in cold areas to shield tissues from freezing damage (Manjarrez et al., 2024).

- **Pathogen Pressure**: Temperature and humidity are two environmental variables that might affect how quickly infections spread. Trees may produce chemicals or physical barriers to prevent infections in response, increasing the expression of genes linked to immune response (Santos et al., 2024).

Trees can adapt to their habitats and survive in harsh situations because of this stress-induced gene expression.

5.5.5 Local Adaptation and Environmental Influence

Local adaptation occurs when populations evolve traits that are specifically suited to their local environmental conditions. Forest trees, which often span large geographic ranges, show significant variation in traits like growth rate, leaf phenology, and drought resistance due to local adaptation (Capblancq et al., 2023).

- **Drought-Prone Regions**: Trees with deeper roots or more effective stomatal control may exhibit greater genetic expression of water-use efficiency-related features in regions where water shortages occur frequently (Gazol et al., 2023).
- **Cold Climates**: Taller latitudes and higher altitudes cause trees to frequently adapt to shorter growing seasons and colder climates. Compared to their counterparts in warmer climates, genes linked to bud burst, dormancy, and frost tolerance are more robustly expressed in these groups.
- **Soil Nutrient Availability**: Because of certain genetic variations that react to nitrogen stress, trees in nutrient-poor soils may display adaptations like greater root surface area or more effective nutrient uptake systems (Leites et al., 2023). The local environment's long-term selection forces, which affect genetic expression patterns and improve survival in particular habitats, are what lead to these adaptations.

5.6 GENETIC TOOLS AND TECHNOLOGIES FOR STUDYING FOREST TREE SPECIES

5.6.1 Quantitative Trait Loci (QTL) Mapping

QTL mapping is one of the most common strategies for linking genetic variation to functional aspects. QTL mapping identifies specific genetic regions associated with the inheritance of complex traits. Chromosome regions containing genes that influence a trait can be found by researchers by observing the co-segregation of genetic markers and phenotypic traits in controlled crosses. QTL mapping has been widely applied to forest trees to study growth, wood quality, and stress tolerance. For instance, QTL mapping of *Populus* (poplar) (Zhu et al., 2023) has identified genetic areas that influence traits such as leaf morphology and wood development.

QTL mapping is still a useful technique for locating large-effect loci influencing critical attributes in populations of forest trees.

5.6.2 Genome-Wide Association Studies (GWAS)

Understanding the genetic foundation of natural diversity in forest trees has become possible with the use of GWAS, an increasingly potent technology. In contrast to QTL mapping, GWAS searches the entire genome for genetic markers connected to specific traits, which are typically single-nucleotide polymorphisms (SNPs). Using this method has shown promising results in discovering several genes and genomic areas linked to complex traits in various environmental settings. GWAS has been used to identify genetic loci linked to characteristics including frost tolerance, water-use efficiency, and disease susceptibility in various forest tree species, including *Pinus sylvestris* (Scots pine) and *Picea abies* (Norway spruce) (Ashwath et al., 2023). GWAS sheds light on the underlying genetic architecture of adaptive characteristics in trees by capturing genetic variation in natural populations.

5.6.3 Next-Generation Sequencing (NGS) Technologies

The rapid and affordable sequencing of entire genomes, transcriptomes, or specific sections of interest has been made possible by NGS, which has completely changed genetic research on forest trees. For tree species with large and complex genomes, NGS has made it possible to gather complete genomic data, providing previously unheard-of resolution to examine genetic diversity and functional trait variation. Reference genomes have been assembled for a number of forest tree species, including *Eucalyptus grandis* and *Quercus Robur* (oak), using whole-genome sequencing. This makes it easier to identify the genes and regulatory components linked to characteristics like wood quality and drought tolerance (Malakar et al., 2024). Furthermore, gene expression patterns in trees are extensively studied using transcriptome sequencing (RNA-seq), which aids in the understanding of how changes in gene expression in response to environmental stressors such as heat or drought occur and how these changes relate to functional attributes.

5.6.4 Single Nucleotide Polymorphism (SNP) Genotyping

One of the most popular techniques for evaluating genetic diversity and locating markers associated with features in forest trees is SNP genotyping. Since SNPs are the most prevalent type of genetic variation found in genomes, they are helpful markers for research on trait associations and evaluations of genetic diversity. Functional characteristics can be finely mapped thanks to high-throughput SNP genotyping systems, which efficiently screen thousands of SNPs across populations (Ousmael et al., 2023). SNP genotyping has been utilized in species such as *Picea sitchensis* (Sitka spruce) and *Pinus taeda* (loblolly pine) to find markers associated with growth rate, disease resistance, and wood qualities. Forest managers can choose individuals with desired genetic profiles for breeding programs with the help of this technology.

5.6.5 Genomic Selection (GS)

GS is a breeding strategy that predicts an individual's breeding values for complex traits by utilizing genetic information throughout the entire genome. In contrast to conventional marker-assisted selection (MAS), which focuses on specific loci, genome-wide selection (GS) captures the combined effects of multiple small-effect genetic markers throughout the entire genome, allowing for more exact performance predictions. GS is being used more and more in forest trees to accelerate breeding cycles, especially for qualities like disease resistance or wood quality that are challenging to quantify. Tree breeders can expedite the improvement of commercially significant features by using genomic data to select individuals (Sharma et al., 2024; Kumar et al., 2020a,b, 2021a–e). GS has been effectively applied to improve growth rate, wood characteristics, and stress tolerance in species such as *Pinus radiata* and *Eucalyptus*.

5.6.6 CRISPR-Cas9 and Genome Editing

The use of CRISPR-Cas9 genome editing technology has created new avenues for studying the functions of forest tree genes and enhancing their characteristics through direct gene manipulation. Through targeted gene editing, CRISPR-Cas9 allows scientists to validate the contribution of individual genes to variance in characteristics and potentially enhance qualities like disease resistance, wood density, and drought resilience (Thapliyal et al., 2023). CRISPR-Cas9 has been used in experiments to knock out genes involved in lignin formation in trees such as *Populus* and *Pinus*, changing the characteristics of the wood. Genome editing may prove to be a critical tool in the future for functionally verifying genes found in forest trees and developing superior tree varieties to address specific environmental concerns or commercial requirements.

5.6.7 Genotyping-by-Sequencing (GBS)

Generating genome-wide SNP data by high-throughput, low-cost GBS is particularly useful for species with big and complex genomes. By focusing on specific genomic regions, GBS reduces genome complexity, making it especially useful for studying genetic diversity and population dynamics in wild tree populations. GBS has been successfully employed to identify genetic markers associated with functional trait variation and local adaptation in species such as *Pseudotsuga menziesii* (Douglas fir) and *Pinus sylvestris* (Scots pine) (Haupt et al., 2024). In addition to providing useful information for understanding how genetic diversity affects qualities like development, stress tolerance, and nutrition usage efficiency, GBS data can also be used to guide conservation efforts that aim to maintain genetic diversity.

5.6.8 Molecular Marker-Assisted Selection (MAS)

In forestry breeding projects, molecular MAS is still a commonly utilized technique. Finding genetic markers that are strongly associated with desired phenotypes is essential to MAS. Even before the features are phenotypically expressed, these markers can be utilized to identify and select individuals that contain advantageous genes. Pitch canker resistance has been selected for using MAS in species like *Pinus radiata*, or Monterey pine. Forest managers can choose trees that possess advantageous genetic variations, enhancing the resilience, productivity, and overall health of forest ecosystems, by incorporating MAS into breeding programs (Pavlenko & Pikunova, 2024). There are still many unanswered questions despite tremendous advancements in our knowledge of the genetic basis of functional features in forest tree species. Unlike many other model organisms, forest trees have enormous genomes, lengthy life cycles, and complex environments, which provide special challenges. Although genetic techniques and technology have opened up new study areas, they are not without restrictions of their own (Watanabe et al., 2024). In the face of climate change and other environmental stresses, better tree breeding, forest management, and conservation activities require an understanding of these issues and the identification of future paths.

5.7 KEY CHALLENGES IN FOREST TREE GENETICS

5.7.1 Complex Genomes and Lack of Reference Genomes

One of the main challenges in investigating the genetics of forest trees is their large and complex genomes. Because of their high levels of polyploidy and repetitive sequences, the genomes of many forest tree species are far larger than those of usual model organisms like rice or *Arabidopsis thaliana*. The genome of *Pinus taeda*, or loblolly pine, is about seven times bigger than the human genome (Gardner et al., 2023). Research may move more slowly as a result of the time and computational resources needed to sequence and assemble such huge genomes.

5.7.2 Long Life Cycles and Delayed Phenotypic Expression

Being long-lived creatures, many species of forest trees require decades to attain reproductive maturity. Because of its lengthy life cycle, phenotypic features of importance, such as disease resistance or wood quality, may not manifest until the tree reaches a particular age, which presents difficulties for genetic research and breeding initiatives (Iler et al., 2023). It is challenging to quickly and precisely correlate genetic markers with phenotypes due to the delayed expression of characteristics. The fact that environmental factors like soil quality and climate influence trait expression across a tree's lifetime complicates the situation. It becomes difficult to distinguish between hereditary and environmental influences as a result, particularly for features like growth rate or drought tolerance that are influenced by environmental factors.

5.7.3 Environmental Variation and Phenotypic Plasticity

The phenotypic features of forest trees are significantly shaped by environmental variation. As a result, it is difficult to distinguish between phenotypic plasticity, or an organism's ability to change its phenotype in response to environmental changes, and genetic variation. The underlying genetic signals can be obscured by phenotypic plasticity, which can lead to incorrect conclusions regarding the genetic basis of trait variation (Kormann et al., 2024). It is challenging to account for every environmental element impacting trait expression in forest ecosystems since the conditions there differ over large geographic areas.

5.7.4 Limited Resources for Genomic Research

Many forest tree species remain poorly understood, despite advances in sequencing technologies, due to a lack of funding, infrastructure, and access to cutting-edge genetic procedures. A small number of commercially valuable species, such as *Eucalyptus, Pinus*, and *Populus*, are the subject of the majority of forest tree genetic studies (Fugeray-Scarbel et al., 2023). We do not fully comprehend the genetic diversity and adaption processes seen in various forest ecosystems due to this imbalance in research focus.

5.8 FUTURE DIRECTIONS IN FOREST TREE GENETICS

5.8.1 Advancements in Sequencing Technologies

The cost of sequencing big genomes is predicted to drop as sequencing technology advances, increasing the viability of sequencing the genomes of non-model tree species. The advancement of long-read sequencing technologies, such as those provided by Oxford Nanopore and PacBio, holds great promise for enhancing the precision of genome assemblies (Espinosa et al., 2024), particularly for species whose sequences are extremely repetitive. Better sequencing technology will also make it easier to build high-quality reference genomes for a larger variety of species of forest trees. These reference genomes will be used to investigate the molecular underpinnings of functional aspects in various forest habitats, as well as genetic diversity and local adaptability.

5.8.2 Integrating Genomics with Environmental Data

To understand how forest trees respond to environmental changes, genetic data must be combined with ecological and environmental data. Advances in remote sensing, climate modeling, and environmental monitoring have made it possible to link genetic variation to specific environmental variables such as temperature, precipitation, and soil qualities (Papin et al., 2024; Kumar et al., 2015, 2016, 2017; Singh et al., 2020). Our knowledge of phenotypic plasticity and local adaptability will increase as a result of this integrative approach, making it possible to forecast how forest trees may react to climate change with more accuracy.

5.8.3 Genomic Selection for Accelerated Breeding

Using genome-wide marker data, GS predicts an individual's genetic potential, offering a viable way to expedite breeding projects. The efficiency of choosing trees with desirable qualities like quicker growth, better wood quality, or more resistance to environmental stress will rise as GS is more commonly used. In conjunction with MAS, the use of GS in forest tree species can aid in the development of new tree kinds that are more suited to upcoming environmental circumstances (McLean et al., 2023).

5.9 CONCLUSION

Understanding the genetic foundation of functional characteristic variation in forest tree species is essential to evaluating their capacity to endure, adapt, and prosper in a changing environment. The field of genomics has advanced our understanding of how genetic variables control traits like growth, stress tolerance, and disease resistance through the use of genomic methods like transcriptome analysis, QTL mapping, and GWAS. Research reveals that intraspecific variation is driven by genetic diversity, gene flow, natural selection, and environmental factors. We can better understand forest trees' responses to environmental changes by fusing genetic studies with ecological and evolutionary viewpoints, which will help with sustainable forest management. As the pressures on ecosystems increase due to climate change, it becomes more important than ever to safeguard genetically robust populations so that forests can continue to provide vital ecosystem services and preserve biodiversity.

REFERENCES

Abdulraheem, M. I., Xiong, Y., Moshood, A. Y., Cadenas-Pliego, G., Zhang, H., & Hu, J. (2024). Mechanisms of plant epigenetic regulation in response to plant stress: Recent discoveries and implications. *Plants*, *13*(2), 163.

Aitken, S. N., Jordan, R., & Tumas, H. R. (2024). Conserving evolutionary potential: Combining landscape genomics with established methods to inform plant conservation. *Annual Review of Plant Biology*, *75*(1), 707–736.

Akram, M. A., Wang, X., Shrestha, N., Zhang, Y., Sun, Y., Yao, S., Li, J., Hou, Q., Hu, W., Ran, J., & Deng, J. (2023). Variations and driving factors of leaf functional traits in the dominant desert plant species along an environmental gradient in the drylands of China. *Science of the Total Environment*, *897*, 165394.

Altieri, S., Niccoli, F., Kabala, J. P., Liyaqat, I., & Battipaglia, G. (2024). Influence of drought and minimum temperature on tree growth and water use efficiency of Mediterranean species. *Dendrochronologia*, *83*, 126162.

Ashwath, M. N., Lavale, S. A., Santhoshkumar, A. V., Mohapatra, S. R., Bhardwaj, A., Dash, U., Shiran, K., Samantara, K., & Wani, S. H. (2023). Genome-wide association studies: An intuitive solution for SNP identification and gene mapping in trees. *Functional & Integrative Genomics*, *23*(4), 297.

Blanca-Reyes, I., Lechuga, V., Llebrés, M. T., Carreira, J. A., Ávila, C., Cánovas, F. M., & Castro-Rodríguez, V. (2024). Under stress: Searching for genes involved in the response of *Abies pinsapo* Boiss to climate change. *International Journal of Molecular Sciences*, *25*(9), 4820.

Blumstein, M. (2024). The drivers of intraspecific trait variation and their implications for future tree productivity and survival. *American Journal of Botany*, *111*(4), e16312.

Bouchard, E., Searle, E. B., Drapeau, P., Liang, J., Gamarra, J. G., Abegg, M., Zambrano, A. A., Alvarez-Davila, E., Alves, L. F., & Paquette, A. (2024). Global patterns and environmental drivers of forest functional composition. *Global Ecology and Biogeography*, *33*(2), 303–324.

Capblancq, T., Lachmuth, S., Fitzpatrick, M. C., & Keller, S. R. (2023). From common gardens to candidate genes: Exploring local adaptation to climate in red spruce. *New Phytologist, 237*(5), 1590–1605.

Climent, J., Alía, R., Karkkainen, K., Bastien, C., Benito-Garzon, M., Bouffier, L., De Dato, G., Delzon, S., Dowkiw, A., Elvira-Recuenco, M., & Cavers, S. (2024). Trade-offs and trait integration in tree phenotypes: Consequences for the sustainable use of genetic resources. *Current Forestry Reports, 10*(3), 196–222.

Ebrahimi Naghani, S., Šmeringai, J., Pleskačová, B., Dobisová, T., Panzarová, K., Pernisová, M., & Robert, H. S. (2024). Integrative phenotyping analyses reveal the relevance of the phyB-PIF4 pathway in Arabidopsis thaliana reproductive organs at high ambient temperature. *BMC Plant Biology, 24*(1), 721.

Espinosa, E., Bautista, R., Larrosa, R., & Plata, O. (2024). Advancements in long-read genome sequencing technologies and algorithms. *Genomics, 116*(3), 110842.

Fugeray-Scarbel, A., Irz, X., & Lemarié, S. (2023). Innovation in forest tree genetics: A comparative economic analysis in the European context. *Forest Policy and Economics, 155*, 103030.

Gardner, S. T., Bertucci, E. M., Sutton, R., Horcher, A., Aubrey, D., & Parrott, B. B. (2023). Development of DNA methylation-based epigenetic age predictors in loblolly pine (*Pinus taeda*). *Molecular Ecology Resources, 23*(1), 131–144.

Gazol, A., Fajardo, A., & Camarero, J. J. (2023). Contributions of intraspecific variation to drought tolerance in trees. *Current Forestry Reports, 9*(6), 461–472.

Giri, K., Mishra, G., Suyal, D. C., Kumar, N., Doley, B., Das, N., Baruah, R. C., Bhattacharyya, R., Bora, N. (2023). Performance evaluation of native plant growth-promoting rhizobacteria for paddy yield enhancement in the jhum fields of Mokokchung, Nagaland, North East India. *Heliyon, 9*, e14588

Hart, S. C., Porter, T. M., Basiliko, N., Venier, L., Hajibabaei, M., & Morris, D. (2023). Fungal community dynamics in coarse woody debris across decay stage, tree species, and stand development stage in northern boreal forests. *Canadian Journal of Forest Research, 54*(1), 12–30.

Haupt, S., Bernhardt, N., Killing, S., Meucci, S., Herzschuh, U., Zakharov, E. S., Harpke, D., Pestryakova, L. A., & Kruse, S. (2024). Biogeography of larches in eastern Siberia–using single nucleotide polymorphisms derived by genotyping by sequencing. *Ecography, 2024*(7), e07092.

Iler, A. M., CaraDonna, P. J., Richardson, L. K., Wu, E. T., Fant, J. B., Pfeiler, K. C., Freymiller, G. A., Godfrey, K. N., Gorman, A. J., Wilson, N., & Jules, E. S. (2023). Genotype accounts for intraspecific variation in the timing and duration of multiple, sequential life-cycle events in a willow species. *American Journal of Botany, 110*(2), e16112.

Kalra, A., Goel, S., & Elias, A. A. (2024). Understanding role of roots in plant response to drought: Way forward to climate-resilient crops. *The Plant Genome, 17*(1), e20395.

Kibria, M. G., Tjoelker, M. G., Marchin, R. M., Arndt, S. K., & Rymer, P. D. (2024). Can species climate niche predict canopy growth, functional traits and phenotypic plasticity in urban trees? *Urban Forestry & Urban Greening, 98*, 128417.

Kormann, J. M., van der Maaten-Theunissen, M., Unterholzner, L., Liesebach, M., Liepe, K. J., & van der Maaten, E. (2024). Variation in vessel traits of northern red oak (*Quercus rubra* L.) provenances revealed high phenotypic plasticity to prevailing environmental conditions. *Trees*, 38(5), 1283–1295.

Kumar, A., Dwivedi, G. K., Tewari, S., Paul, J., Anand, R., Kumar, N., Kumar, P., Singh, H., & Kaushal, R. (2020a). Carbon mineralization and inorganic nitrogen pools under *Terminalia chebula* Retz.-Based agroforestry system in Himalayan foothills, India. *Forest Science, 66*(5), 634–643.

Kumar, A., Dwivedi, G. K., Tewari, S., Paul, J., Sah, V. K., Singh, H., Kumar, P., Kumar, N., & Kaushal, R. (2020b). Soil organic carbon pools under *Terminalia chebula* Retz. based agroforestry system in Himalayan foothills, India", *Current Science, 118*(7), 1098–1103.

Kumar, A., Kumar, P., Singh, H., Bisht, S., & Kumar, N. (2021a). Relationship of physiological plant functional traits with soil carbon stock in temperate forest of Garhwal Himalaya. *Current Science*, *120*(8), 1368–1373

Kumar, A., Kumar, P., Singh, H., & Kumar, N. (2021b). Modulation of plant functional traits under essential plant nutrients during seasonal regime in natural forests of Garhwal Himalayas. *Plant and Soil*, *465*, 197–212.

Kumar, A., Kumar, P., Singh, H., & Kumar, N. (2021d). Impact of plant functional traits on near saturated hydraulic conductivity of soil under different forests of Kempty watershed in Garhwal Himalaya. *Indian Journal of Soil Conservation*, *49*(1), 38–44.

Kumar, A., Kumar, P., Singh, H., & Kumar, N. (2021e). Adaptation and mitigation potential of roadside trees with bio-extraction of heavy metals under vehicular emissions and their impact on physiological traits during seasonal regimes. *Urban Forestry & Urban Greening*, *58*, 126900.

Kumar, A., Mushtaq, M., Kumar, P., Sharma, D. P., & Gahlaut, V. (2024). Insights into flowering mechanisms in apple (Malus× domestica Borkh.) amidst climate change: An exploration of genetic and epigenetic factors. *Biochimica et Biophysica Acta (BBA)-General Subjects*, *1868*(5), 130593.

Kumar, A., Tewari, S., Singh, H., Kumar, P., Kumar, N., Bisht, S., Kushwaha, S., Tamta, N., & Kaushal, R. (2021c). Biomass accumulation and carbon stocks in different agroforestry system prevalent in Himalayan foothills, India. *Current Science*. *120*(6), 1083–1088.

Kumar, N., Jeena, N., & Singh, H. (2019). Elevated temperature modulates rice pollen structure: A study from foothill Himalayan agro-ecosystem in India. *3 Biotech*, 9, 175.

Kumar, N., Jeena, N., Kumar, A., Khairakpam, R., & Singh, H. (2021f). Comparative response of rice cultivars to elevated air temperature in Bhabar region of Indian Himalaya: Status on yield attributes. *Heliyon*, 7, e07474.

Kumar, N., Kumar, N., Shukla, A., Shankhdhar, S. C., & Shankhdhar, D. (2015). Impact of terminal heat stress on pollen viability and yield attributes of rice (*Oryza sativa* L.). *Cereal Research Communications*, *43*(4), 616–626.

Kumar, N., Shankhdhar, S. C., & Shankhdhar, D. (2016). Impact of elevated temperature on antioxidant activity and membrane stability in different genotypes of rice (*Oryza sativa* L.). *Indian Journal of Plant Physiology (Now Plant Physiology Report)*. *21*(1), 37–43.

Kumar, N., Suyal, D. C., Sharma, I. P., Verma, A., & Singh, H. (2017). Elucidating stress proteins in rice (*Oryza sativa* L.) genotype under elevated temperature: A proteomic approach to understand heat stress response. *3 Biotech*, *7*, 205.

Kumar, S., Wang, S., Wang, M., Zeb, S., Khan, M. N., Chen, Y., Zhu, G., & Zhu, Z. (2024). Enhancement of sweetpotato tolerance to chromium stress through melatonin and glutathione: Insights into photosynthetic efficiency, oxidative defense, and growth parameters. *Plant Physiology and Biochemistry*, *208*, 108509.

Lagattuta, K. A., Park, H. L., Rumker, L., Ishigaki, K., Nathan, A., & Raychaudhuri, S. (2024). The genetic basis of autoimmunity seen through the lens of T cell functional traits. *Nature communications*, *15*(1), 1204.

Larocque, G. R., Shugart, H. H., Xi, W., & Holm, J. A. (2024). Forest succession models. In: Larocque, G. R. (ed.) *Ecological Forest Management Handbook* (pp. 166–208). CRC Press, Boca Raton.

Leites, L., & Benito Garzón, M. (2023). Forest tree species adaptation to climate across biomes: Building on the legacy of ecological genetics to anticipate responses to climate change. *Global Change Biology*, *29*(17), 4711–4730.

Li, Y., Yuan, Y., Hu, Z., Liu, S., & Zhang, X. (2024). Genetic transformation of forest trees and its research advances in stress tolerance. *Forests*, *15*(3), 441.

Lindh, A., Sundqvist, M. K., Axelsson, E. P., Hasselquist, N. J., Aguilar, F. X., Alloysius, D., & Ilstedt, U. (2024). Functional traits to predict financial value of enrichment planting in degraded tropical forests. *New Forests*, 55(5), 1283–1310.

Love, J. M., & Ferris, K. G. (2024). Local adaptation to an altitudinal gradient: The interplay between mean phenotypic trait variation and phenotypic plasticity in *Mimulus laciniatus. Perspectives in Plant Ecology, Evolution and Systematics*, *63*, 125795.

Lu, J. Z., Bluhm, C., Foltran, E., Pérez, C. A. R., Ammer, C., Caruso, T., Glatthorn, J., Lamersdorf, N., Polle, A., Sandmann, D., &, Scheu, S. (2024). Functional traits in soil-living oribatid mites unveil trophic reorganization in belowground communities by introduced tree species. *Geoderma*, *448*, 116947.

Malakar, A., Chandra, G., & Barthwal, S. (2024). Tree genome databases: A new era in the development of cyber-infrastructures for forest trees. In *Genomics Data Analysis for Crop Improvement* (pp. 285–308). Singapore: Springer Nature Singapore.

Manjarrez, L. F., Guevara, M. Á., de María, N., Vélez, M. D., Cobo-Simón, I., López-Hinojosa, M., Cabezas, J. A., Mancha, J. A., Pizarro, A., Díaz-Sala, M. C. & Cervera, M. T. (2024). Maritime pine rootstock genotype modulates gene expression associated with stress tolerance in grafted stems. *Plants*, *13*(12), 1644.

Maple, R., Zhu, P., Hepworth, J., Wang, J. W., & Dean, C. (2024). Flowering time: From physiology, through genetics to mechanism. *Plant Physiology*, *195*(1), 190–212.

Mattioli, W., Ferrara, C., Colonico, M., Gentile, C., Lombardo, E., Presutti Saba, E., & Portoghesi, L. (2024). Assessing forest accessibility for the multifunctional management of protected areas in Central Italy. *Journal of Environmental Planning and Management*, *67*(1), 197–216.

McLean, D., Apiolaza, L., Paget, M., & Klápště, J. (2023). Simulating deployment of genetic gain in a radiata pine breeding program with genomic selection. *Tree Genetics & Genomes*, *19*(4), 33.

Meger, J., Ulaszewski, B., Chmura, D. J., & Burczyk, J. (2024). Signatures of local adaptation to current and future climate in phenology-related genes in natural populations of *Quercus robur*. *BMC Genomics*, *25*(1), 78.

Miryeganeh, M., & Armitage, D. W. (2024). Epigenetic responses of trees to environmental stress in the context of climate change. *Biological Reviews*, *100*(1), 131–148.

Navarro-Ramos, M. J., van Leeuwen, C. H., Olsson, C., Elmberg, J., Månsson, J., Martín-Vélez, V., Lovas-Kiss, Á., & Green, A. J. (2024). Seed dispersal between aquatic and agricultural habitats by greylag geese. *Agriculture, Ecosystems & Environment*, *359*, 108741.

Ousmael, K., Whetten, R. W., Xu, J., Nielsen, U. B., Lamour, K., & Hansen, O. K. (2023). Identification and high-throughput genotyping of single nucleotide polymorphism markers in a non-model conifer (*Abies nordmanniana* (Steven) Spach). *Scientific Reports*, *13*(1), 22488.

Papin, V., Bosc, A., Sanchez, L., & Bouffier, L. (2024). Integrating environmental gradients into breeding: Application of genomic reactions norms in a perennial species. *Heredity*, *133*(3), 160–172.

Pavlenko, A. A., & Pikunova, A. V. (2024). Mapping of genetic factors that determine economically important traits of pear (Pyrus) and marker-assisted selection techniques. *Russian Journal of Genetics*, *60*(5), 551–571.

Peck, L. D., & Sork, V. L. (2024). Can DNA methylation shape climate response in trees? *Trends in Plant Science, 29*(10), 1089–1102.

Pena, R., & Tibbett, M. (2024). Mycorrhizal symbiosis and the nitrogen nutrition of forest trees. *Applied Microbiology and Biotechnology*, *108*(1), 461.

Qin, Z., Stubbings, W. A., Chen, M., Li, F., Wu, F., & Wang, S. (2024). Co-exposure with copper alters the uptake, accumulation, subcellular distribution, and biotransformation of organophosphate triesters in rice (*Oryza sativa* L.). *Journal of Agricultural and Food Chemistry, 72*(35), 19312–19322.

Rahman, M. A., Arndt, S., Bravo, F., Cheung, P. K., van Doorn, N., Franceschi, E., del Río, M., Livesley, S. J., Moser-Reischl, A., Pattnaik, N., & Torquato, P. R. (2024). More than a canopy cover metric: Influence of canopy quality, water-use strategies and site climate on urban forest cooling potential. *Landscape and Urban Planning*, *248*, 105089.

Rojas, B. E., Tonetti, T., & Figueroa, C. M. (2023). Trehalose 6-phosphate metabolism in C4 species. *Current Opinion in Plant Biology*, *72*, 102347.

Santos, B., Pinto, G., Berenguer, H., Correia, B., Amaral, J., Gómez-Cadenas, A., & Alves, A. (2024). Physiological and hormonal responses of drought-stressed *Eucalyptus* seedlings infected with *Neofusicoccum kwambonambiense*. *Plant Pathology*, *73*(4), 810–823.

Scotti-Saintagne, C., de Sousa Rodrigues, A., Roig, A., & Fady, B. (2024). A comprehensive strategy for the conservation of forest tree genetic diversity: An example with the protected *Pinus nigra* subsp. *salzmannii* (Dunal) Franco in France. *Conservation Genetics*, *25*(2), 469–480.

Shah, I. H., Manzoor, M. A., Jinhui, W., Li, X., Hameed, M. K., Rehaman, A., Li, P., Zhang, Y., Niu, Q., & Chang, L. (2024). Comprehensive review: Effects of climate change and greenhouse gases emission relevance to environmental stress on horticultural crops and management. *Journal of Environmental Management*, *351*, 119978.

Sharma, U., Sankhyan, H. P., Kumari, A., Thakur, S., Thakur, L., Mehta, D., Sharma, S., Sharma, S., & Sankhyan, N. (2024). Genomic selection: A revolutionary approach for forest tree improvement in the wake of climate change. *Euphytica*, *220*(1), 9.

Singh H, Yadav M, Kumar N, Kumar A, & Kumar M. (2020). Assessing adaptation and mitigation potential of roadside trees under the influence of vehicular emissions: A case study of *Grevillea robusta* and *Mangifera indica* planted in an urban city of India. *PLoS One*, 15(1), e0227380.

Stuke, M., Yun, K., & Kim, S. H. (2024). Adapting a process-oriented cold hardiness model to conifers. *Forest Ecology and Management*, *553*, 121611.

Sullivan, M. K., Fayolle, A., Bush, E., Ofosu-Bamfo, B., Vleminckx, J., Metz, M. R., & Queenborough, S. A. (2024). Cascading effects of climate change: New advances in drivers and shifts of tropical reproductive phenology. *Plant Ecology*, *225*(3), 175–187.

Susila, H., Gawarecka, K., Youn, G., Jurić, S., Jeong, H., & Ahn, J. H. (2024). THYLAKOID Formation 1 interacts with FLOWERING LOCUS T and modulates temperature-responsive flowering in Arabidopsis. *The Plant Journal*, *120*(1), 60–75.

Thapliyal, G., Bhandari, M. S., Vemanna, R. S., Pandey, S., Meena, R. K., & Barthwal, S. (2023). Engineering traits through CRISPR/cas genome editing in woody species to improve forest diversity and yield. *Critical Reviews in Biotechnology*, *43*(6), 884–903.

Wang, X., Zhou, X., Zhang, M., Ge, W., Yang, G., Zhou, H., Ma, L., Liu, K., Qi, W., & Bu, H. (2024). Influence of seed mass and shape on light plasticity of germination of alpine plants on the Tibetan Plateau: The role of photoblastic taxa, dispersal ability, and life history. *Global Ecology and Conservation*, *51*, e02896.

Watanabe, M., Ueno, S., Hasegawa, Y., Hirayama, S., Iwai, J., Kakui, H., & Moriguchi, Y. (2024). Development and application of a KASP marker for marker-assisted selection against the male-sterile gene MALE STERILITY 4 (MS4) in Japanese cedar (*Cryptomeria japonica* D. Don). *New Forests*, *55*(5), 1349–1362.

Ye, Z., Mu, Y., Van Duzen, S., & Ryser, P. (2024). Root and shoot phenology, architecture, and organ properties: An integrated trait network among 44 herbaceous wetland species. *New Phytologist*, *244*(2), 436–450.

Zhang, B., Wang, Z., Dai, X., Gao, J., Zhao, J., Ma, R., Chen, Y., Sun, Y., Ma, H., Li, S., & Li, W. (2024a). A COMPASS histone H3K4 trimethyltransferase pentamer transactivates drought tolerance and growth/biomass production in *Populus trichocarpa*. *New Phytologist*, *241*(5), 1950–1972.

Zhang, H., Liu, S., Yu, J., Li, J., Shangguan, Z., & Deng, L. (2024b). Thinning increases forest ecosystem carbon stocks. *Forest Ecology and Management*, *555*, 121702.

Zhu, X., Weng, Q., Bush, D., Zhou, C., Zhao, H., Wang, P., & Li, F. (2023). High-density genetic linkage mapping reveals low stability of QTLs across environments for economic traits in *Eucalyptus*. *Frontiers in Plant Science*, *13*, 1099705.

6 Intraspecific Trait Variation and Tree Improvement Program

Ram Gopal, Naresh Singh, Pankaj Lal, and Neha

6.1 INTRODUCTION

People within the same species exhibit intraspecific trait variability (ITV), also known as intraspecific functional variability, when they exhibit different trait values and trait syndromes. This concept shows how adaptation and genetic variation are two important ways that ITV can show how well a species can adapt to changes in its environment (Geber and Griffen, 2003). Genetic diversity represents the extent of physical variation among individual genotypes (Hughes et al., 2008; Kumar et al., 2020a,b). Species can experience genetic variability—a topic that has been widely explored in evolutionary literature (Fisher, 1930), and this variability plays a significant role in the evolution of species over time. Furthermore, adaptability, or plasticity, refers to the ability of each individual genotype to generate multiple phenotypes in response to varying environmental conditions (DeWitt et al., 1998).

6.1.1 Explanation of Intraspecific Variation: Genetic and Phenotypic Diversity within a Species

Intraspecific variation within a species refers to the genetic and phenotypic diversity among individuals, which arises from various factors. Key sources of genetic diversity include mutations, gene flow, and sexual reproduction, all of which contribute to a species' ability to adapt to its local environment (Schradin, 2013). The survival and evolutionary potential of populations depends on this genetic variation, which plays a vital role in determining traits like disease resistance and adaptability (Des Roches et al., 2018; Kumar et al., 2021a,b). On the other hand, phenotypic variation involves observable traits influenced by both genetic and environmental factors. By exhibiting unique physical characteristics, behaviors, and physiological responses, individuals can effectively occupy different ecological niches. For instance, in species such as finches, variations in seed size illustrate how individuals can optimize resource use (Lichtenstein et al., 2016). Recognizing the importance of diverse traits within a population is essential for maintaining ecological stability and mitigating the impacts of environmental changes (Schoen& Brown, 1991; Kumar et al., 2021c). Therefore, conserving this variation is crucial for biodiversity and the ongoing provision of ecosystem services that benefit both nature and human communities (Lichtenstein et al., 2016).

DOI: 10.1201/9781003545231-6

6.1.2 Importance of Trait Variability in Natural Populations for Adaptation and Survival

The ability of forest trees to adapt and survive in changing environmental conditions depends on the variability of traits within their natural populations. This is important because having a range of functional traits allows trees to better cope with environmental stresses such as climate change, pests, and diseases. Species that show significant intraspecific trait variability tend to be more resilient to habitat changes, as they can adjust their physiological and morphological characteristics to make the most of available resources in different conditions. Through phenotypic plasticity, individual trees can respond to immediate environmental changes, enhancing their chances of survival and reproduction in shifting environments. Moreover, genetic diversity within populations increases trait variability and serves as a reservoir of adaptive potential, which is vital for long-term resilience against environmental changes. Given the growing pressures on forest ecosystems from human activities and climate change, preserving this trait variability is essential for the survival of individual species and the overall health of forest ecosystems (Gailinget al., 2009; Kumar et al., 2015).

6.1.3 Role in Tree Improvement

ITV can significantly benefit species by enhancing their adaptability to changing environmental conditions. This variability allows trees to display a range of functional traits, enabling them to better manage stressors like drought, pests, and diseases within their species. For instance, trees with different levels of drought tolerance can thrive in various moisture conditions, supporting their survival and growth across diverse habitats (Westerband et al., 2021). Using ITV helps in phenotypic plasticity, which means that trees can change their traits in response to changes in their environment. This helps populations grow and keep the ecosystem stable. Tree breeding programs need to understand the idea of ITV because it helps them find traits that are useful in certain situations. This helps them choose genotypes that are better able to handle climate change. Researchers can make adaptive management plans better by looking into the genetic and phenotypic diversity within tree populations (Kumar et al., 2022; Gazol et al., 2023). This will help keep forest ecosystems strong and productive even when the environment is changing all the time.

6.2 OVERVIEW OF SELECTION PROCESSES

Artificial selection and natural selection serve different roles in tree improvement initiatives. Natural selection occurs when environmental factors favor the survival and reproduction of the best-fit individuals in a population, leading to adaptations that improve fitness over time. Through this gradual process, trees can develop traits that are beneficial for their environments, such as resistance to drought or pests. In contrast, selective breeding is a human-driven process where breeders intentionally choose trees with specific desirable traits for propagation, like quicker growth or superior wood quality. This method can produce favourable traits more rapidly than natural selection by focusing on characteristics that are beneficial for commercial forestry or ecological restoration.

By leveraging both natural selection and artificial selection, tree improvement projects can enhance forest ecosystems, making them more resilient and productive (Kumar et al., 2017, 2019; Klisz et al., 2023; Gazol et al., 2023).

6.2.1 Genetic Basis of Trait Variation

Understanding the genetic basis of trait variation in forest trees is crucial for grasping how these species adapt to their environments and enhance their resilience to changing conditions. This variation is significantly shaped by the intricate interactions of genetic factors, including multiple alleles at different loci that influence phenotypic traits such as growth rate, disease resistance, and phenology. Research has demonstrated that many adaptive traits in forest trees follow a polygenic inheritance pattern, meaning they are governed by numerous genes, each contributing a small effect to the overall phenotype. For example, specific genes linked to traits like drought resistance or growth have been identified in species such as *Pinuspinaster* and *Pinustaeda*, underscoring the importance of genetic diversity in breeding programs aimed at boosting forest productivity and resilience (Oluwajuwon et al., 2022). Additionally, common garden experiments have shown significant genetic differentiation among tree populations along environmental gradients, suggesting local adaptation to different climatic conditions (Hamrick et al., 1981).

6.2.2 Understanding Genetic Variation

Understanding genetic variation in forest trees is essential for developing effective conservation and management strategies, as it plays a key role in the adaptability and resilience of these species to environmental changes. Forest trees possess high levels of genetic diversity due to their long life spans, out-crossing mating systems, and large effective population sizes, which contribute to significant standing genetic variation. This genetic diversity often manifests in phenotypic traits critical for survival, such as growth rates, disease resistance, and responses to climate conditions. Research has demonstrated that tree populations can show considerable genetic differentiation along environmental gradients, indicating local adaptation to specific habitats. For example, common garden experiments have illustrated how different populations react to uniform environmental conditions, offering insights into their adaptive traits and genetic composition. Additionally, identifying major genes linked to key traits, along with the polygenic nature of most adaptive characteristics, underscores the complexity of genetic variation in forest trees. This information not only helps programs that breed trees to get better qualities, but it also helps conservation efforts to keep the genetic diversity that is important for the long-term health of forest ecosystems (Hurel et al., 2021; Cruz-Salazar et al., 2021; Zytynska et al., 2011).

TABLE 6.1
Differences Between Heritable and Nonheritable Traits

Aspect	Heritable Traits	NonheritableTraits
Definition	Traits that can be passed from parents to offspring through genetic mechanisms	Traits influenced primarily by environmental factors rather than genetic inheritance
Examples	Growth rate, disease resistance, and flowering time in response to climate	Leaf size variations due to light availability or tree height variations caused by soil quality
Genetic basis	Controlled by specific genes and often show significant heritability, allowing for selection	Largely determined by environmental conditions, exhibiting phenotypic plasticity rather than genetic stability
Adaptation potential	Can evolve over generations through natural or artificial selection, enhancing survival in changing environments	Limited adaptive potential since these traits do not contribute to the genetic makeup of future generations

6.2.3 Overview of Genetic Sources of Variation

The genetic variation in forest trees comes from several important mechanisms, including mutation, recombination, and gene flow. Mutation adds new alleles to a population's gene pool through heritable changes in DNA, potentially creating entirely new traits that can provide adaptive advantages. During sexual reproduction, recombination occurs, mixing genetic material between parental genotypes to create offspring with unique trait combinations. This process boosts genetic diversity by producing new allele combinations that may be advantageous in different environmental situations. Furthermore, gene flow, which is the transfer of alleles between populations through pollen and seed dispersal, enhances genetic variation by bringing in a new genetic material to local populations, thus increasing their ability to adapt. Together, these mechanisms help ensure that forest trees maintain strong genetic diversity, which is vital for their survival and adaptability to changing environmental conditions and stresses, such as climate change and diseases. Understanding these genetic sources is crucial for effective tree breeding and conservation strategies aimed at improving the resilience of forest ecosystems (Cruz-Salazar et al., 2021) (Table 6.1).

6.2.4 Molecular Markers and Genetic Mapping

Molecular markers and genetic mapping are essential for comprehending ITV in tree improvement initiatives. Microsatellites and single-nucleotide polymorphisms (SNPs) are genetic markers that help scientists find quantitative trait loci (QTLs) that are linked to specific traits. This helps breeders choose individuals with the traits they want more efficiently. Marker-assisted selection (MAS) utilizes these molecular markers to improve traditional breeding techniques, greatly speeding up the creation of enhanced tree varieties while preserving genetic diversity (Kumar et al., 2017, 2019; Cortés et al., 2020). By identifying the genetic factors that influence traits such as growth rate or disease resistance, researchers can determine the specific alleles linked to these characteristics. Breeding programs can then apply this knowledge to develop hybrids that are more resilient to environmental challenges. Furthermore, utilizing molecular markers allows for the evaluation of genetic diversity within populations, which is crucial for effective conservation and sustainable management in forestry (Hasan et al., 2021).

6.2.5 Molecular Tools for Assessing Intraspecific Variation and Genetic Mapping of QTLs for Trait Discovery

Molecular tools, such as SNPs and microsatellites, are very important for studying differences within the same species in forest trees. They give us important information about genetic diversity and population structure. SNPs, which indicate variations at a single nucleotide position in the genome, enable high-resolution genetic mapping and help identify QTLs linked to key traits such as growth rate and disease resistance (Fischer et al., 2017). Their wide distribution throughout the genome makes them advantageous, enabling thorough evaluations of genetic variation among different populations. Microsatellites, which are also called simple sequence repeats, on the other hand, are highly polymorphic markers that can show how different populations are genetically because they mutate so quickly. Research has demonstrated that while SNPs offer a broader view of the genome, microsatellites are effective in capturing particular population dynamics and historical patterns of gene flow. For example, research shows that estimates of genetic differentiation from SNP data and microsatellite markers are very similar. By combining these molecular tools, researchers can deepen their understanding of ITV in forest trees, which can lead to more effective breeding strategies focused on enhancing resilience and adaptability in changing environments (Caignard et al., 2019; Kumar et al., 2021d; Klisz et al., 2023).

6.3 PHENOTYPIC VARIATION IN FOREST TREES

Phenotypic variation in forest trees plays a vital role in their adaptability and survival, shaped by both genetic and environmental influences. This variation is evident in various traits, including growth rates, leaf shapes, and reproductive features, which can vary widely among individuals of the same species. Research indicates that forest trees show significant phenotypic differences along environmental gradients, suggesting local adaptation to particular habitats. For example, common garden experiments have demonstrated that trees from different populations can display unique growth patterns and seasonal traits when cultivated in the same conditions, emphasizing the impact of genetic variation on these characteristics. Moreover, phenotypic plasticity enables trees to modify their traits in response to environmental shifts, enhancing their resilience. The way genetic diversity and environmental factors interact shows how important it is to understand phenotypic variation in forest trees in order to protect biodiversity and ecosystem health through effective management and conservation efforts (Hurel et al., 2021).

6.4 METHODS FOR MEASURING INTRASPECIFIC VARIATION

Intraspecific variation between forest trees is a key part of looking at ecosystems as a whole to understand biodiversity, adaptation, and resilience. Researchers have developed several approaches based on genetic, functional, and phenotypic traits for the estimation of such variation.

6.4.1 Genetic Analysis

6.4.1.1 Airborne Imaging Spectroscopy

Recent studies show that airborne imaging spectroscopy is effective for evaluating genetic variation in tree populations. This technique enables large-scale monitoring without the need for extensive in situ sampling, which can be resource-heavy. By examining spectral reflectance data, researchers can forecast genetic structure by focusing on specific wavelength regions linked to leaf characteristics such as water content and pigments. This method offers a systematic and spatially continuous way to monitor genetic diversity at the canopy level.

6.4.1.2 Molecular Techniques

Traditional genetic analysis involves collecting samples from individual trees and assessing allele frequencies through molecular markers. Techniques such as microsatellite analysis and SNP genotyping can effectively gauge genetic diversity within populations. While these methods can be labor-intensive, they provide valuable insights into the genetic makeup of tree species.

6.4.2 Functional Trait Measurement

6.4.2.1 Intraspecific Trait Variation (ITV)

Functional traits like leaf thickness, specific leaf area, and nutrient concentrations have been studied to understand their variability within species across various environmental gradients. Research indicates that ITV plays a significant role in tree performance under stressful conditions, such as drought (Czyżet al., 2020).

6.4.2.2 Dendrometer Bands

We use dendrometer bands to track tree growth over time, offering valuable insights into how individual trees react to environmental changes. This method makes it possible to precisely track growth rates and can help link functional traits to performance metrics in a range of environmental conditions (Maracahipes-Santos et al., 2023).

6.4.2.3 Remote Sensing Technologies

6.4.2.3.1 LiDAR and Satellite Imagery

Light Detection and Ranging (LiDAR) technology is useful for evaluating forest structure and biomass, allowing for indirect inferences about intraspecific variation through the analysis of canopy structure. Satellite imagery enhances this by offering wider spatial coverage, which helps in assessing biodiversity on a larger scale.

6.5 EXPERIMENTAL TRIALS

6.5.1 Provenance Trials

Provenance trials consist of planting seeds from various populations in a shared garden environment to examine differences in growth and adaptability under consistent environmental conditions. This approach aids in recognizing local adaptations and the degree of intraspecific variation (ITV) across diverse geographic areas.

6.5.2 Field Trials and Provenance Testing

Field trials and provenance testing are essential methods in forestry that evaluate how well different tree species adapt and perform in various environmental conditions. Provenance trials involve planting seeds from different geographical origins (provenances) at multiple test sites to assess their growth, survival, and other characteristics. This method helps determine which provenances are most suitable for specific planting locations, especially considering the impacts of climate change. The process starts with collecting seeds from various locations within a species' range, followed by organized planting in controlled experimental setups. Over time, researchers track important performance indicators such as height, survival rates, and phenology. This information is crucial for creating seed transfer guidelines, allowing forest managers to choose the best seed sources for reforestation projects. Although setting up these trials can be resource-heavy and time-consuming, they offer vital insights into genetic adaptation and phenotypic plasticity among tree populations. The results can show how plants have changed in response to their environment and help predict how different types of plants will do in new environments. This will ultimately lead to more sustainable forest management (Risk et al., 2021).

6.6 QUANTITATIVE VS. QUALITATIVE TRAITS

In the study of forest trees, traits are generally divided into two primary categories: quantitative and qualitative. Recognizing these differences is crucial for effective tree breeding, conservation efforts, and management strategies.

6.6.1 Quantitative Traits

Quantitative traits are characteristics that can be measured and show a range of variations. They are influenced by multiple genes, making them polygenic. Because these traits can be quantified and represented with numbers, they are ideal for statistical analysis.

Examples: *Growth Rate*: Height, diameter at breast height (DBH), and biomass are common quantitative traits measured in forest trees. For instance, studies have shown significant variation in growth rates among different tree species and provenances, which can be linked to environmental factors such as soil quality and moisture availability.

Wood Properties: Traits such as wood density, fiber length, and chemical composition (e.g., cellulose and lignin content) are also quantitative. These traits are crucial for determining the economic value of timber and the ecological role of trees in carbon sequestration.

6.6.2 Qualitative Traits

Qualitative traits, on the other hand, are specific characteristics that generally adhere to a Mendelian inheritance pattern. These traits are often governed by a single gene or a small number of genes and can be classified into distinct categories.

Examples: *Disease Resistance*: The presence or absence of resistance to certain pathogens is a qualitative trait. For instance, some tree species may show resistance to pests or diseases because of particular genetic factors.

Leaf Morphology: Traits like leaf shape (e.g., lobed vs. unlobed) or color (such as green compared to variegated) are included in this category. These characteristics can affect how trees interact with their surrounding environment.

Differentiating between traits measured quantitatively vs. qualitatively

Trait Type	Quantitative Traits	Qualitative Traits
Definition	Measurable characteristics exhibiting continuous variation, influenced by multiple genes	Discrete characteristics typically controlled by one or a few genes
Examples	• Height (e.g., maximum tree height) • Diameter at breast height (DBH) • Wood density	• Leaf color (e.g., green vs. variegated) • Disease symptoms (e.g., presence or absence of resistance) • Leaf shape (e.g., lobed vs. unlobed)
Measurement	Quantified using statistical methods; can be assessed through field trials and growth metrics	Assessed through observational studies and genetic analyses
Analysis methods	QTL mapping and regression analysis to identify genetic markers associated with traits	Genetic mapping to identify alleles linked to specific traits
Importance	Critical for understanding growth potential, adaptability, and economic value of timber	Essential for conservation strategies and breeding programs focused on specific characteristics

6.7 COMMON STATISTICAL APPROACHES

In tree improvement, a range of statistical methods is used to analyze traits and improve breeding programs. Important techniques include estimating heritability, conducting variance analysis, and studying phenotypic correlations.

Statistical Approach	Description	Applications
Heritability estimation	Measures the proportion of observed variation in a trait that can be attributed to genetic factors. It is typically estimated using methods such as ANOVA or mixed models	Helps in understanding the genetic potential for traits, guiding selection in breeding programs

(Continued)

(Continued)

Statistical Approach	Description	Applications
Variance analysis	Involves partitioning the total variance observed in traits into genetic, environmental, and interaction components. Techniques like ANOVA are commonly used	Identifies significant sources of variation, allowing for targeted breeding strategies and improved trait selection
Phenotypic correlation studies	Examines the relationship between different phenotypic traits to understand how they influence each other. This can be done using correlation coefficients or regression analysis	Useful for identifying traits that may be selected together, aiding in the development of multitrait selection strategies

6.8 CHALLENGES IN MANAGING TRAIT VARIATION

Managing trait variation in forest trees poses several challenges that can greatly affect tree improvement, conservation efforts, and ecosystem management. Below are some key challenges identified from the search results:

6.8.1 Complexity of Intraspecific Variation

ITV can be just as significant as interspecific variation, making it challenging to evaluate the adaptive capacity of trees. This complexity complicates predictions regarding how individual trees will respond to changing environmental conditions, including heightened aridity and drought stress. The connection between ITV and tree performance is frequently ambiguous and varies greatly among species, which can impede the development of effective management strategies.

6.8.2 Long Breeding Cycles

Tree breeding in forests usually requires lengthy cycles because many species take a long time to flower and mature. This prolonged process can slow down the release of genetically enhanced materials into the market, making it difficult to effectively address new pests, diseases, and the impacts of climate change.

6.8.3 Variable Juvenile–Mature Correlations

There is frequently a weak connection between the traits shown in juvenile stages and those seen in mature trees. This variability makes the selection process more challenging, as traits that seem appealing in young trees might not carry over to their adult performance, which could result in the mismanagement of genetic resources (Grattapaglia et al., 2018).

6.8.4 Emerging Pests and Diseases

The emergence of new pests and diseases can quickly change the way we manage trees. It's essential to keep an eye on trait variation since some genotypes might be more vulnerable than others. This requires continuous monitoring and adjustments to breeding programs to maintain resilience against these challenges (Grattapaglia et al., 2018).

6.8.5 Environmental Changes

Rapid changes in climate and environmental conditions present major challenges for managing trait variation. Trees need to adapt swiftly to survive, yet grasping how various traits contribute to resilience under stress is complicated. While phenotypic plasticity can facilitate short-term adaptation, depending on this mechanism may result in trade-offs that affect long-term fitness (Gazol et al., 2023).

6.8.6 Genetic Maladaptation Risks

Gene flow between populations can help reduce extinction risks, but it can also result in genetic maladaptation, especially when central populations interbreed with those on the periphery. This mixing can weaken local adaptations that are essential for survival in particular environments (Gazol et al., 2023).

6.8.7 Limitations of Current Genetic Approaches

Traditional genetic dissection methods, such as quantitative trait mapping, have encountered difficulties in practical applications because of the complex multifactorial inheritance of traits in forest trees. Even with progress in genomics and MAS, it is still challenging to convert research findings into effective breeding strategies (Grattapaglia et al., 2018).

6.8.8 Integration of Functional Traits

Understanding the interaction of various traits within a single species is crucial for effective management, yet it can be complicated by factors like environmental filtering and phylogenetic influences. To effectively integrate functional traits in response to environmental gradients, thorough studies are needed that connect both interspecific and intraspecific variations (Homeier et al., 2021).

6.9 CASE STUDIES ON INTRASPECIFIC TRAIT VARIATION IN FOREST TREES

Case Study	Focus	Key Findings	Reference
Intraspecific variation and drought tolerance	Examines functional traits along aridity gradients in trees	Intraspecific variation in functional traits significantly impacts tree performance under drought conditions, with specific traits varying by species and environmental context. This variability can influence growth and survival rates	Gazol et al. (2023)

(*Continued*)

(*Continued*)

Case Study	Focus	Key Findings	Reference
Functional trait variability in riparian forests	Investigates how intraspecific variability affects tree species persistence in modified environments	Higher intraspecific variability in traits like leaf thickness and phosphorus content enhances species persistence in riparian forests impacted by agricultural expansion, indicating adaptability to altered conditions	Maracahipes-Santos et al. (2023)
Genetic variation in tropical tree species	Reviews genetic diversity among tropical tree species in plantations	Genetic variation within populations is crucial for restoration efforts, with some species exhibiting high levels of diversity while others show low diversity, impacting their adaptability and resilience to environmental changes	Cruz-Salazar et al. (2021)
Intraspecific variation in response to elevated CO	Analyzes tree productivity responses to elevated CO_2 levels across different genotypes	Significant intraspecific variation was found in photosynthesis and growth responses to elevated CO_2, indicating potential for breeding programs focused on enhancing productivity under changing atmospheric conditions	Poorter (1993)
Airborne imaging spectroscopy for genetic variation detection	Utilizes imaging spectroscopy to detect genetic variation within a *Fagussylvatica* population	Airborne imaging spectroscopy can effectively reveal genotype-specific phenotypic features, providing a cost-efficient method for assessing genetic variation across large spatial scales, crucial for biodiversity monitoring efforts	Czyż et al. (2020)

6.10 FUTURE PERSPECTIVES IN TRAIT VARIATION AND TREE IMPROVEMENT

6.10.1 Biotechnological Advances

6.10.1.1 Genomic Insights

The incorporation of genomics into tree improvement provides a thorough insight into genetic diversity and trait variation among species. Advanced sequencing technologies can pinpoint genetic markers associated with desirable traits, making MAS more efficient.

6.10.1.2 CRISPR Technology

The CRISPR/Cas9 system is transforming tree breeding through its ability to perform precise gene editing. This innovative technology facilitates targeted changes in the genomes of forest trees, improving traits like growth rate, disease resistance, and stress tolerance. For example, CRISPR has been effectively utilized to develop genetically modified poplar trees that exhibit enhanced qualities for biomass production and greater resistance to diseases (Xie et al., 2023).

6.10.1.3 Targeted Trait Improvement

By using CRISPR to disable or alter specific genes linked to unwanted traits, researchers can create tree varieties that are more adapted to particular environmental conditions. This focused strategy can greatly speed up the breeding process when compared to conventional methods.

6.11 CLIMATE-RESILIENT TREE BREEDING

6.11.1 Identifying Resilient Genotypes

Intraspecific variation offers a rich source of genetic diversity that can be utilized to create tree varieties that are resilient to climate change. By examining how various genotypes react to environmental stressors (such as drought and temperature changes), breeders can identify individuals that show exceptional adaptability.

6.11.2 Multienvironment Trials

Testing different genotypes in various environmental conditions will help pinpoint those that thrive under various climate scenarios. This strategy guarantees that the chosen varieties are both high-yielding and capable of withstanding future climate challenges.

6.11.3 Functional Trait Assessment

Focusing on functional traits that enhance stress tolerance, like root depth, leaf structure, and water-use efficiency, will be essential for creating tree species that can withstand climate challenges. Gaining insights into the ecological effects of these trait variations will help steer breeding initiatives aimed at building stronger forest ecosystems.

6.12 CONSERVATION OF GENETIC RESOURCES—IMPORTANCE OF GENETIC RESOURCE CONSERVATION

6.12.1 Preserving Intraspecific Diversity

Conserving genetic resources within species is essential for maintaining the adaptive potential of tree populations. The genetic diversity present within species enhances their ability to adapt to changing environments and resist pests and diseases.

6.12.2 Seed Banks and Genetic Reserves

Preserving genetic resources within species is essential for maintaining the adaptive potential of tree populations. The genetic diversity present within species enhances their ability to adapt to changing environments and resist pests and diseases.

6.12.3 Integrating Conservation with Breeding Programs

Future breeding programs need to include conservation strategies aimed at preserving genetic diversity while also prioritizing desirable traits. This combined approach will promote sustainable forest management practices that enhance both productivity and biodiversity.

6.13 CONCLUSION

In conclusion, ITV is very important for the success of tree improvement programs because it helps tree species become more resilient, adaptable, and productive in the face of environmental problems. By gaining insights into and utilizing the genetic and phenotypic diversity within species, researchers and forest managers can create targeted breeding strategies that not only enhance growth rates and disease resistance but also promote the long-term sustainability of forest ecosystems. Using cutting-edge biotechnological tools like genomics and CRISPR, along with a focus on traits that are resistant to climate change, will make these programs even better at dealing with the urgent effects of climate change. It is also important to protect intraspecific genetic resources to keep biodiversity high and make sure that future generations will be able to access the genetic diversity they need to adapt. In the end, it will take a comprehensive approach that combines scientific progress with environmental protection in order to move tree improvement projects forward and create strong forests that can thrive in an uncertain future.

REFERENCES

Caignard, T., Delzon, S., Bodénès, C., Dencausse, B., & Kremer, A. (2019). Heritability and genetic architecture of reproduction-related traits in a temperate oak species. *Tree Genetics & Genomes*, *15*, 1–12.

Cortés, A. J., Restrepo-Montoya, M., & Bedoya-Canas, L. E. (2020). Modern strategies to assess and breed forest tree adaptation to changing climate. *Frontiers in Plant Science*, *11*, 583323.

Cruz-Salazar, B., Ruiz-Montoya, L., Ramírez-Marcial, N., & García-Bautista, M. (2021). Relationship between genetic variation and diversity of tree species in tropical forests in the El Ocote Biosphere Reserve, Chiapas, Mexico. *Tropical Conservation Science*, *14*, 1–14. https://doi.org/10.1177/1940082920978143

Czyż, E. A., Guillén Escribà, C., Wulf, H., Tedder, A., Schuman, M. C., Schneider, F. D., & Schaepman, M. E. (2020). Intraspecific genetic variation of a Fagussylvatica population in a temperate forest derived from airborne imaging spectroscopy time series. *Ecology and Evolution*, *10*(14), 7419–7430.

Des Roches, S., Post, D. M., Turley, N. E., Bailey, J. K., Hendry, A. P., Kinnison, M. T., & Palkovacs, E. P. (2018). The ecological importance of intraspecific variation. *Nature Ecology & Evolution*, *2*(1), 57–64.

DeWitt, T. J., Sih, A., & Wilson, D. S. (1998). Costs and limits of phenotypic plasticity. *Trends in ecology & evolution*, *13*(2), 77–81.

Fischer, M. C., Rellstab, C., Leuzinger, M., Roumet, M., Gugerli, F., Shimizu, K. K., & Widmer, A. (2017). Estimating genomic diversity and population differentiation–an empirical comparison of microsatellite and SNP variation in *Arabidopsis halleri*. *BMC Genomics*, *18*, 1–15.

Gailing, O., Vornam, B., Leinemann, L., & Finkeldey, R. (2009). Genetic and genomic approaches to assess adaptive genetic variation in plants: forest trees as a model. *Physiologia Plantarum*, *137*(4), 509–519.

Gazol, A., Fajardo, A., & Camarero, J. J. (2023). Contributions of intraspecific variation to drought tolerance in trees. *Current Forestry Reports*, *9*(6), 461–472.

Geber, M. A., & Griffen, L. R. (2003). Inheritance and natural selection on functional traits. *International Journal of Plant Sciences*, *164*(S3), S21–S42.

Grattapaglia, D., Silva-Junior, O. B., Resende, R. T., Cappa, E. P., Müller, B. S., Tan, B., & El-Kassaby, Y. A. (2018). Quantitative genetics and genomics converge to accelerate forest tree breeding. *Frontiers in Plant Science*, *9*, 1693.

Hamrick, J. L., Mitton, J. B., & Linhart, Y. B. (1981). *Levels of Genetic Variation in Trees: Influence of Life History Characteristics* (pp. 35–41). Berkeley, CA: Gen. Tech. Rep. PSW-48.

Hasan, N., Choudhary, S., Naaz, N., Sharma, N., & Laskar, R. A. (2021). Recent advancements in molecular marker-assisted selection and applications in plant breeding programmes. *Journal of Genetic Engineering and Biotechnology*, *19*(1), 128.

Homeier, J., Seeler, T., Pierick, K., & Leuschner, C. (2021). Leaf trait variation in species-rich tropical Andean forests. *Scientific Reports*, *11*(1), 9993.

Hughes, A. R., Inouye, B. D., Johnson, M. T., Underwood, N., & Vellend, M. (2008). Ecological consequences of genetic diversity. *Ecology Letters*, *11*(6), 609–623.

Hurel, A., de Miguel, M., Dutech, C., Desprez-Loustau, M. L., Plomion, C., Rodríguez-Quilón, I., & Budde, K. B. (2021). Genetic basis of growth, spring phenology, and susceptibility to biotic stressors in maritime pine. *Evolutionary Applications*, *14*(12), 2750–2772.

Klisz, M., Chakraborty, D., Cvjetković, B., Grabner, M., Lintunen, A., Mayer, K., & Rossi, S. (2023). Functional traits of boreal species and adaptation to local conditions. In: Girona, M. M., Morin, H., Gauthier, S., & Bergeron, Y. (eds.) *Boreal Forests in the Face of Climate Change: Sustainable Management* (pp. 323–355). Cham: Springer International Publishing.

Kumar, A., Dwivedi, G. K., Tewari, S., Paul, J., Anand, R., Kumar, N., Kumar, P., Singh, H., & Kaushal, R. (2020a). Carbon mineralization and inorganic nitrogen pools under *Terminalia chebula* Retz.-Based agroforestry system in Himalayan foothills, India. *Forest Science*, *66*(5), 634–643.

Kumar, A., Dwivedi, G. K., Tewari, S., Paul, J., Sah, V. K., Singh, H., Kumar, P., Kumar, N., & Kaushal, R. (2020b). Soil organic carbon pools under *Terminalia chebula* Retz. based agroforestry system in Himalayan foothills, India. *Current Science*, *118*(7), 1098–1103.

Kumar, A., Kumar, P., Singh, H., Bisht, S., & Kumar, N. (2021a). Relationship of physiological plant functional traits with soil carbon stock in temperate forest of Garhwal Himalaya. *Current Science*, *120*(8), 1368–1373.

Kumar, A., Kumar, P., Singh, H., & Kumar, N. (2021b). Modulation of plant functional traits under essential plant nutrients during seasonal regime in natural forests of Garhwal Himalayas. *Plant and Soil*, *465*, 197–212.

Kumar, A., Kumar, P., Singh, H., & Kumar, N. (2021c). Impact of plant functional traits on near saturated hydraulic conductivity of soil under different forests of Kempty watershed in Garhwal Himalaya. *Indian Journal of Soil Conservation*, *49*(1), 38–44.

Kumar, A., Singh, H., Kumari, G., Bisht, S., Malik, A., Kumar, N., Singh, M., Raturi, A., Barthwal, S., Thakur, A., & Kaushal, R. (2022). Adaptive resilience of roadside trees to vehicular emissions via leaf enzymatic, physiological, and anatomical trait modulations. *Environmental Pollution*, *313*, 120191.

Kumar, N., Jeena, N., & Singh, H. (2019). Elevated temperature modulates rice pollen structure: a study from foothill Himalayan agro-ecosystem in India. *3 Biotech*, *9*: 175.

Kumar, N., Jeena, N., Kumar, A., Khairakpam, R., & Singh, H. (2021d). Comparative response of rice cultivars to elevated air temperature in Bhabar region of Indian Himalaya: Status on yield attributes. *Heliyon*, *7*, e07474.

Kumar, N., Kumar, N., Shukla, A., Shankhdhar, S. C., & Shankhdhar, D. (2015). Impact of terminal heat stress on pollen viability and yield attributes of rice (*Oryza sativa* L.). *Cereal Research Communications*, *43*(4), 616–626.

Kumar, N., Shankhdhar, S. C., & Shankhdhar, D. (2016). Impact of elevated temperature on antioxidant activity and membrane stability in different genotypes of rice (*Oryza sativa* L.). *Indian Journal of Plant Physiology (Now Plant Physiology Report)*, *21*(1), 37–43.

Kumar, N., Suyal, D. C., Sharma, I. P., Verma, A., & Singh, H. (2017). Elucidating stress proteins in rice (*Oryza sativa* L.) genotype under elevated temperature: a proteomic approach to understand heat stress response. *3 Biotech*, *7*, 205.

Lichtenstein, J. L., Pruitt, J. N., & Modlmeier, A. P. (2016). Intraspecific variation in collective behaviors drives interspecific contests in acorn ants. *Behavioral Ecology*, *27*(2), 553–559.

Maracahipes-Santos, L., Silvério, D. V., Maracahipes, L., Macedo, M. N., Lenza, E., Jankowski, K. J., & Brando, P. M. (2023). Intraspecific trait variability facilitates tree species persistence along riparian forest edges in Southern Amazonia. *Scientific Reports*, *13*(1), 12454.

Oluwajuwon, T. V., Attafuah, R., Offiah, C. J., & Krabel, D. (2022). Genetic variation in tropical tree species and plantations: a review. *Open Journal of Forestry*, *12*(3), 350–366.

Poorter, H. (1993). Interspecific variation in the growth response of plants to an elevated ambient CO_2 concentration. *CO_2 and Biosphere*, *104*(1), 77–97.

Risk, C., McKenney, D. W., Pedlar, J., & Lu, P. (2021). A compilation of North American tree provenance trials and relevant historical climate data for seven species. *Scientific Data*, *8*(1), 29.

Schoen, D. J., & Brown, A. H. (1991). Intraspecific variation in population gene diversity and effective population size correlates with the mating system in plants. *Proceedings of the National Academy of Sciences*, *88*(10), 4494–4497.

Schradin, C. (2013). Intraspecific variation in social organization by genetic variation, developmental plasticity, social flexibility or entirely extrinsic factors. *Philosophical Transactions of the Royal Society B: Biological Sciences*, *368*(1618), 20120346.

Westerband, A. C., Funk, J. L., & Barton, K. E. (2021). Intraspecific trait variation in plants: a renewed focus on its role in ecological processes. *Annals of Botany*, *127*(4), 397–410.

Xie, D., Gu, B., Liu, Y., Ye, P., Ma, Y., Wen, T., & Zhao, Z. (2023). Efficient targeted recombination with CRISPR/Cas9 in hybrids of *Caenorhabditis nematodes* with suppressed recombination. *BMC Biology*, *21*(1), 203.

Zytynska, S. E., Fay, M. F., Penney, D., & Preziosi, R. F. (2011). Genetic variation in a tropical tree species influences the associated epiphytic plant and invertebrate communities in a complex forest ecosystem. *Philosophical Transactions of the Royal Society B: Biological Sciences*, *366*(1569), 1329–1336.

7 Understanding Functional Trait Variation in Different Forest Ecosystems

Nada Tadi

7.1 INTRODUCTION

Forests consist of a broad and extensive range of biodiversity, which form dynamic systems and provide various critical ecosystem services. Trees and plants in forests exhibit specialized adaptations in the form of functional traits that enhance survival, growth, and efficiency in these unique ecosystems characterized by high biodiversity, intense competition, and varied microclimatic conditions. Functional traits are morphological, biochemical, physiological, structural, phenological, or behavioral characteristics that are expressed in phenotypes of individual organisms and are considered relevant to the response of such organisms to the environment and/or their effects on ecosystem properties and influence performance or fitness (Nock et al., 2016). Functional traits directly or indirectly influence the role of organisms in an ecosystem.

7.2 TYPES OF FUNCTIONAL TRAITS

Several factors affect the development and expression of functional traits, ranging from genetic to environmental influences. Therefore, there are large numbers of functional traits, and different combinations of these functional traits can be observed in different forest ecosystems. Functional traits can be categorized into three broad types which are as follows:

7.2.1 Morphological Trait

Morphological functional traits are the physical characteristics that directly affect the survival, growth, and reproduction ability of organisms in different environments. These traits help plants in obtaining resources such as light, water, and nutrients from the environment; provide mechanical support; and play an important role in reproduction and in resistance to factors that cause stress such as herbivory, drought, and wind. The examples of key morphological traits are as follows:

DOI: 10.1201/9781003545231-7

7.2.1.1 Leaf Traits

- **Leaf Size and Shape**: Leaf size and shape influence the ability of a plant to capture sunlight, regulate water loss, and resist stress caused by environmental factors. Larger leaves have higher efficiency in capturing light (Lusk et al., 2019). Compound leaves have greater water conductivity and higher photosynthetic rates, which allow them to acquire resources more efficiently and grow at a faster rate. (Yang et al., 2019).
- **Leaf Thickness**: Thicker leaves typically have more mesophyll tissue and larger cells with bigger vacuoles, which aid in water storage and help maintain hydration during drought. Leaf thickness is positively associated with increased hydraulic conductivity and elastic modulus (Scoffoni et al., 2014). Several studies have found a strong relationship between leaf thickness and leaf water potential (Scoffoni et al., 2014).
- **Leaf Surface Characteristics**: Many plants have leaves that may have various structures such as trichomes, cuticles, or spines, which can give protection from feeding by herbivores, help in reducing water loss, and mitigate damage from ultraviolet radiation.

7.2.1.2 Plant Height

Plant height plays an important role in competing for light in forest ecosystems. Taller trees usually dominate the canopy and have better access to sunlight, while shorter plants in the understory are adapted to capturing diffused light. Various traits such as crown area, total leaf area, leaf area index (Sterck & Bongers, 2001), tree allometry (Osada, 2012), biomass allocation (Claveau et al., 2005), and trunk diameter growth (Claveau et al., 2002) are strongly influenced by plant height.

7.2.1.3 Wood Density and Flexibility

- **Wood Density**: Wood density is an important trait that affects mechanical and physiological properties (Kraft et al., 2010) of a tree. It supports the structure, influences carbon storage, and helps in resisting damage from wind, herbivores, and other environmental factors. It influences the competitiveness of tree species and shapes the composition, structure, and function of forest ecosystems, which affect the mortality of trees and wood decomposition. Also, a strong link between wood density and biomass production (Poorter et al., 2019) makes it crucial for assessing carbon uptake and storage (Bouchard et al., 2024).
- **Stem Flexibility**: Plants with thick and flexible stems show resistance to physical damage by wind or snow. The flexibility of the stem is found to help in reducing the resistance to the airflow, which minimizes the risk of failure (Sellier & Fourcaud, 2009). In areas affected by heavy snowfall, species often develop stems, which are narrow and flexible, that can bend without breaking.

7.2.1.4 Root Architecture

- **Root Depth and Spread**: Root traits influence a plant's ability to access water and nutrients. In many plants, sometimes, roots can grow beyond the soil layers and reach weathered and/or fractured hard rock and bedrock

(Schwinning, 2010) and access water and nutrients, which are unavailable to surface roots (Estrada-Medina et al., 2012).

- **Root-to-Shoot Ratio**: The root-to-shoot ratio reflects the allocation of resources between aboveground growth (light capture) and belowground growth (water and nutrient acquisition). Plants in dry or nutrient-limited environments may invest more in root growth relative to shoot growth. The allocation of photosynthetically derived materials between roots and shoots depends on plant species, environmental conditions, and plant phenology (Klepper, 1991). Generally, the root-to-shoot ratio is influenced by (deficiency in) light, water, and major mineral nutrients (Cakmak et al., 1994).

7.2.1.5 Seed Size and Dispersal Mechanisms

- **Seed Size**: Larger seeds typically have more energy reserves in the form of food storage, which helps in seedling establishment in competitive or resource-limited environments, such as the shaded understory of tropical forests, by supporting embryo growth and enabling strong seedling growth before weeds can create competition. On the other hand, smaller seeds can be dispersed over longer distances by wind or animals. Smaller seeds also mean more seeds per plant, increasing the likelihood of accidental ingestion and dispersal, enhancing their chances of colonizing new or disturbed areas (Ambika et al., 2014).
- **Dispersal Traits**: Various plant species have developed morphological adaptations such as wings, fleshy fruits, or spines on seeds, which help in the dispersal of seeds more effectively. Wind-dispersed species often have lightweight seeds with structures like wings or feathery pappus while animal-dispersed species may produce fleshy fruits that attract frugivores or have spiny seeds, which stick on the hairs of the animals.

7.2.1.6 Bark Thickness and Texture

- **Bark Thickness**: Bark thickness varies greatly among forest species and is an important trait for protecting trees from environmental stresses such as fire, herbivory, and temperature fluctuations. Research suggests that thicker bark correlates with higher fire resistance (Bold et al., 2010).
- **Bark Texture**: Smooth bark reduces the ability of insects to grip tree surfaces. In some tree species with exfoliating bark, which regularly renews, are better at shedding lianas, leading to fewer liana infestations (Talley et al., 1996). Rough bark, however, facilitates liana attachment, increasing infestation. Additionally, thorns and spines on the stem bark help protect against bark-feeding and climbing animals (Sfair et al., 2016).

7.2.1.7 Crown Architecture

The architecture of crown (branching patterns, shape, and density) affects both the allocation of light to leaf area and the arrangement and display of the leaf area (Valladares & Pugnaire, 1999). It affects light-capturing efficiency and distribution of resources to support growth. Species that dominate the canopy in tropical and temperate forests often have wide, spreading crowns that maximize light interception, while species in the understory may have narrow, compact crowns that allow them to optimize light capture in shaded conditions.

7.2.2 Physiological Traits

Physiological functional traits refer to the internal processes that regulate metabolic activities in an organism and influence the ability of the organism to acquire resources, grow, reproduce, and survive. These traits are related to the efficiency of plants in using water, light, nutrients, and the ability to cope with environmental stress. Key physiological functional traits are as follows:

7.2.2.1 Photosynthetic Capacity

- **Leaf Nitrogen Content**: Nitrogen is a key component of chlorophyll and enzymes involved in photosynthesis. Leaves with higher nitrogen content typically have higher photosynthetic rates (Evans, 1989).
- **Stomatal Density and Conductance**: The density and opening of stomata control the amount of CO_2 entering the leaf and the rate of water vapor leaving it. Plants with higher stomatal conductance not only have higher photosynthetic rates due to increased CO_2 uptake but also have increased water loss (Cochard et al., 1996).
- **Chlorophyll Concentration**: In low-light environments such as forest understories, the photopigment content of leaves, especially chlorophyll b, can effectively absorb low-light and enhance the ability of leaves to capture light (Hui et al., 2023) and help plants to perform photosynthesis under low-light conditions.

7.2.2.2 Water-Use Efficiency (WUE)

- **Stomatal Regulation**: Plants balance CO_2 uptake for photosynthesis with water loss through transpiration by regulating the opening and closing of stomata. There are two main types of stomatal regulation: isohydric and anisohydric (Tardieu & Simonneau, 1998). Isohydric plants decrease stomatal conductance as soil water potential drops, maintaining a stable midday leaf water potential even during drought. In contrast, anisohydric species allow their midday leaf water potential to decrease as soil moisture declines during drought.
- **Leaf Structure**: Traits such as leaf thickness, cuticle thickness (waxy layer), and hairiness can reduce water loss and increase water-use efficiency by limiting evaporation. These traits are common in species from environments with limited water availability, such as temperate forests with seasonal droughts or tropical dry forests.
- **Leaf Size**: Second, leaf size indirectly affects water-use efficiency by influencing several water-use traits, including vein density (Sack et al., 2012), stomatal density (Conesa et al., 2020), specific leaf area (SLA) (Milla & Reich, 2007), tree height, and wood density (Malhado et al., 2009).

7.2.2.3 Nutrient-Use Efficiency

- **Leaf Longevity**: Nutrient-use efficiency is better related to leaf lifespan (Ankila, 2000). Species with leaves having a long lifespan often have higher nutrient-use efficiency, as they conserve and recycle nutrients within the plant over extended periods. This phenomenon can be seen in nutrient-poor environments of tropical rainforests, where rapid nutrient cycling occurs but soil nutrients are often scarce.

- **Nutrient Resorption**: Before leaves are shed, some species can withdraw nutrients such as nitrogen and phosphorus from old leaves and store them in other parts of the plant. This mechanism for nutrient conservation can be seen in environments with a limited nutrient availability in boreal and temperate forests prior to leaf senescence in forest tree species (Hagen-Thorn et al., 2006).
- **Mycorrhizal Associations**: In nutrient-deficient soils, it is important for many forest species to form symbiotic relationships with mycorrhizal fungi, which help plants access nutrients like phosphorus from the soil by increasing the surface-absorbing capacity of the roots of host plants (Kayama & Yamanaka, 2014). Usually, plants with efficient mycorrhizal relationships can outcompete others.

7.2.2.4 Respiration Rate

- **Temperature Sensitivity**: Respiration rates in plants rise with increasing temperature, so plants in warmer environments tend to have higher baseline respiration rates. Leaf respiration is particularly sensitive to temperature and reaches its peak at 45°C or between 48°C and 60°C, beyond which it declines sharply (Weerasinghe et al., 2014).
- **Leaf Mass per Area (LMA)**: In nutrient-poor or drought-prone environments where conserving energy is important, species have developed leaves with high LMA, which have lower respiration rates, as they are designed for long-term energy storage rather than quick turnover. In contrast, species with low LMA usually have higher rates of photosynthesis and respiration per unit leaf dry mass (Poorter et al., 2009).
- **Root Respiration**: Root respiration is an important trait in plants growing in environments where the availability of water or nutrients is limited and roots must maintain energy for nutrient uptake. Root respiration includes the respiration by living root cells, symbiotic mycorrhizal fungi and related microorganisms, decomposers acting on root secretions and dead root tissues in rhizosphere (Moyano et al., 2009). A significant amount of root respiration occurs in the fine roots, which can make up more than 95% of the root respiration of pine and birch stands (Kumar et al., 2019, 2021f, 2022; Giri et al., 2023).

7.2.2.5 Cold and Heat Tolerance

- **Antifreeze Proteins**: Some plant species in cold regions produce antifreeze proteins that bind to and prevent the growth of ice crystals in the apoplast (Gupta & Deswal, 2014) during subzero temperatures in winter and protect tissues from damage, thus providing tolerance to freezing.
- **Heat-Shock Proteins**: Some plant species produce heat-shock proteins to protect cellular functions during extreme heat conditions while ensuring survival and recovery after heat stress. Thus, the production of heat-shock

proteins develops thermotolerance. Also, it has been found that heat-shock proteins are formed under other environmental stresses such as water stress (Vásquez-Robinet et al., 2010).

- **Thermoregulation by Transpiration and Wilting**: Transpiration keeps the leaf cooler by several degrees as compared to the surrounding air temperature. Water stress can cause leaves to wilt, which reduces water loss and lowers the amount of light hitting the leaf surface, which in turn decreases the energy load on the leaf (Ranawana et al., 2023).

7.2.2.6 Drought Tolerance

- **Water Storage**: Some plant species can store water in specialized tissues and sapwood, which helps them to survive water-scarce conditions. This stored water helps in maintaining water balance without affecting the ability of the plant to transport water efficiently, which is especially important in water-limited environments (Stratton et al., 2000).
- **Xeromorphic Traits**: Plants in drought-prone areas have developed traits like small, thick leaves, sunken stomata, and waxy cuticles that reduce to water loss and make plants more resilient to water stress. Also, such plants can have structures that limit the loss of water and prevent cells from collapsing by providing structural support during dry periods or store water in photosynthetic tissues.
- **Root Adaptations**: Deep root systems and extensive lateral roots assist plants in accessing water from deeper and wider soil layers during droughts. Deep roots can be critical in alleviating water stress in many crops (Gewin, 2010).

7.2.3 Phenological Traits

Phenological functional traits refer to the timing of events related to life cycles in organisms, such as the onset of leafing, flowering, fruiting, and senescence (leaf shedding) in plants. These traits are influenced by environmental factors such as temperature, light, and water availability, and they play an important role in the response of plants and ecosystems to seasonal and climatic changes. They directly impact plant fitness and ecological interactions such as competition for resources, pollination, seed dispersal, and interactions with herbivores. Key phenological functional traits are as follows:

7.2.3.1 Leaf Phenology

- **Leafing Out (budburst)**: Temperature, photoperiod, and moisture are factors that influence the emergence of leaf in spring. The species that leaf out early can get access to sunlight and resources before others but also face risks of frost damage or herbivory. Air temperature is the key factor regulating budburst and leaf-out in temperate and boreal plants, with winter chilling and spring warmth being crucial for bud development (Linkosalo et al., 2006).
- **Leaf Longevity**: Generally, leaves are retained by evergreen tree species for multiple years, which enables them to continue photosynthesis throughout

the year, even in environments where the availability of nutrients is poor in environments like boreal or tropical rainforests. In contrast, leaves are shed annually by deciduous species to avoid damage during winter as in temperate forests or dry seasons as in tropical dry forests (Holbrook et al., 1995).

- **Senescence**: Many species of trees shed leaves in autumn in response to decreasing light and temperature to conserve water and nutrients during winter. Also, there are tree species that shed leaves during dry seasons to reduce water loss. For example, leaf senescence occurs in several species of Rosacea family due to the influence of temperature (Heide, 2011).

7.2.3.2 Flowering Phenology

- **Seasonal Flowering**: Light intensity and temperature affect the timing of flowering. Generally, high temperatures accelerate flowering, whereas low temperatures delay it (Lee et al., 2023). For example, in many temperate species, flowering takes place in the spring or early summer, which provides optimal temperatures and light conditions for pollinator activity, whereas in tropical species, flowering may occur year-round or be timed to coincide with specific environmental triggers, such as the onset of the rainy season.
- **Mass Flowering Events**: Some tropical and temperate tree species may exhibit simultaneous flowering of large numbers of individuals. This strategy attracts a large number of pollinators, which helps wind-pollinated plants with pollination and overwhelms seed-eating animals, so some seeds survive.
- **Pollinator Interactions**: Most of the plant species are fully or partially dependent on pollinators for reproduction. Therefore, flowering phenology is closely related to the availability of pollinators and synchronization between flowering time and peak activity of pollinators (e.g., bees, birds, and bats) ensures successful reproduction due to which a wide range of flowering times across species can be seen in tropical forests caused by the diversity of pollinators.

7.2.3.3 Fruit and Seed Phenology

- **Fruit Maturation**: In tropical forests, fruiting times may be staggered throughout the year, allowing species to take advantage of different seed dispersers and reduce competition for dispersal services. In temperate forests, fruiting typically occurs in late summer or early autumn, allowing seeds to mature before winter dormancy. In tropical forests, fruit maturation and ripening can be also observed throughout the year and peak activity of fruit maturation has been recorded from March to May. However, the time required for fruit maturation varied considerably with species (Singh & Kushwaha, 2006).
- **Seed Dispersal Timing**: The timing of seed release is critical for successful colonization. Wind-dispersed species often release seeds during dry periods or when wind speeds are high, increasing the chances of seeds being carried long distances. Animal-dispersed species may time their fruiting to coincide with the peak activity of fruit-eating animals, such as birds or mammals.

- **Dormancy and Germination**: Many species in temperate and alpine forests produce seeds that enter a period of dormancy, delaying germination until conditions are favorable (e.g., spring). This ensures that seedlings emerge during optimal growing conditions. In contrast, species in tropical forests with more consistent growing conditions may not require dormancy, allowing seeds to germinate soon after dispersal. Dormancy in temperate-zone deciduous fruit trees is a phase of development that allows trees to survive unfavorable conditions during the winter (Campoy et al., 2011).

7.2.3.4 Reproductive Phenology

The timing of reproductive events (flowering, seed production, and fruiting) is often synchronized with environmental factors such as temperature, light availability, and moisture. Species with different reproductive strategies may stagger their reproductive efforts to avoid competition for pollinators or seed dispersers. Species in temperate and alpine forests tend to have distinct reproductive seasons, with most flowering and seed production occurring during the warmer months. In tropical forests, some species reproduce continuously throughout the year, while others may have one or two distinct reproductive periods.

7.2.3.5 Phenological Adaptations to Climate and Environmental Stress

- **Temperature**: In temperate and alpine forests, many species rely on temperature changes as a cue for phenological events like leafing out, flowering, and fruiting. Air temperature was found to be the most crucial cue in determining the seasonal onset and senescence. Temperature directly regulates the dynamics of plant development via the processes of chilling and forcing requirements during dormancy (Luedeling et al., 2013; Kumar et al., 2015, 2016, 2017).
- **Moisture**: In tropical and temperate dry forests, rainfall is a key driver of phenological events. The number of tree species that flush leaves on a specific date is closely linked to rainfall in the two weeks prior (Lieberman & Lieberman, 1984). Many species synchronize their leafing, flowering, and fruiting with the onset of the rainy season to take advantage of increased water availability.
- **Photoperiod (Day Length)**: Photoperiod is an important cue for phenology in many temperate and alpine species, especially for flowering and leafing. Photoperiod ensures proper timing, especially if winter chilling is insufficient. Some species rely more on photoperiod, affecting how they respond to warming, with potential impacts on species interactions and ecosystems (Crabbe et al., 2016).

7.2.3.6 Phenological Synchrony and Asynchrony

- **Synchrony**: In some ecosystems, species exhibit synchronized phenological events to increase reproductive success or survival. For example, masting, or mast seeding, is the highly variable and synchronized production of seeds by plant populations that is very common in perennial plants (Crone and Rapp, 2014).

- **Asynchrony**: In diverse tropical ecosystems, phenological asynchrony is common, with different species flowering and fruiting at different times throughout the year. This reduces competition for pollinators and seed dispersers and helps maintain high biodiversity. In temperate forests, species with staggered leaf-out times may reduce competition for light and nutrients, promoting coexistence.

7.3 KEY FUNCTIONAL TRAITS UNDER DIFFERENT FOREST ECOSYSTEMS

Forests are some of the most complex and diverse ecosystems on Earth, with plant species exhibiting a range of functional traits that allow them to thrive in a highly dynamic environment. These forests are influenced by temperature, precipitation, canopy structure, density, soil, and competition for light and other resources. Below are some of the key functional traits of trees and plants that enable them to adapt and flourish in different forest environments or ecosystems.

7.3.1 Leaf Trait

Many tropical trees, such as *Shorea robusta* and *Tectona grandis*, have large leaves that enhance light capture efficiency, particularly in the low-light conditions found in dense canopies (Lusk et al., 2019; Kumar et al., 2020a,b, 2021a–e). This adaptation enables rapid growth and efficient energy production in the competitive forest environment. Additionally, many tropical plants have pointed or drip-tip leaves that allow rainwater to runoff quickly, preventing fungal growth and minimizing leaf damage. Trees like *Butea monosperma* feature thick cuticles and hydrophobic layers on their leaves, which help reduce water loss in the humid climate and conserve moisture during short dry spells.

Subtropical species, such as *Eucalyptus* spp. and *Pinus* spp., have leathery (sclerophyllous) leaves with thick cuticles, which help in minimizing water loss by reducing transpiration and helping the plants survive during dry periods. In deciduous species like *Shorea robusta*, shedding of leaves during the dry season to avoid damage (Holbrook et al., 1995) reduces water loss through transpiration and conserves moisture.

In temperate forests, deciduous trees (e.g., *Acer oblongum*) shed their leaves in autumn to conserve water and energy during the winter, when sunlight is limited. In contrast, evergreen trees (e.g., *Pinus* sp. and *Abies* sp.) retain needle-like leaves year-round, reducing surface area to minimize water loss, which helps them survive in nutrient-poor soils. Deciduous species like *Betula utilis* have broad leaves that maximize light capture during spring and summer, supporting efficient photosynthesis and rapid growth. Trees with a high SLA, such as *Salix* sp. and *Betula* sp., have larger leaves relative to their mass, which enhances photosynthesis and encourages fast growth under optimal conditions.

In alpine forests, plants typically have thick, small leaves with a waxy cuticle that helps reduce water loss and provides insulation against cold temperatures. These adaptations are crucial for surviving freezing conditions and conserving moisture.

Coniferous trees, such as pines and spruces, have needle-like leaves that further minimize water loss by reducing surface area, having thick cuticles, and fewer stomata. This enables photosynthesis to continue year-round, even during the colder months. Alpine plants often exhibit high LMA, which indicates thicker, denser leaves that are more resistant to frost damage and desiccation. High LMA also enhances energy efficiency, allowing these plants to store nutrients for the brief growing season.

7.3.2 The Root Systems

Many tropical trees, such as *Ceiba pentandra,* have shallow roots that spread widely near the soil surface to efficiently absorb nutrients from the organic-rich upper layers, whereas various tropical trees like *Tamarindus indica* have deep roots to access groundwater in dry periods. Tall trees such as *Ceiba pentandra, Ficus benghalensis,* and *Dipterocarpus* sp. often develop large, surface-level buttress roots for stability in the shallow, unstable soils, preventing toppling during storms. Epiphytes, like orchids and bromeliads, grow on other plants and use aerial roots to absorb water and nutrients from the air, rainwater, and organic debris.

Subtropical forests have trees like *Pinus roxburghii* with deep taproots in drier areas to access groundwater during droughts for providing water and nutrients, which are unavailable to surface roots and for anchoring the plant in loose soils. Also, shallow, wide-spreading roots of trees like *Morus* spp. and *Punica granatum* capture water from rains, helping plants store moisture for dry periods. Many subtropical plants and trees like *Pinus roxburghii* form symbiotic relationships with mycorrhizal fungi, enhancing nutrient uptake, particularly phosphorus, and improving water absorption and stress tolerance in plants by increasing the surface-absorbing capacity of the roots of host plants (Kayama & Yamanaka, 2014).

In temperate forests, there are also trees with deep roots (e.g., *Cedrus deodara*) which allow them to access groundwater during dry periods and help in alleviating water stress (Gewin, 2010) and provide stability in windy conditions. Also, dense networks of fine roots enable efficient nutrient and water uptake, particularly in heterogeneous soils with varying nutrient availability. Symbiotic fungi extend the root systems of temperate trees like *Abies pindrow*, which helps in improving access to nutrients like phosphorus and nitrogen, especially in nutrient-poor soils.

Plants of alpine forests also develop deep roots for stability in rocky, shallow soils and to access water and nutrients. Fine roots spread near the soil surface to maximize nutrient and water absorption, particularly during the short growing season when resources are more abundant from snowmelt or rain.

7.3.3 Water-Use Efficiency

Tropical plants often exhibit high photosynthetic rates due to abundant water and warmth. They regulate stomatal openings to balance water loss with carbon intake, particularly during drought conditions. Some tropical trees like *Acacia tortilis* redistribute water from deeper soil layers to the surface roots, benefiting surrounding vegetation during dry spells and promoting forest water cycling (Horton & Hart, 1998), and it is known as hydraulic lift. Certain plants like Bromeliads (*Aechmea* sp.) and cacti store rainwater in specialized structures, providing a reserve during dry periods and supporting other organisms in the ecosystem.

Many subtropical species (e.g., *Olea europaea*) regulate stomatal openings to conserve water during droughts, minimizing transpiration while maintaining essential moisture (Tardieu & Simonneau, 1998). Some subtropical species like *Taxodium* spp. also conduct hydraulic lift. Temperate trees balance their transpiration and photosynthetic needs, using water efficiently while ensuring consistent growth despite seasonal fluctuations in rainfall. During periods of sufficient moisture, temperate trees also open their stomata for photosynthesis and close them during droughts to conserve water. This regulation optimizes water use while maintaining growth.

In alpine forests, due to limited water in soils, the alpine plants have evolved high water-use efficiency, reduced transpiration, and maximized available moisture during the growing season. Alpine plants maintain stable leaf water potential despite fluctuating water availability, such as water from melting of snow or drought. This strategy (Isohydric Water Regulation) helps them survive in water-scarce environments.

7.3.4 Reproductive Strategies

Many tropical plants such as *Artocarpus heterophyllus, Musa* spp., and *Ficus* spp. flower and fruit year-round, ensuring continuous resources for pollinators and seed dispersers. This intermittent or continuous flowering and fruiting maintains a steady food supply. Tropical plants have evolved specialized pollination mechanisms such as large, colourful flowers that attract birds and bats, while small, nocturnal flowers appeal to moths and bats. Tropical plants often rely on animals for seed dispersal, with birds, bats, and mammals consuming fruit and dispersing seeds. Some species also use wind or water dispersal, particularly those near riverbanks or forest edges, with seeds adapted for flotation or wind travel.

Subtropical plants such as *Celtis australis*, *Myrica esculenta*, and *Prunus cerasoides* depend heavily on animals for seed dispersal. Birds, bats, and mammals disperse seeds after consuming fruit, benefiting both the plants and the animals. Some subtropical plants producing small or winged seeds (e.g., *Pinus roxburghii* and *Alnus nepalensis*) use wind for dispersal. Also, seeds of some subtropical plants (e.g., *Alnus nepalensis* and *Juglans regia*) are dispersed by water.

Temperate plants typically flower and fruit during the growing season when conditions are optimal. Changes in temperature and day length trigger flowering, ensuring that pollination occurs during favorable conditions. Wind dispersal is common in temperate forests, with species like oaks and maples producing winged seeds that travel long distances. Animal-mediated dispersal is also widespread, with animals eating fleshy fruits and depositing seeds in new locations, promoting genetic diversity and expanding plant populations.

Alpine plants, particularly conifers, also depend on wind for pollination. Reproductive structures like pollen cones allow for large-scale pollen dispersal, enabling fertilization in low-density populations. Seeds in alpine forests are adapted for dispersal. Many alpine plants, such as *Pinus* spp. and *Picea* sp., produce winged seeds carried by the wind. Animal-mediated dispersal is also common, with rodents and birds transporting seeds to new areas.

7.3.5 Heat and Cold Adaptations

Tropical plants face the challenges of high heat and excessive water loss and have developed several mechanisms to cope with these conditions. Many tropical plants including *Mangifera indica* and *Carica papaya* produce heat-shock proteins, which develop thermotolerance that prevents cellular damage.

Efficient photosynthesis is regulated to balance heat and water use, with thick, waxy cuticles on leaves reducing water loss and protecting against intense sunlight. Leaf hairs (trichomes) help trap moisture and reflect sunlight, and stomata are carefully regulated to minimize water loss while supporting photosynthesis.

In case of temperate plants, which experience seasonal temperature fluctuations, they have developed adaptations to survive extreme cold. Temperate trees like *Quercus* spp. and *Carya* spp. have thick bark that insulates living tissues, protecting them from freezing temperatures and physical damage. Deciduous species shed leaves in winter, reducing water loss and preventing frost damage. This also minimizes the risk of freezing in winter when water availability is low. Leaf senescence in *Prunus* spp. and *Pyrus* spp. is attributed to temperature (Heide, 2011).

Also, many species enter deep dormancy during winter to survive unfavorable conditions during the winter (Campoy et al., 2011) halting metabolic processes and protecting sensitive tissues from cold.

Alpine plants endure extreme cold, high winds, and short growing seasons and have developed specialized adaptations. Some alpine plants produce antifreeze proteins that prevent ice crystal formation in cells, protecting tissues from freeze damage (Gupta & Deswal, 2014).

In addition to that, many alpine plants enter dormancy during winter, conserving energy and resources until warmer conditions return. They also develop dark pigmentation in their leaves or stems, which absorbs solar radiation, helping the plant warm up during cold mornings and protecting it from harmful UV radiation at high altitudes.

7.3.6 Defensive Strategies

Many tropical plants such as *Azadirachta indica* produce toxic secondary metabolites (e.g., alkaloids, terpenoids, and tannins) that make their tissues unpalatable or harmful to herbivores. Tropical plants like *Acacia* spp. and *Zizyphus* spp. have spines, thorns, or tough leaves that deter herbivores. Some species also have sticky secretions or fine hairs that trap pests or reduce feeding by other animals. Certain plants form mutualistic relationships with ants (e.g., *Acacia* spp.), offering nectar or shelter in exchange for protection against herbivores.

In some subtropical as well as in temperate plants, thorns, spines (e.g., *Berberis aristata*), and leathery, tough, fibrous, and lignin-rich leaves (*Magnolia* spp.) help reduce herbivory by making plants difficult to consume. Trichomes (plant hairs) on leaves also deter herbivores by making them less palatable. Some plants also produce tough seeds with hard coats (e.g., *Juglans regia*) to prevent damage during dispersal. Some trees (e.g., *Pinus roxburghii*) produce organic compounds like alkaloids and terpenoids, along with essential oils and resins, which make plants unpalatable or toxic. Also, the timing of flowering and seed production is adjusted to avoid peak herbivore activity. Seed dormancy or overproduction of seeds helps ensure plant survival and reproduction.

In alpine forests, plants have low, cushion-like growth forms, tough woody stems, and spines or prickles that provide physical protection. Dense mats of plants also limit herbivore access.

Dense trichomes can deter herbivores and reflect sunlight to reduce grazing. Also, they often produce organic compounds (e.g., alkaloids and terpenoids) and bitter phenolic compounds that deter herbivores. Essential oils and Volatile Organic Compounds (VOCs) serve to repel grazers or signal neighboring plants to ramp up defenses. Alpine plants often flower and set seeds early to avoid herbivore pressure, use seed dormancy for protection, and regenerate vegetatively to recover from grazing. Cold and desiccation tolerance, along with slow growth and energy allocation toward defenses rather than rapid growth, enable alpine plants to survive both herbivory and harsh environmental conditions.

7.4 CONCLUSION

In conclusion, plant functional traits across forest ecosystems are finely tuned to their specific environmental challenges. Tropical forests feature large leaves and rapid growth to exploit abundant sunlight and water. Subtropical forests have drought-resistant leaves and deep roots to cope with seasonal water scarcity. Temperate forests balance deciduous and evergreen strategies, with dense wood and specialized leaves, allowing species to thrive in variable temperatures and water availability. Alpine forests, adapted to extreme cold, have small, thick leaves and low-growth forms to minimize exposure to harsh conditions. Across all ecosystems, traits like water-use efficiency, reproductive strategies, and defenses are crucial for survival and supporting resilience through strategies like rapid growth in tropical and temperate forests or delayed senescence in alpine species. These functional traits not only ensure survival but also maintain ecosystem processes, biodiversity, and the stability of forests.

REFERENCES

Ambika, S., Manonmani, V., and Somasundaram, G. (2014). Review on effect of seed size on seedling vigour and seed yield. *Research Journal of Seed Science*, *7*(2), 31–38.

Ankila, J. H. (2000). Photosynthetic nutrient-use efficiency in tree fast-growing tropical trees with differing leaf longevities. *Tree Physiology*, *20*(14), 937–944. https://doi.org/10.1093/treephys/20.14.937

Bold, G., Speck, T., Blömer, J., Bertling, J., and Speck, O. (2010). Insulation capability of the bark of trees with different fire adaptation. *Journal of Materials Science*, *45*, 5950–5959. https://doi.org/10.1007/s10853-010-4680-4.

Bouchard, E., Searle, E. B., Drapeau, P., Liang, J., Gamarra, J. G., Abegg, M., Alberti, G., Zambrano, A. A., Alvarez-Davila, E., Alves, L. F., and Avitabile, V. (2024). Global patterns and environmental drivers of forest functional composition. *Global Ecology and Biogeography*, *33*, 303–324.

Cakmak, I., Hengeler, C., and Marschner, H. (1994) Partitioning of shoot and root dry matter and carbohydrates in bean plants suffering from phosphorus, potassium and magnesium deficiency. *Journal of Experimental Botany*, *45*, 1245–1250.

Campoy, J. A., Ruiz, D., and Egea, J. (2011). Dormancy in temperate fruit trees in a global warming context: a review. *Scientia Horticulturae*, *130*(2), 357–372.

Claveau, Y., C. Messier, P.G. Comeau and K.D. Coates. (2002). Growth and crown morphological responses of boreal conifer seedlings and saplings with contrasting shade tolerance to a gradient of light and height. *Canadian Journal of Foresr Research*, 32:458–468.

Claveau Y., Messier C., and Comeau, P. G. (2005). Interacting influence of light and size on aboveground biomass distribution in sub-boreal conifer saplings with contrasting shade tolerance. *Tree Physiology*, *25*, 373–384.

Cochard, H., Bréda, N., and Granier, A. (1996). Whole tree hydraulic conductance and water loss regulation in Quercus during drought: evidence for stomatal control of embolism? *Annales des Sciences Forestieres*, *53*(2–3), 197–206.

Conesa, M. À., Muir, C. D., Molins, A., and Galmés, J. (2020). Stomatal anatomy coordinates leaf size with *Rubisco kinetics* in the Balearic *Limonium, AoB Plants,* 12. https://doi.org/10.1093/aobpla/plz050.

Crabbe, R. A., Dash, J., Rodriguez-Galiano, V. F., Janous, D., Pavelka, M., and Marek, M. V. (2016). Extreme warm temperatures alter forest phenology and productivity in Europe. *Science of the Total Environment*, *563*, 486–495.

Crone, E. E., and Rapp, J. M., (2014). Resource depletion, pollen coupling, and the ecology of mast seeding. *Annals of the New York Academy of Sciences, 1322*, 21–34.

Estrada-Medina, H., Graham, R. C., Allen, M. F., Jiménez-Osornio, J. J., and Robles-Casolco, S. (2012). The importance of limestone bedrock and dissolution karst features on tree root distribution in northern Yucatán, México. *Plant and Soil*, *362*, 37–50.

Evans, J. R. (1989). Photosynthesis and nitrogen relationships in leaves of C3 plants. *Oecologia*, *78*, 9–19.

Gewin V. (2010). An underground revolution. *Nature*, *466*, 552–553.

Giri, K., Mishra, G., Suyal, D. C., Kumar, N., Doley, B., Das, N., Baruah, R. C., Bhattacharyya, R., and Bora, N. (2023). Performance evaluation of native plant growth-promoting rhizobacteria for paddy yield enhancement in the jhum fields of Mokokchung, Nagaland, North East India. *Heliyon*, *9*, e14588.

Gupta, R., and Deswal, R. (2014). Antifreeze proteins enable plants to survive in freezing conditions. *Journal of Biosciences*, *39*, 931–944.

Hagen-Thorn, A., Varnagiryte, I., Nihlgård, B., and Armolaitis, K. (2006). Autumn nutrient resorption and losses in four deciduous forest tree species. *Forest Ecology and Management*, *228*(1–3), 33–39.

Heide, O.M. (2011). Temperature rather than photoperiod controls growth cessation and dormancy in Sorbus species. *Journal of Experimental Botany*, *62*(15), 5397–5404.

Holbrook, N. M., Whitbeck, J. L. and Mooney, H. A. (1995). Drought responses of neotropical dry forests trees. In: Bullock, S. H., Medina, E., and Mooney, H. A., Eds., *Seasonally Dry Tropical Forests*, Cambridge University Press, Cambridge, pp. 243–276. https://doi.org/10.1017/CBO9780511753398.010

Horton, J., and Hart, S. (1998). Hydraulic lift: a potentially important ecosystem process. *Trends in Ecology and Evolution*, *13*, 232–235. https://doi.org/10.1016/S0169-5347(98)01328-7.

Hui, J., Wu, C., Li, X., Huang, L., Jiang, Y., and Zhang, B. (2023). The effect of light availability on photosynthetic responses of four *Aglaonema commutatum* cultivars with contrasting leaf pigment. *Applied Sciences*, *13*(5), 3021. https://doi.org/10.3390/app13053021

Kayama, M., and Yamanaka, T. (2014) Growth characteristics of ectomycorrhizal seedlings of *Quercus glauca*, *Quercus salicina*, and *Castanopsis cuspidata* planted on acidic soil. *Trees: Structure and Function, 28*, 569–583.

Klepper, B. (1991). Root-shoot relationships. In Waisel, Y., Eshel, A., and Kafkafi, U., Eds., *Plant Roots: The Hidden Half*, Marcel Dekker Inc., New York, pp. 265–286.

Kraft, N. J. B., Metz, M. R., Condit, R. S., and Chave, J. (2010). The relationship between wood density and mortality in a global tropical forest data set. *New Phytologist*, *188*, 1124–1136

Kumar, A., Dwivedi, G. K., Tewari, S., Paul, J., Anand, R., Kumar, N., Kumar, P., Singh, H., and Kaushal, R. (2020a). Carbon mineralization and inorganic nitrogen pools under *Terminalia chebula* Retz.-Based agroforestry system in Himalayan foothills, India. *Forest Science*, *66*(5), 634–643.

Kumar, A., Dwivedi, G. K., Tewari, S., Paul, J., Sah, V. K., Singh, H., Kumar, P., Kumar, N., Kaushal, R. (2020b) Soil organic carbon pools under *Terminalia chebula* Retz. based agroforestry system in Himalayan foothills, India. *Current Science*, *118*(7), 1098–1103.

Kumar, A., Kumar, P., Singh, H., Bisht, S., and Kumar, N. (2021a). Relationship of physiological plant functional traits with soil carbon stock in temperate forest of Garhwal Himalaya. *Current Science*, *120*(8), 1368–1373.

Kumar, A., Kumar, P., Singh, H., and Kumar, N. (2021b). Modulation of plant functional traits under essential plant nutrients during seasonal regime in natural forests of Garhwal Himalayas. *Plant and Soil*, *465*, 197–212.

Kumar, A., Kumar, P., Singh, H., and Kumar, N. (2021d). Impact of plant functional traits on near saturated hydraulic conductivity of soil under different forests of Kempty watershed in Garhwal Himalaya. *Indian Journal of Soil Conservation*, *49*(1), 38–44.

Kumar, A., Kumar, P., Singh, H., and Kumar, N. (2021e). Adaptation and mitigation potential of roadside trees with bio-extraction of heavy metals under vehicular emissions and their impact on physiological traits during seasonal regimes. *Urban Forestry & Urban Greening*, *58*, 126900.

Kumar, A., Singh, H., Kumari, G., Bisht, S., Malik, A., Kumar, N., Singh, M., Raturi, A., Barthwal, S., Thakur, A., and Kaushal, R. (2022). Adaptive resilience of roadside trees to vehicular emissions via leaf enzymatic, physiological, and anatomical trait modulations. *Environmental Pollution*, *313*, 120191.

Kumar, A., Tewari, S., Singh, H., Kumar, P., Kumar, N., Bisht, S., Kushwaha, S., Tamta, N., and Kaushal, R. (2021c). Biomass accumulation and carbon stocks in different agroforestry system prevalent in Himalayan foothills, India. *Current Science*, *120*(6), 1083–1088.

Kumar, N., Jeena, N., and Singh, H. (2019). Elevated temperature modulates rice pollen structure: a study from foothill Himalayan agro-ecosystem in India. *3 Biotech*, *9*, 175.

Kumar, N., Jeena, N., Kumar, A., Khairakpam, R., and Singh, H. (2021f). Comparative response of rice cultivars to elevated air temperature in Bhabar region of Indian Himalaya: status on yield attributes. *Heliyon*, *7*, e07474.

Kumar, N., Kumar, N., Shukla, A., Shankhdhar, S. C., and Shankhdhar, D. (2015). Impact of terminal heat stress on pollen viability and yield attributes of rice (*Oryza sativa* L.). *Cereal Research Communications*, *43*(4), 616–626.

Kumar, N., Shankhdhar, S. C., and Shankhdhar, D. (2016). Impact of elevated temperature on antioxidant activity and membrane stability in different genotypes of rice (*Oryza sativa* L.). *Indian Journal of Plant Physiology (Now Plant Physiology Report)*, *21*(1), 37–43.

Kumar, N., Suyal, D. C., Sharma, I. P., Verma, A., and Singh, H. (2017). Elucidating stress proteins in rice (*Oryza sativa* L.) genotype under elevated temperature: a proteomic approach to understand heat stress response. *3 Biotech*, *7*, 205.

Lee, Z., Kim, S., Choi, S. J., Joung, E., Kwon, M., Park, H. J., and Shim, J. S. (2023). Regulation of flowering time by environmental factors in plants. *Plants*, *12*, 3680. https://doi.org/10.3390/plants12213680

Lieberman, D., and Lieberman, M. (1984). The causes and consequences of synchronous flushing in a dry tropical forest. *Biotropica*, *16*, 193–201.

Linkosalo, T., Hakkinen, R., and Hanninen H. (2006). Models of the spring phenology of boreal and temperate trees: is there something missing? *Tree Physiology*, *26*, 1165–1172.

Luedeling, E., Guo, L., Dai, J., Leslie, C., and Blanke, M. M. (2013). Differential responses of trees to temperature variation during the chilling and forcing phases, *Agricultural and Forest Meteorology*, *181*, 33–42, ISSN 0168-1923. https://doi.org/10.1016/j.agrformet.2013.06.018

Lusk, C. H., Grierson, E. R. P., and Laughlin, D. C. (2019). Large leaves in warm, moist environments confer an advantage in seedling light interception efficiency. *New Phytologist*, *223*(3), 1319–1327.

Malhado, A. C. M., Malhi, Y., Whittaker, R. J., Ladle, R. J. Ter Steege, H., Phillips, O. L., Butt, N., Aragao, L., Quesada, C. A., Araujo- Murakami, A., Arroyo, L., Peacock, J., Lopez-Gonzalez, G., Baker, T. R., and Anderson, L. O. (2009). Spatial trends in leaf size of Amazonian rainforest trees. *Biogeosciences*, *6*, 1563–1576. https://doi.org/10.5194/bg-6-1563-2009

Milla, R., and Reich, P. B. (2007). The scaling of leaf area and mass: the cost of light interception increases with leaf size. *Proceedings of the Royal Society: Biological Sciences*, *274*, 2109–2115. https://doi.org/10.1098/rspb.2007.0417

Moyano, F. E., Atkin, O. K., Bahn, M., Bruhn, D., Burton, A. J., Heinemeyer, A., and Wieser, G. (2009). Respiration from roots and the mycorrhizosphere. In Kutsch, W. L., Bahn, M., and Heinemeyer, A., Eds., *Soil Carbon Dynamics–An Integrated Methodology*, Cambridge University Press, Cambridge, UK.

Nock, C. A., Vogt, R. J., and Beisner, B. E. (2016). Functional traits. *eLS*, *1*, 1–8.

Osada, N. (2012). Crown exposure to light and tree allometry of 11 tree species in a snowy cool-temperate forest in Japan. *Plant Ecology*, *213*,783–794.

Poorter, H., Niinemets, Ü., Poorter, L., Wright, I. J., and Villar, R. (2009). Causes and consequences of variation in leaf mass per area (LMA): a meta-analysis. *New Phytologist*, *182*(3), 565–588.

Poorter, L., Rozendaal, D., Sanaphre-Villanueva, L., Bongers, F., Almeida, J., Andrade, J., Dupuy, J., Hernandez-Stefanoni, J., Lohbeck, M., van der Sande, M., Álvarez, F., Villa, P., Moreno, V., Massoca, P., Junqueira, A., Utrera, L., Kennard, D., Dent, D., Piotto, D., and Colletta, G. (2019). Wet and dry tropical forests show opposite successional pathways in wood density but converge over time. *Nature Ecology and Evolution*, *3*. 928–934. https://doi.org/10.1038/s41559-019-0882-6

Ranawana, S. R. W. M. C. J. K., Bramley, H., Palta, J. A., and Siddique, K. H. (2023). Role of Transpiration in Regulating Leaf Temperature and its Application in Physiological Breeding. In *Translating Physiological Tools to Augment Crop Breeding*. Springer Nature, Singapore, pp. 91–119

Schwinning, S. (2010). The ecohydrology of roots in rocks, *Ecohydrology*, *245*, 238–245

Scoffoni, C., Vuong, C., Diep, S., Cochard, H., and Sack, L. (2014). Leaf shrinkage with dehydration: coordination with hydraulic vulnerability and drought tolerance. *Plant Physiology*, *164*(4), 1772–1788.

Sellier, D., and Fourcaud, T. (2009). Crown structure and wood properties: influence on tree sway and response to high winds. *American Journal of Botany*, *96*(5), 885–896.

Sfair, J. C., Rochelle, A. L. C., Rezende, A. A., and Martins, R. O. B. E. R. T. O. (2016). Liana avoidance strategies in trees: combined attributes increase efficiency. *Tropical Ecology*, *57*(3), 559–566.

Singh, K. P., and Kushwaha, C. P. (2006). Diversity of flowering and fruiting phenology of trees in a tropical deciduous forest in India. *Annals of Botany*, *97*(2), 265–276.

Sterck, F. J., and Bongers, F. (2001) Crown development in tropical rain forest trees: patterns with tree height and light availability. *Journal of Ecology*, *89*, 1–13.

Stratton, L., Goldstein, G., and Meinzer, F. C. (2000). Stem water storage capacity and efficiency of water transport: their functional significance in a Hawaiian dry forest. *Plant, Cell and Environment*, *23*(1), 99–106.

Sack, L., Scoffoni, C., McKown, A. D., Frole, K., Rawls, M., Havran, J. C., Tran, H. and Tran, T. (2012). Developmentally based scaling of leaf venation architecture explains global ecological patterns. *Nature Communications*, *3*(1), 837.

Talley, S.M., W. N. Setzer and B. R. Jackes. (1996). Host associations of two adventitious-root-climbing vines in a north Queensland tropical rain forest. *Biotropica*, *28*, 361–366.

Tardieu, F., and Simonneau, T. (1998). Variability of species among stomatal control under fluctuating soil water status and evaporative demand: modeling isohydric and anisohydric behaviours. *Journal of Experimental Botany*, *49*, 419–432.

Valladares, F., and Pugnaire, F. I. (1999). Tradeoffs between irradiance capture and avoidance in semiarid environments simulated with a crown architecture model. *Annals of Botany*, *83*, 459–470

Vásquez-Robinet, C., Watkinson, J. I., Sioson, A. A., Ramakrishnan, N., Heath, L. S., and Grene, R. (2010). Differential expression of heat shock protein genes in preconditioning for photosynthetic acclimation in water-stressed loblolly pine. *Plant Physiology and Biochemistry*, *48*(4), 256–264.

Weerasinghe, L. K, Creek, D., Crous, K. Y., Xiang, S., Liddell, M. J., Turnbull, M. H., and Atkin, O. K. (2014). Canopy position affects the relationships between leaf respiration and associated traits in a tropical rainforest in Far North Queensland. *Tree Physiology*, *34*, 564–584.

Yang, D., Zhang, Y. J., Song, J., Niu, C. Y., and Hao, G. Y. (2019). Compound leaves are associated with high hydraulic conductance and photosynthetic capacity: evidence from trees in Northeast China. *Tree Physiology*, *39*, 729–739. https://doi.org/10.1093/treephys/tpy147

8 Functional Trait Relationships with Tree Growth and Productivity

Divya Chaudhary, Anjali Joshi, and Arushi Arora

8.1 INTRODUCTION

Functional traits are fundamental characteristics that influence plant fitness and adaptability, shaping how trees interact with their environment. These traits encompass morphological, physiological, and phenological attributes that allow trees to optimize resource use and enhance functional efficiency. By understanding how these traits vary across different environmental conditions, researchers can gain critical insights into plant performance, ecological strategies, and overall forest dynamics (Violle et al., 2007). As environmental conditions change, particularly due to climate change, the study of functional traits becomes increasingly relevant for predicting the resilience and productivity of tree species. For instance, in water-scarce environments, trees often exhibit traits that promote drought avoidance. Traits such as increased wood density and higher leaf mass per area (LMA) help reduce the risk of hydraulic failure and enhance structural defence (Correia et al., 2018). However, these adaptations come with trade-offs. Trees with such traits tend to have slower growth rates and reduced biomass production, as they prioritize survival over rapid growth (Pérez-Ramos et al., 2019). These trade-offs underscore the importance of understanding how functional traits influence tree growth and productivity across various ecosystems.

This chapter explores the intricate relationships between functional traits, tree growth, and productivity. It delves into key concepts of tree growth patterns and productivity measures, and then examines how different functional traits ranging from morphological to physiological directly influence these outcomes. The chapter also discusses how trees use several traits to navigate trade-offs, emphasizing the role of environmental influences in shaping trait expression. Case studies will provide real-world examples of how functional traits impact tree growth and productivity, offering valuable insights into forest management and conservation efforts in the context of global environmental change. This chapter aims to give an extensive understanding of the processes influencing tree development and productivity by concentrating on the function of functional traits. This will help readers gain a greater knowledge of how trees adapt to their constantly changing environment.

DOI: 10.1201/9781003545231-8

8.2 TREE GROWTH AND PRODUCTIVITY

8.2.1 Understanding Tree Growth Patterns

Tree growth is a complex process influenced by various genetic, environmental, and physiological variables. It primarily occurs through two mechanisms: primary growth, which causes shoot and root elongation, and secondary growth, which increases the girth of the tree through vascular cambium activity. Primary growth takes place in small regions known as apical meristems, where new leaves, as well as the height and length of branches and roots, are developed. One or more leaves are formed at the node, a location at the apical meristems, followed by the internode of the stem. Environmental factors like light and nutrient availability affect how many leaves a tree species produces throughout a growing season; under favourable conditions, more leaves are produced. However, the availability of water during the maturation process has the biggest impact on the length of the internodes and the size of the leaves. For example, a spring drought or root damage can reduce water supply, leading to smaller leaves and a sparse canopy appearance.

Apart from the elongation of their roots and shoots, trees undergo secondary growth, which expands their diameter. This process is essential for the formation of woody tissue, which provides structural support as well as storage capacity. Peeling off a branch reveals the vascular cambium, the soft inner layer that resides near the wood. Every year, the vascular cambium generates new inner bark (phloem) on the outside and new wood (xylem) on the inside. Sugars from the leaves are moved downward by the phloem, while water and nutrients are moved upward by the xylem. In temperate climates, the vascular cambium stops growing throughout the winter, resulting in observable annual growth rings at the end of the growing season. These annual rings might not exist in tropical regions where growth happens all year round (Larcher, 2003). Certain species, particularly those found in temperate areas, have prominent rings because earlywood, or spring wood, is less dense than latewood, or wood formed later in the summer and autumn. Hardwood trees that exhibit distinct earlywood and latewood characteristics are referred to as "ring-porous," whereas trees that maintain the same wood structure throughout the growing season are referred to as "diffuse-porous." However, coniferous and softwood species are not referred to by this terminology (Tyree and Zimmermann, 2013).

Cork, the outer layer of bark, is formed by the cork cambium, which accounts for a small but significant proportion of diameter growth. Mature trees maintain a consistent thickness of bark because the outermost layer sheds even as new cork is generated periodically. Consequently, the diameter growth of adult trees is primarily due to the vascular cambium's production of wood, but cork cambium growth plays a substantial role in diameter expansion in saplings (Tyree and Zimmermann, 2013). Growth patterns in trees differ significantly based on species and environmental variables. For instance, trees in tropical rainforests experience continuous growth throughout the year due to the favourable climate, while temperate trees display seasonal growth, with rapid expansion during the growing season and dormancy during the winter months (Larcher, 2003).

8.2.2 Productivity Measures in Forest Ecosystems

Productivity in forest ecosystems refers to the rate at which trees and other plants create biomass, which is crucial for carbon cycling and energy transfer within ecosystems. Tree productivity is commonly measured using two key metrics: net primary productivity (NPP) and gross primary productivity (GPP). NPP is the amount of carbon stored in plant tissues after autotrophic respiration is taken into consideration, while GPP represents the total carbon captured through photosynthesis (Chapin et al., 2002). A number of variables, including soil fertility, water availability, light intensity, and climate, have a major impact on productivity, which varies greatly throughout forest types and geographic regions. For instance, tropical forests tend to have high productivity due to plentiful sunlight and rainfall, whereas boreal forests show lower productivity because of colder temperatures and shorter growing seasons (Luyssaert et al., 2007).

Forest productivity is frequently quantified in the field by measuring both GPP (the overall rate of carbon capture and storage) and NPP (the rate of carbon intake after accounting for autotrophic respiration). According to Malhi et al. (2017), the Global Ecosystems Monitoring protocol offers a thorough framework for monthly or seasonal measurements of all the main components of NPP and autotrophic respiration. To calculate NPP, several characteristics of forest productivity are monitored, such as canopy litterfall, leaf loss from herbivory, aboveground woody productivity for medium and large trees (diameter >10 cm, measured every three months), and annual wood productivity for smaller trees (diameter 2–10 cm). Additional factors include coarse root productivity, which is deduced from aboveground productivity data; fine root productivity, which is assessed using ingrowth cores every three months; and branch turnover in live trees.

Autotrophic respiration (Ra) is estimated by aggregating several measurements: monthly measurements of rhizosphere respiration, aboveground woody respiration (scaled with surface area), and belowground coarse root and bole respiration (using a fixed multiplier of stem respiration). Additionally, leaf dark respiration is assessed from measurements taken across several leaves during two different seasons. From these data, GPP is calculated by summing NPP and Ra, with the equation $GPP = NPP + Ra$. Carbon-use efficiency (c), a key indicator of how efficiently plants use captured carbon for growth, is represented by the ratio $c = NPP/GPP$.

Notably, measures based on tree diameter at breast height only account for a minor portion of GPP and NPP. While wood production usually only contributes around 10% of GPP and 25% of NPP, larger contributions are made by factors including respiration, fine root turnover, and litterfall (Malhi et al., 2017). Furthermore, there is typically little correlation between wood production and overall NPP or GPP (Malhi et al., 2011). Despite the inherent assumptions in these estimates, they provide a robust approximation of stand-level GPP and NPP, helping researchers better understand forest productivity and carbon dynamics.

8.2.3 Factors Influencing Tree Growth and Productivity

Determining how ecosystems react to environmental changes and how taxonomic and functional diversity affect these responses is one of the main concerns in contemporary ecosystem science (Lavorel and Garnier, 2002). Environmental gradients

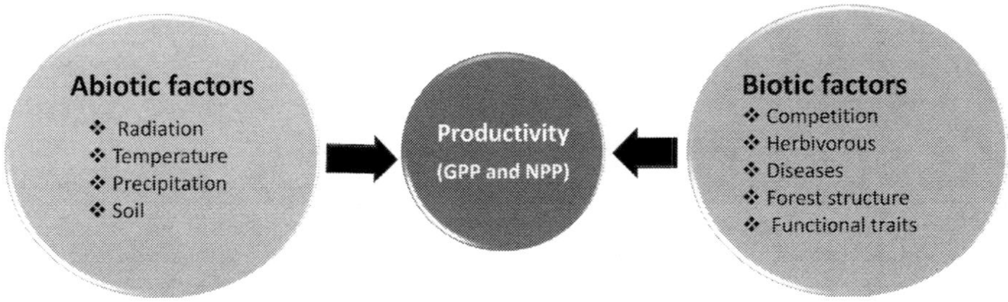

FIGURE 8.1 Key factors influencing tree productivity.

can provide valuable insights into the variables governing ecosystem function since environmental conditions vary both geographically and temporally. Tropical forests, with their high levels of biodiversity, provide particularly valuable systems for studying these dynamics (Wright, 2002). Because of these forests' richness, scientists can investigate the association between diversity and function without the confounding effects of specific dominant species. Because of their steady soil moisture levels and year-round growing seasons, tropical elevation gradients are particularly helpful among these systems because they provide "natural laboratories" for researching how temperature affects ecosystem function without being impacted by seasonal fluctuations or dormant periods (Malhi et al., 2010; Sundqvist et al., 2013).

Direct impacts of the environment on ecosystem production should be distinguished from indirect controls that work through species composition and forest structure. This distinction makes it clearer how much productivity can be deduced from evaluating the composition of ecosystems. Environmental factors, such as temperature, rainfall, and soil fertility, are typically viewed as direct drivers of productivity (Figure 8.1). Some studies indicate that higher average temperatures within the typical tropical climate range may promote tree growth, even though temperature is typically not a limiting factor in most tropical places (Cleveland et al., 2011). Rainfall has been associated with faster tree growth rates in seasonal tropical forests (Brienen and Zuidema, 2005). However, other studies show that solar radiation is a key determinant of productivity throughout the Amazon, especially during the rainy season (Nemani et al., 2003; Graham et al., 2003). Another aspect is soil fertility: in lowland tropical forests, the availability of phosphorus (P) often serves as the limiting element, whereas in colder, montane regions with younger soils, nitrogen (N) is typically the limiting factor (Quesada et al., 2012). As long as no other resources are restricting, increasing any of these elements can generally improve tree growth, demonstrating the direct and immediate impact of the environment on production.

8.3 RELATIONSHIP BETWEEN FUNCTIONAL TRAITS AND TREE GROWTH

The functional traits in plants refer to the set of morphological, phenological, or physiological features that majorly influence the growth, survival, and/or reproduction of the plant, thereby ascertaining the region and extent of a plant species' distribution (Violle et al., 2007; Kumar et al., 2019, 2021f, 2022; Giri et al., 2023).

These functional characteristics vary greatly between species as well as within them. The distribution of different species in a given area is significantly affected by local environmental conditions, demonstrating the presence of environmental filtering in nature, which selects a subset of species from a larger species pool, resulting in correlations between specific traits and regional and local environmental conditions (Boet et al., 2020). Moreover, the way resources are partitioned amongst different species creates a pattern of trade-off between various traits in these species affecting the plant survival strategies leading to differential success of the species under different environmental conditions. Plant functional characteristics are therefore important in determining plant performance and can be utilized for predicting species dispersion across environmental gradients (Sterck et al., 2014).

8.3.1 Morphological Traits and Growth Patterns

A wide range of morphological traits acts as functional characteristics determining the species distribution, survival, and growth (Figure 8.2). In recent decades, researchers have investigated the relationships between wood density traits in trees and various plant functions. Wood density has a significant role in converting assimilated carbon into biomass for stems and branches, thereby affecting the height of the tree as well as crown development. Wood density is more closely related to tree height than diameter. Generally, tree species with denser wood exhibit slower growth rates,

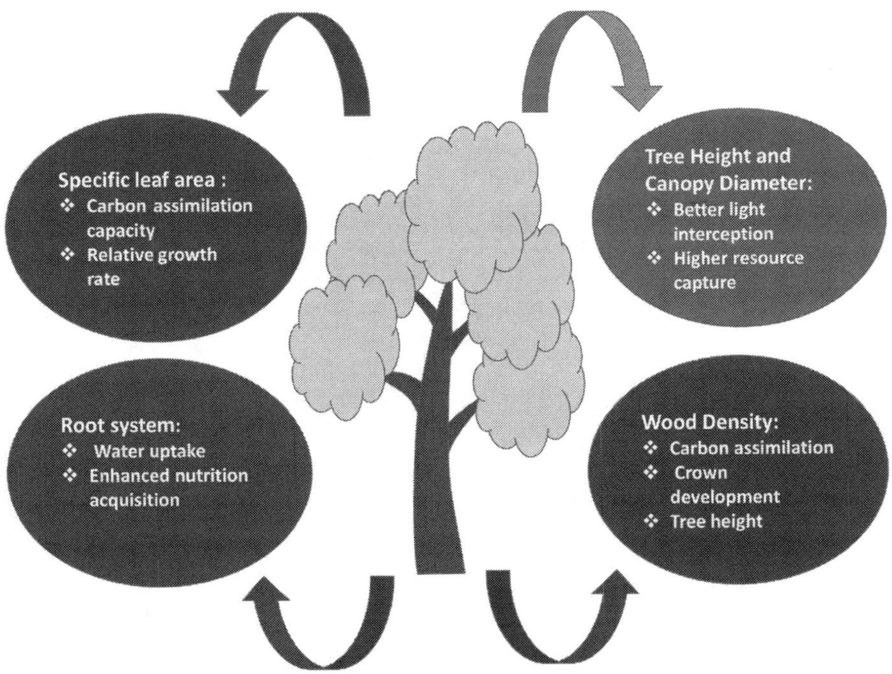

FIGURE 8.2 Functional traits determining tree growth and productivity.

later reproduction, and enhanced strength. Furthermore, wood density influences plant water dynamics, including internal water balance and water transportation capacity, and has a substantial impact on plant death. Higher wood density is linked with enhanced energy requirement for wood development and reduced hydraulic conductance, thereby negatively affecting the plant growth (Chave et al., 2009). In the vast majority of tree species, higher wood density is frequently associated with lower mortality and increased survival. It has also been discovered that the total carbon absorption capacity and, thus, the relative growth rates in trees are influenced by the photosynthetic area of the tree species, which is governed by specific leaf area, tree height, and overall crown size (Figure 8.2). Higher specific leaf area leads to significant surplus assimilation of carbon and therefore significantly enhances growth rates (Rees et al., 2010). Similarly, tree species with greater height and larger canopy diameter, on account of having better light interception, have higher resource capture capacity and thus are associated with enhanced growth rates. Turnbull et al. (2012) also examined the influence of mass of the seeds on growth rate and seedling survival and reported that seed mass had a positive impact on the likelihood of seedling survival in tree species; however, species with smaller seeds demonstrated higher relative growth rates in terms of height, though this trend persisted only for initial months after planting and dissipates when growth rates are compared at later stages. According to these findings, seed mass is a valuable metric for forecasting a tree's ability to establish itself during forest restoration initiatives.

8.3.2 Physiological Traits and Their Impact on Growth

Tree physiological traits including drought tolerance, hydraulic conductivity, and carbon assimilation capacity have a direct effect on the performance and fitness of tree species under a natural environment (Zhu et al., 2018). Generally, individuals possessing greater hydraulic conductivity and better gaseous exchange support efficient water and nutrient movement in the plant, facilitating carbon assimilation and thereby speeding up the volumetric growth. Physiological activities such as gas exchange and xylem hydraulic conductivity are directly related and influenced by wood density. In dry forest ecosystems, trees with low wood density can achieve higher growth rates due to increased water storage, facilitating earlier seasonal growth. Associations have also been observed between wood density of species and phenophases, depending on the growth limiting factors, with the life cycle of high wood density tree species, which are less water efficient, being more influenced by water availability, whereas those with low wood density possessing surplus water being more influenced by photoperiod. Poor hydraulic conductivity is associated with reduced water transport efficiency, but higher wood density improves tree species' resilience to environmental stresses such as pathogen attack, drought-induced embolism or cavitation, and mechanical damage, allowing for adaptation to drier and stressed conditions. Tree species having wood density values between 0.5 and 0.65 g cm^{-3} are considered most efficient in water transport. Under extreme drought conditions, in spite of the stomatal closure, water is also lost from the cuticle layer, which is measured as leaf minimum conductance, a critical drought tolerance trait that helps predict mortality in plants (Duursma et al., 2019). Therefore, under drought stress conditions, tree species with higher leaf minimum conductance possess better survival and growth. The water potential at which the leaves lose their turgor, which causes stomata to close and consequently lowers gas exchange and plant growth, is known as the leaf turgor loss point. It is used to evaluate a tree species' ability to withstand drought. Plant species with a greater negative leaf turgor loss point can absorb water from the soil at much lower soil water concentrations, allowing them to maintain hydraulic conductance, gas exchange, and growth even when water availability is limited, increasing their ability to withstand water stress. High evaporative demand may limit photosynthetic carbon assimilation due to stomatal closure intended to reduce water stress, which would adversely affect plant growth. Temperature also has a significant impact on the rates of photosynthesis and maintenance respiration (Keenan et al., 2013).

8.3.3 Phenological Traits and Growth Cycles

Phenological traits of plants need to be assessed to have a better insight into various ecosystem processes such as biomass production, pollinator competition, light, temperature, precipitation, and photoperiod, as well as plant growth patterns (Kikuzawa and Lechowicz, 2011). The seasonality of phenology of plants is particularly affected by environmental factors such as precipitation and temperature. Leaves, being the predominant organ of photosynthate assimilation, act as a key trait for plant growth and competitive success; therefore, the environmental and evolutionary selective pressure affecting leaves and ultimately plant growth needs to be studied in detail (Givnish, 1987). Reich et al. (2004) reported the influence of the environment on leaf phenology, substantially affecting plant functioning traits such as shoot and leaf growth. Leaf emergence patterns in trees have associations with both shoot growth and the leaves' survival period (Kikuzawa et al., 1996). Leaf phenology and leaf emergence is a strong determinant of species demography, species interaction, and tree biomass production (Diez et al., 2012). Deciduous plants usually finish their growth earlier than evergreens, while they may or may not start it earlier, according to Dhaila et al. (1995). Evergreen and deciduous trees respond similarly to different climatic circumstances in terms of functional features such as branch extension and leaf growth. However, deciduous species tend to enter leaf fall shortly after the rainy season or at the advent of dry season, evergreens, on the other hand, keep their leaves longer throughout the winter and shed them in the summer, when new shoots begin to grow (Garkoti and Singh, 1994). Under same climatic conditions, deciduous tree species produce and expand their leaves simultaneously and in flush during the peak growing season while evergreens typically display sequential unfolding of the leaves and shoot growth over a longer period of time expanded through seasons (Kikuzawa and Lechowicz, 2011; Kumar et al., 2015, 2016, 2017; Singh et al., 2020).

8.3.4 Growth Response to Environmental Factors

Relative growth rates in tree species are affected by key functional traits, most of which are significantly influenced by abiotic factors (Violle et al., 2007). Montane landscapes frequently exhibit predictable changes in numerous abiotic variables as elevation increases, altering plant performance. For instance, temperature decrease is studied to lower the transpiration and metabolism in various tree species (Körner, 2016). Also, the typical reduction of gaseous exchange and hence the photosynthesis rate and trees growth are found under lower partial pressure of carbon dioxide and low atmospheric pressure, which is the characteristic of elevated landscapes. However, in montane landscapes, these effects are effectively counterbalanced by the corresponding increase in radiation intensity and carbon dioxide and water vapour diffusion coefficients restoring the photosynthesis rate and plant growth (Wang et al., 2017). Soil conditions also influence the tree growth and regeneration with less developed and less fertile soils negatively affecting the species regeneration and growth (FAO, 2015; Müller et al., 2016). Due to the more favourable climatic variables experienced by plants at lower elevation, tree species growing at lower altitudes are characterized by taller stems to compete for light (Mao et al., 2018), better leaf nutrition leading to enhanced photosynthetic capacity (Drollinger et al., 2017), thinner and larger leaves for effective light capture (Liu et al., 2020), and tree architecture supporting efficient gaseous exchange and water transport (Yang et al., 2020) showcasing an acquisitive strategy that enhances resource acquisition and overall growth. Contrastingly, trees species at higher elevation are exposed to harsher climatic conditions and therefore functions as per the conservative strategy to enhance conservation of available resources and ensure plant survival by featuring thick and small leaves protecting against high UV-B radiation (Turunen and Latola, 2005), shorter stems to reduce vulnerability to strong winds (Mao et al., 2018), and greater dry matter in leaves coupled with lower specific leaf area enhancing leaf lifespan and durability and stems resistant to cavitation facilitating gaseous exchange and water transport under freezing conditions (Yang et al., 2020). Temperature and precipitation also influence the growth rates. Elevated temperatures are associated with lower precipitation and water stress conditions leading to stomatal closure limiting photosynthetic carbon assimilation and growth (Keenan et al., 2013; Paine et al., 2015). Excessive heat and light characteristic of arid ecosystems are navigated by certain tree species by self-shading involving modification of their branching pattern and allocating resources to the longer axis versus the shorter axis helping in the maintenance of survival and efficient growth rates (Paquette et al., 2007).

8.4 FUNCTIONAL TRAITS AND TREE PRODUCTIVITY

The relationship between tree production and functional features is an important research subject because it influences how biodiversity affects the overall ecosystem function. A recent research study depicts that the diversity of the functional traits within the tree communities can significantly enhance ecosystem productivity. By analyzing the complementary traits of different tree species, Zhang et al. (2021) demonstrated through a global meta-analysis that both functional diversity and species identity significantly boost ecosystem productivity. The interplay between the environmental factors and functional traits is important for unveiling the tree growth dynamics. Functional traits associated with tree size and competition for light were identified by Maynard et al. (2021) as critical for predicting tree performance across varied environmental conditions. The overall interactions between environmental

TABLE 8.1
Functional Traits with Their Description and Impact on Tree Productivity

S. No.	Functional Trait	Description	Impact on Productivity	References
1	Leaf mass/area	Indicative of leaf thickness and water-retention capacity	Drought resistance and increased productivity under stress conditions	Correia et al. (2018)
2	Xylem pressure (Ψ50)	Measures water potential at which the conductivity of xylem is lost	Lower xylem pressure indicates better tolerance to drought	Correia et al. (2018)
3	Leaf area to sapwood ratio	Efficiency of capturing light and allocation of resources	High ratios are indicative of increased photosynthetic growth and capacity	Correia et al. (2018)
4	Neighbourhood diversity	Defines the diversity of species surrounding a focal tree	Enhances individual tree growth and reduces competition	Fichtner et al. (2017); Fichtner et al. (2020)
5	Functional diversity	Describes various functional traits within a community	Enhances the total biomass production through niche complementarity	Zhang et al. (2021); Wan et al. (2023)

conditions, neighbourhood diversity, and species traits often create a complex framework that influences the tree communities and their response to ecological pressures and forest productivity. Examples of functional traits with their details are presented in Table 8.1.

8.4.1 Trait–Productivity Relationships in Different Forest Types

The trait–productivity correlations in various forest types are multidimensional and complex, influenced by a number of factors such as species composition, management tactics, and environmental circumstances. Different forest types have different relationships between functional traits and tree productivity. Because light and nutrients are abundant in tropical forests, traits that improve resource acquisition, such as large leaf area and high photosynthetic rates, are strongly linked to high production (Poorter et al., 2008). These traits allow tropical trees to capitalize on favourable conditions and grow rapidly, contributing to the high productivity of these ecosystems.

In contrast, in temperate and boreal forests, where resources can be more limited, traits that enhance resource-use efficiency, such as high wood density and conservative water use, are vital for productivity (Reich, 2014). Trees in such areas may grow slowly, but they are better adapted to survive and maintain productivity under harsh conditions such as cold winters and nutrient-poor soils. As seen in the boreal forests, coniferous species such as spruce and pine are often dominant. Such species exhibit traits such as needle-like leaves with low surface area that reduce water loss and allow them to maintain productivity in environments where nutrients and water are limited (Givnish, 2002).

The mixed broadleaf conifer forests show a negative association between stand productivity and precipitation and temperature, demonstrating the importance of environmental factors in determining forest productivity (Wu et al. 2019). Several intraspecific variations in plant traits like leaf area influence the community productivity and assembly, which in turn have huge implications for carbon cycling and resilience to climate extremes (Sigdel et al., 2022). This study is further strengthened by the fact that specific traits have the capability to enhance the ecosystem including soil fertility and biomass production, critical for maintaining forest productivity and health (Hua et al., 2021).

The biomass productivity approach can potentially evaluate different forest site qualities, revealing that mature forests often have a lower potential than younger forest types (Yan et al., 2023). Zhang et al. (2019) emphasized that forest productivity is influenced significantly by anthropogenic and natural disturbances, highlighting the need for management strategies to maintain tree productivity under changing climatic conditions. Additionally, the multifunctionality of forests with low productivity underscores the critical evaluation of consideration of parameters such as forest age, species richness, and site type while evaluating ecosystem services. Thus, the low-productivity forests essentially maintain high multi-functionality, which is critical for conservation and management strategies (Jönsson and Snäll, 2020; Kumar et al., 2020a,b, 2021a–e). This evaluation is needed for the management of different forest types to optimize both forest productivity and biodiversity. In summary, the relationships between traits and productivity in several forest types are influenced and shaped by combinations of climatic, ecological, and management factors, where high-productivity forests often support greater biomass and biodiversity and low-productivity forests play a vital role in species conservation and ecosystem services. An illustrative understanding of such dynamics is vital for effective management and conservation of forests.

8.4.2 Influence of Environmental Conditions on Trait Productivity

The impact of environmental variables on tree trait productivity is a multifaceted topic that encompasses soil features, biotic interactions, and climatic considerations. Numerous tree species and ecosystems are impacted by these environmental factors in terms of tree growth, survival, and reproductive success. It was clearly stated by Sánchez-Salguero et al. (2016) that global warming conditions can cause an increased vulnerability in forest ecosystems, especially at the rear edges of species distributions, where the trees struggle to adapt to rapid disturbances in precipitation and temperature patterns. Additionally, some research showed that long-lived tree species cannot adapt to abrupt changes in climate, which has a major effect on forest ecosystems around the world (Butt et al., 2013). Nevertheless, trees in urban areas face intense stress from climate change as well as pollution, ultimately diminishing their vitality and growth (Moser-Reischl et al., 2019). These studies depict the vital role of certain climatic factors that influence the productivity and health of trees.

Another critical component influencing tree productivity and growth encompasses the soil characteristics and topography. Factors such as topography and soil fertility are vital in the determination of richness and productivity in tree species

(Wei and Xiaoxian, 2016). Tree growth and distribution are also significantly influenced by environmental factors including water availability and soil quality (Zhang et al., 2023). The interaction between climatic conditions and soil properties creates a varied growth response amongst the tree species, complicating predictions related to forest productivity under unpredictable environmental conditions. Furthermore, the growth of tree species is significantly influenced by biotic interactions, such as competition and facilitation between the species. The intraspecific variation in the growth responses to fluctuations in the environment influences the species diversity in forests, particularly in mixed-species forests where the competition for resources can enhance or inhibit the growth of trees (Aussenac et al., 2017). These interactions help in understanding the complexities of forest ecosystems and the need to emphasize both biotic and abiotic factors while assessing tree productivity.

Overall, the interplay of environmental factors such as climatic conditions, soil parameters, and biotic interactions influences tree productivity traits. Further, understanding such relationships is vital for the prediction of the response of forests to the unpredictable changes in the environment for developing an effective management and an effective conservation approach.

8.5 CASE STUDIES ON THE RELATIONSHIP BETWEEN FUNCTIONAL TRAITS AND TREE GROWTH AND PRODUCTIVITY

Plant functional traits are key to predict growth and survival in numerous tree species. Research on harvested trees has highlighted the significance of wood density in predicting tree biomass through allometric relationships (Chave et al., 2005). Incorporating wood density reveals that the largest diameter trees do not always correspond to the highest biomass. Wood density contributed about 33% variation in growth rates amongst the fastest-growing juvenile rain forest tree species in Panama, although the correlations tend to weaken in slow-growing species, adult trees, and other functional traits (Rüger et al., 2012). By analysing the national forest inventory data throughout Spain, Martínez-Vilalta et al. (2010) similarly found an inverse association between growth rate and wood density. Larger-scale research by Poorter et al. (2008) showed that wood density accounted for 11% of the variation in growth rate across five neotropical forest locations. They attributed this consistency to the comparable trade-offs that all rainforest trees faced. According to Yan et al. (2013), species abundance in eastern China was positively correlated with wood density, whereas species abundance was negatively correlated with leaf nitrogen content.

Studies have indicated that in tropical forests, seedling survival during severe drought largely depends on the xylem's resistance to embolism formation, with substantial variability in embolism resistance among tropical tree species (Tyree et al., 2003; Markesteijn et al., 2010). Species in these forests have also been shown to have significant variations in how they respond to drought in terms of leaf shedding (Pineda-García et al., 2011). In a three-year study, Devi and Garkoti (2013) examined phenological traits such as leaf fall, leaf growth, shoot growth, and leaf pool size in a total of six deciduous and thirteen evergreen tree species in the tropical forest of Assam, India and observed considerable interspecific variation in traits such as time of leaf initiation, leaf area, rate of leaf expansion, leaf lifespan, and shoot extension in evergreen tree species compared to their deciduous counterparts providing them capabilities of better resource utilization and survival. The deciduous species in the study began producing leaves before the advent of the monsoons allowing them to ripe plants take advantage of the monsoon season with full foliage by generating drier (Negi, 2006). This early boost in net primary production allows for increased biomass allocation to photosynthetic systems, hence increasing productivity. Deciduous species generally exhibited faster growth rates, initiating growth before evergreen species with faster shoot growth and rapid leaf development and expansion, thus gaining competitive advantages for forest resources over evergreen trees. The evergreen species were characterized by lesser leaf number, less shoot elongation, greater leaf lifespan/retention, and lower leaf pool size than the deciduous species. Owing to their greater leaf lifespan, the evergreens spend less resources on the generation of their photosynthetic machinery; however, deciduous, on the other hand, possess shorter leaf lifespan and higher photosynthetic capacity which allow them to optimize growth during favourable seasons.

The global forests are facing drought-induced mortality, with the physiological adaptations to drought varying significantly, leading to diverse mortality patterns in different tree species and changes in community structure. Wu et al. (2022) examined the functional traits, including carbon dynamics, water relations, stomatal regulation, and tree growth in four subtropical tree species to evaluate their mechanism of survival under water-limited conditions. Stomatal closure, reduction in turgor loss point and leaf minimum conductance, enhanced leaf dry matter, and LMA as well as stabilization of light-saturated photosynthetic rates, electron transport, and RuBisCO carboxylation were the mechanisms employed by *Michelia macclurei*, *Ormosia pinnata*, Castanopsis *fissa*, and *Dalbergia odorifera* for reducing water loss and imparting drought tolerance. *M. macclurei* displayed maximum tolerance to drought and hence better survival and growth.

8.6 CONCLUSION

The association of functional traits with tree growth and productivity highlights the delicate balance that trees must strike between survival and development, especially under changing environmental conditions. Although some characteristics, like LMA and higher wood density, help with structural defence and drought tolerance, they frequently come at the expense of slower growth rates and biomass production. These trade-offs demonstrate the importance of trait integration and the necessity for trees to prioritize different functions depending on their environmental conditions. In light of global climate change, the case studies covered in this chapter demonstrate how these dynamics manifest in ecosystems and offer insightful information for conservation and forest management strategies. By comprehending the role of functional traits, scientists and forest managers may more accurately predict how trees and forests will respond to continuous environmental changes, ensuring resilience and sustainable forest productivity.

REFERENCES

Aussenac, R., Bergeron, Y., Mekontchou, C. G., Gravel, D., Pilch, K., and Drobyshev, I. 2017. Intraspecific variability in growth response to environmental fluctuations modulates the stabilizing effect of species diversity on forest growth. *Journal of Ecology* 105: 1010–1020. https://doi.org/10.1111/1365-2745.12728

Boet, O., Arnan, X., and J. Retana. 2020. The role of environmental vs. biotic filtering in the structure of European ant communities: a matter of trait type and spatial scale. *PLoS One* 15: 1–18. https://doi.org/10.1371/journal.pone.0228625

Brienen, R. J., and Zuidema, P. A. 2005. Relating tree growth to rainfall in Bolivian rain forests: a test for six species using tree ring analysis. *Oecologia* 146(1): 1–12.

Butt, N., Pollock, L. J., & McAlpine, C. 2013. Eucalypts face increasing climate stress. *Ecology and Evolution* 3: 5011–5022. https://doi.org/10.1002/ece3.873

Chapin, F. S., Matson, P. A., Mooney, H. A., and Vitousek, P. M. 2002. *Principles of terrestrial ecosystem ecology*. Springer, pp. 1–447.

Chave, C., Andalo, S., Brown, M. A., Cairns, J. Q., Chambers, D., Eamus, H., Fo¨lster, F., Fromard, N., Higuchi, T., Kira, J. P., Lescure, B. W., Nelson, H., Ogawa, H., Puig, B., Rie´ra, T., and Yamakura. 2005. Tree allometry and improved estimation of carbon stocks and balance in tropical forests. *Ecosystem Ecology*. https://doi.org/10.1007/s00442-005-0100-x

Chave, J., Coomes, D., Jansen, S., Lewis, S.L., Swenson, N.G. and Zanne, A.E. (2009). Towards a worldwide wood economics spectrum. *Ecology Letters*, *12*(4): 351-366.

Cleveland, C. C., Townsend, A. R., Taylor, P., Alvarez-Clare, S., Bustamante, M. M., Chuyong, G., Dobrowski, S. Z., Grierson, P., Harms, K. E., Houlton, B. Z., and Marklein, A. 2011. Relationships among net primary productivity, nutrients and climate in tropical rain forest: a pan-tropical analysis. *Ecology Letters* 14: 939–947. https://doi.org/10.1111/j.1461-0248.2011.01658.x

Correia, D., Bouchard, M., Filotas, É., and Raulier, F. 2018. Disentangling the effect of drought on stand mortality and productivity in northern temperate and boreal forests. *Journal of Applied Ecology* 56: 758–768. https://doi.org/10.1111/1365-2664.13305

Devi, A. F., and Garkoti, S. C. 2013. Variation in evergreen and deciduous species leaf phenology in Assam, India. *Trees* 27: 985–997. https://doi.org/10.1007/s00468-013-0850-8

Dhaila, S., Singh, S. P., Negi, G. C. S., and Rawat, Y. S. 1995. Shoot growth phenology of co-existing evergreen and deciduous species in Anoak forest. *Ecological Research* 10: 151–159.

Diez, J. M., Inbenz, I., Miller Rushing, A. J., Mazer, S. J., Crimmins, T. M., Crimmins, M. A., Bertelsen, C. D., and Inouye, D. W. 2012. Forecasting phenology: from species variability to community patterns. *Ecology Letters*. https://doi.org/10.1111/j.146-0248

Drollinger, S., Müller, M., Kobl, T., Schwab, N., Böhner, J., Schickhoff, U., and Scholten, T. 2017. Decreasing nutrient concentrations in soils and trees with increasing elevation across a treeline ecotone in Rolwaling Himal, Nepal. *Journal of Mountain Science* 14: 843–858. https://doi.org/10.1007/s11629-016-4228-4

Duursma, R. A., Blackman, C. J., Lopéz, R., Martin-StPaul, N. K., Cochard, H., and Medlyn, B. E. 2019. On the minimum leaf conductance: its role in models of plant water use, and ecological and environmental controls. *New Phytologist* 221: 693–705. https://doi.org/10.1111/nph.15395

FAO. 2015. Understanding mountain soils: a contribution from mountain areas to the International Year of Soils 2015. Food and Agriculture Organization of the United Nations.

Fichtner, A., Härdtle, W., Liu, Y., Bruelheide, H., Kunz, M., and von Oheimb, G. 2017. From competition to facilitation: how tree species respond to neighbourhood diversity. *Ecology Letters* 20: 892–900. https://doi.org/10.1111/ele.12786

Fichtner, A., Schnabel, F., Bruelheide, H., Kunz, M., Mausolf, K., Schuldt, A., Härdtle, W., and von Oheimb, G. 2020. Neighbourhood diversity mitigates drought impacts on tree growth. *Journal of Ecology* 108: 865–875. https://doi.org/10.1111/1365-2745.13353

Garkoti, S. C., and Singh, S. P. 1994. Nutrient cycling in the three central Himalayan forests ranging from close canopied to open canopied treeline forests, India. *Arctic and Alpine Research* 26: 339–348.

Giri, K., Mishra, G., Suyal, D. C., Kumar, N., Doley, B., Das, N., Baruah, R. C., Bhattacharyya, R., and Bora, N. 2023. Performance evaluation of native plant growth-promoting rhizobacteria for paddy yield enhancement in the jhum fields of Mokokchung, Nagaland, North East India. *Heliyon* 9: e14588.

Givnish, T. J. 2002. Adaptive significance of evergreen vs. deciduous leaves: solving the triple paradox. *Silva Fennica* 36: 703–743.

Givnish, T. J. 1987. Comparative studies of leaf form: assessing the relative roles of selective pressures and phylogenetic constraints. *New Phytologist* 106: 131–160. https://doi.org/10.1111/j.1469-8137.1987.tb04687.x

Graham, E. A., Mulkey, S. S., Kitajima, K., Phillips, N. G., and Wright S. J. 2003. Cloud cover limits net CO_2 uptake and growth of a rainforest tree during tropical rainy seasons. *Proceedings of the National Academy of Sciences* 100: 572–576. https://doi.org/10.1073/pnas.0133045100

Hua, Z., Wang, Z., Wen, Z., and Yang, Y. 2021. Effects of plant functional traits on ecosystem services: a review. *Chinese Journal of Plant Ecology* 45: 1140–1153. https://doi.org/10.17521/cjpe.2020.0142

Jönsson, M., and Snäll, T. 2020. Ecosystem service multifunctionality of low-productivity forests and implications for conservation and management. *Journal of Applied Ecology* 57: 695–706. https://doi.org/10.1111/1365-2664.13569

Keenan, T. F., Hollinger, D. Y., Bohrer, G., Dragoni, D., Munger, J. W., Schmid, H. P., and Richardson, A. D. 2013. Increase in forest water-use efficiency as atmospheric carbon dioxide concentrations rise. *Nature* 499: 324–327.

Kikuzawa, K., Koyama, H., Umeki, K., and Lechowicz, M. J. 1996. Some evidence for an adaptive linkage between leaf phenology and shoot architecture in sapling trees. *Functional Ecology* 10: 252–257.

Kikuzawa, K., and Lechowicz, M. J. 2011. *Ecology of leaf longevity. Ecological Research Monographs*. Springer, Berlin.

Körner, C. 2016. Plant adaptation to cold climates. *F1000 Research* 5: 15. https://doi.org/10.12688/f1000research.9107.1

Kumar, A., Dwivedi, G. K., Tewari, S., Paul, J., Anand, R., Kumar, N., Kumar, P., Singh, H., and Kaushal, R. 2020a. Carbon mineralization and inorganic nitrogen pools under *Terminalia chebula* Retz.-Based agroforestry system in Himalayan foothills, India. *Forest Science* 66(5): 634–643.

Kumar, A., Dwivedi, G. K., Tewari, S., Paul, J., Sah, V. K., Singh, H., Kumar, P., Kumar, N., and Kaushal, R. 2020b. Soil organic carbon pools under *Terminalia chebula* Retz. based agroforestry system in Himalayan foothills, India. *Current Science* 118(7): 1098–1103.

Kumar, A., Kumar, P., Singh, H., Bisht, S., and Kumar, N. 2021a. Relationship of physiological plant functional traits with soil carbon stock in temperate forest of Garhwal Himalaya. *Current Science* 120(8): 1368–1373.

Kumar, A., Kumar, P., Singh, H., and Kumar, N. 2021b. Modulation of plant functional traits under essential plant nutrients during seasonal regime in natural forests of Garhwal Himalayas. *Plant and Soil* 465: 197–212.

Kumar, A., Kumar, P., Singh, H., and Kumar, N. 2021d. Impact of plant functional traits on near saturated hydraulic conductivity of soil under different forests of Kempty watershed in Garhwal Himalaya. *Indian Journal of Soil Conservation* 49(1): 38–44.

Kumar, A., Kumar, P., Singh, H., and Kumar, N. 2021e. Adaptation and mitigation potential of roadside trees with bio-extraction of heavy metals under vehicular emissions and their impact on physiological traits during seasonal regimes. *Urban Forestry & Urban Greening* 58: 126900.

Kumar, A., Singh, H., Kumari, G., Bisht, S., Malik, A., Kumar, N., Singh, M., Raturi, A., Barthwal, S., Thakur, A., and Kaushal, R. (2022). Adaptive resilience of roadside trees to vehicular emissions via leaf enzymatic, physiological, and anatomical trait modulations. *Environmental Pollution* 313: 120191.

Kumar, A., Tewari, S., Singh, H., Kumar, P., Kumar, N., Bisht, S., Kushwaha, S., Tamta, N., and Kaushal, R. 2021c. Biomass accumulation and carbon stocks in different agroforestry system prevalent in Himalayan foothills, India. *Current Science* 120(6): 1083–1088.

Kumar, N., Jeena, N., Kumar, A., Khairakpam, R., and Singh, H. 2021f. Comparative response of rice cultivars to elevated air temperature in Bhabar region of Indian Himalaya: status on yield attributes. *Heliyon* 7: e07474.

Kumar, N., Jeena, N., and Singh, H. 2019. Elevated temperature modulates rice pollen structure: a study from foothill Himalayan Agro-ecosystem in India. *3 Biotech* 9: 175.

Kumar, N., Kumar, N., Shukla, A., Shankhdhar, S. C., and Shankhdhar, D. 2015. Impact of terminal heat stress on pollen viability and yield attributes of rice (*Oryza sativa* L.). *Cereal Research Communications* 43(4): 616–626.

Kumar, N., Shankhdhar, S. C., and Shankhdhar, D. 2016. Impact of elevated temperature on antioxidant activity and membrane stability in different genotypes of rice (*Oryza sativa* L.). *Indian Journal of Plant Physiology (Now Plant Physiology Report)* 21(1): 37–43.

Kumar, N., Suyal, D. C., Sharma, I. P., Verma, A., and Singh, H. 2017. Elucidating stress proteins in rice (*Oryza sativa* L.) genotype under elevated temperature: a proteomic approach to understand heat stress response. *3 Biotech* 7: 205.

Larcher, W. 2003. *Physiological Plant Ecology: Ecophysiology and Stress Physiology of Functional Groups*. Springer Science & Business Media, Germany. pp. 1–505.

Lavorel, S., and Garnier, E. 2002. Predicting changes in community composition and ecosystem functioning from plant traits: revisiting the Holy Grail. *Functional Ecology* 16: 545–556. https://doi.org/10.1046/j.1365-2435.2002.00664.x

Liu, W., Zheng, L., and Qi, D. 2020. Variation in leaf traits at different altitudes reflects the adaptive strategy of plants to environmental changes. *Ecology and Evolution* 10: 8166–8175. https://doi.org/10.1002/ece3.6519

Luyssaert, S., Inglima, I., Jung, M., Richardson, A. D., Reichstein, M., Papale, D., Piao, S. L., Schulze, E. D., Wingate, L., Matteucci, G., and Aragao, L. E. O. C. 2007. CO_2 balance of boreal, temperate, and tropical forests derived from a global database. *Global Change Biology* 13: 2509–2537. https://doi.org/10.1111/j.1365-2486.2007.01439.x

Malhi, Y., Doughty, C., and Galbraith, D. 2011. The allocation of ecosystem net primary productivity in tropical forests. *Philosophical Transactions of the Royal Society B: Biological Sciences* 366: 3225–3245. https://doi.org/10.1098/rstb.2011.0062

Malhi, Y., Girardin, C. A., Goldsmith, G. R., Doughty, C. E., Salinas, N., Metcalfe, D. B., Huaraca Huasco, W., Silva-Espejo, J. E., del Aguilla-Pasquell, J., Farfán Amézquita, F., and Aragão, L. E. 2017. The variation of productivity and its allocation along a tropical elevation gradient: a whole carbon budget perspective. *New Phytologist* 214: 1019–1032. https://doi.org/10.1111/nph.14189

Malhi, Y., Silman, M., Salinas, N., Bush, M., Meir, P., and Saatchi, S. 2010. Introduction: elevation gradients in the tropics: laboratories for ecosystem ecology and global change research Y. MALHI et al. ELEVATION GRADIENTS IN THE TROPICS. *Global Change Biology* 16(12): 3171–3175.

Mao, L., Chen, S., Zhang, J., and Zhou, G. 2018. Altitudinal patterns of maximum plant height on the Tibetan Plateau. *Journal of Plant Ecology* 11: 85–91. https://doi.org/10.1093/jpe/rtw128

Markesteijn, L., Poorter, L., Bongers, F., Paz, H., and Sack, L. 2010. Hydraulics and life history of tropical dry forest tree species: coordination of species' drought and shade tolerance. *New Phytologist* 191: 480–495.

Martínez-Vilalta, J., Mencuccini, M., Vayreda, J., and Retana, J. 2010. Interspecific variation in functional traits, not climatic differences among species ranges, determines demographic rates across 44 temperate and Mediterranean tree species. *Journal of Ecology* 98: 1462–1475.

Maynard, D.S., Bialic-Murphy, L., Zohner, C.M., Averill, C., van den Hoogen, J., Ma, H., Mo, L., Smith, G.R., Aubin, I., Berenguer, E. and Boonman, C.C. (2021). Global trade-offs in tree functional traits. *bioRxiv: 2021-09*. https://doi.org/10.1101/2021.09.16.458157.

Moser-Reischl, A., Rötzer, T., Biber, P., Ulbricht, M., Uhl, E., Qu, L., Koike, T., and Pretzsch, H. 2019. Growth of Abies sachalinensis along an urban gradient affected by environmental pollution in Sapporo, Japan. *Forests* 10: 707. https://doi.org/10.3390/f10080707

Müller, M., Schickhoff, U., Scholten, T., Drollinger, S., Böhner, J., and Chaudhary, R. P. 2016. How do soil properties affect alpine treelines? General principles in a global perspective and novel findings from Rolwaling Himal, Nepal. *Progress in Physical Geography* 40: 135–160. https://doi.org/10.1177/0309133315615802

Negi, G. C. S. 2006. Leaf and bud demography and shoot growth in evergreen and deciduous trees of central Himalaya, India. *Tree* 20: 416–429.

Nemani, R. R., Keeling, C. D., Hashimoto, H., Jolly, W. M., Piper, S. C., Tucker, C. J., Myneni, R. B., and Running, S. W. 2003. Climate-driven increases in global terrestrial net primary production from 1982 to 1999. *Science* 300: 1560–1563.

Paine, C. E. T., Amissah, L., Auge, H., Baraloto, C., Baruffol, M., Bourland, N., Bruelheide, H., Daïnou, K., de Gouvenain, R. C., Doucet, J. L., and Doust, S. 2015. Globally, functional traits are weak predictors of juvenile tree growth, and we do not know why. *Journal of Ecology* 103: 978–989. https://doi.org/10.1111/1365-2745.12401

Paquette, A., Bouchard, A., and A. Cogliastro. 2007. Morphological plasticity in seedlings of three deciduous species under shelterwood under-planting management does not correspond to shade tolerance ranks. *Forest Ecology and Management* 241: 278–287.

Pérez-Ramos, I.M., Matías, L., Gómez-Aparicio, L. and Godoy, Ó. (2019). Functional traits and phenotypic plasticity modulate species coexistence across contrasting climatic conditions. *Nature Communications*, 10(1): 2555.

Pineda-García, F., Paz, H., and Tinoco-Ojanguren, C. 2011. Morphological and physiological differentiation of seedlings between dry and wet habitats in a tropical dry forest. *Plant, Cell & Environment* 34: 1536–1547.

Poorter, L., Wright, S. J., Paz, H., Ackerly, D. D., Condit, R., Ibarra-Manríquez, G., Harms, K. E., Licona, J. C., Martinez-Ramos, M., Mazer, S. J., and Muller-Landau, H. C. 2008. Are functional traits good predictors of demographic rates? Evidence from five neotropical forests. *Ecology* 89: 1908–1920. https://doi.org/10.1890/07-0207.1

Quesada, C. A., Phillips, O. L., Schwarz, M., Czimczik, C. I., Baker, T. R., Patiño, S., Fyllas, N. M., Hodnett, M. G., Herrera, R., Almeida, S., and Alvarez Dávila, E. 2012. Basin-wide variations in Amazon forest structure and function are mediated by both soils and climate. *Biogeosciences* 9: 2203–2246.

Rees, M., Osborne, C. P., Woodward, F. I., Hulme, S. P., Turnbull, L. A., and Taylor, S. H. 2010. Partitioning the components of relative growth rate: how important is plant size variation? *The American Naturalist* 176: E152–E161.

Reich, P. B. 2014. The world-wide 'fast–slow'plant economics spectrum: a traits manifesto. *Journal of Ecology* 102: 275–301.

Reich, P. B., Uhl, C., Walters, M. B., Prugh, L., and Ellsworth, D. S. 2004. Leaf demography and phenology in Amazonian rain forest: a census of 40000 leaves of 23 tree species. *Ecological Monographs* 74: 3–23. https://doi.org/10.1890/02-4047

Rüger, N., Wirth, C., Wright, S. J., and Condit, R. S. 2012. Functional traits explain light and size response of growth rates in tropical tree species. *Ecology* 93: 2626–2636.

Sánchez-Salguero, R., Camarero, J. J., Gutiérrez, E., González-Rouco, J. F., Gazol, A., Sangüesa-Barreda, G., Andreu-Hayles, L., Linares, J. C., and Seftigen, K. 2016. Assessing forest vulnerability to climate warming using a process-based model of tree growth: bad prospects for rear-edges. *Global Change Biology* 23: 2705–2719. https://doi.org/10.1111/gcb.13541

Schnabel, F., Schwarz, J., Dănescu, A., Fichtner, A., Nock, C. A., Bauhus, J., and Potvin, C. 2019. Drivers of productivity and its temporal stability in a tropical tree diversity experiment. *Global Change Biology* 25: 4257–4272. https://doi.org/10.1111/gcb.14792

Sigdel, S. R., Liang, E., Rokaya, M. B., Rai, S., Dyola, N., Sun, J., Zhang, L., Zhu, H., Chettri, N., Chaudhary, R. P., and Peñuelas, J. 2022. Functional traits of a plant species fingerprint ecosystem productivity along broad elevational gradients in the Himalayas. *Functional Ecology* 37: 383–394. https://doi.org/10.1111/1365-2435.14226

Singh, H, Yadav, M, Kumar, N, Kumar, A, and Kumar, M. 2020. Assessing adaptation and mitigation potential of roadside trees under the influence of vehicular emissions: a case study of *Grevillea robusta* and *Mangifera indica* planted in an urban city of India. *PLoS One* 15(1): e0227380.

Sterck, F., Markesteijn, L., Toledo, M., Schieving, F., and Poorter, L. 2014. Sapling performance along resource gradients drives tree species distributions within and across tropical forests. *Ecology* 95: 2514–2525. https://doi.org/10.1890/13-2377.1

Sundqvist, M. K., Sanders, N. J., & Wardle, D. A. 2013. Community and ecosystem responses to elevational gradients: processes, mechanisms, and insights for global change. *Annual Review of Ecology, Evolution, and Systematics* 44(1): 261–280. https://doi.org/10.1146/annurev-ecolsys-110512-135750

Turnbull, L. A., Philipson, C. D., Purves, D. W., Atkinson, R. L., Cunniff, J., Goodenough, A., Hautier, Y., Houghton, J., Marthews, T. R., Osborne, C. P., and Paul-Victor, C. 2012. Plant growth rates and seed size: a re-evaluation. *Ecology* 93: 1283–1289.

Turunen, M. and Latola, K. 2005). UV-B radiation and acclimation in timberline plants. *Environmental Pollution*, 137(3): 390–403.

Tyree, M. T., Engelbrecht. B. M. J., Vargas. G., and Kursar, T. 2003. Desiccation tolerance of five tropical seedlings in Panama. Relationship of field assessment of drought performance. *Plant Physiology* 132: 1439–1447.

Tyree, M. T., and Zimmermann, M. H. 2013. *Xylem Structure and the Ascent of Sap*. Springer Science & Business Media, Germany.

Violle, C., Navas, M. L., Vile, D., Kazakou, E., Fortunel, C., Hummel, I., and Garnier, E. 2007. Let the concept of trait be functional! *Oikos* 116: 882–892. https://doi.org/10.1111/j.2007.0030-1299.15559.x

Wan, X., Joly, F., Jia, H., Zhu, M., Fu, Y., and Huang, Z. 2023. Functional identity drives tree species richness-induced increases in litterfall production and forest floor mass in young tree communities. *New Phytologist* 240(3): 1003–1014. https://doi.org/10.1111/nph.19216

Wang, H., Prentice, I. C., Davis, T. W., Keenan, T. F., Wright, I. J., and Peng, C. 2017. Photosynthetic responses to altitude: an explanation based on optimality principles. *New Phytologist* 213: 976–982. https://doi.org/10.1111/nph.14332

Wei, C., and Xiaoxian, Z. 2016. Partitioning tree species diversity and developmental changes in habitat associations in a subtropical evergreen broadleaf secondary forest in Southern China. *Forests* 7: 228. https://doi.org/10.3390/f7100228

Wright, J. S. 2002. Plant diversity in tropical forests: a review of mechanisms of species coexistence. *Oecologia* 130: 1–14. https://doi.org/10.1007/s004420100809

Wu, T., Tan, N., Tissue, D. T., Huang, J., Duan, H., Su, W., Song, Y., Liu, X., Liu, Y., Li, X., and Lie, Z. 2022. Physiological traits and response strategies of four subtropical tree species exposed to drought. *Environmental and Experimental Botany* 203. https://doi.org/10.1016/j.envexpbot.2022.105046

Wu, Z., Zhang, Z., and Wang, J. 2019. Estimating the productive potential of five natural forest types in northeastern china. *Forest Ecosystems* 6(1). https://doi.org/10.1186/s40663-019-0204-0

Yan, E. R., Yang, X. D., Chang, S. X., and Wang, X. H. 2013. Plant trait-species abundance relationships vary with environmental properties in subtropical forests in eastern China. *PLoS One* 8: e61113. https://doi.org/10.1371/journal.pone.0061113

Yang, D., Wang, A. Y., Zhang, J. L., Bradshaw, C. J. A., and Hao, G. Y. 2020. Variation in stem xylem traits is related to differentiation of upper limits of tree species along an elevational gradient. *Forests* 11: 1–14. https://doi.org/10.3390/f11030349

Yuen, J. Q., Fung, T., and Ziegler, A. D. 2015. Review of allometric equations for major land covers in SE Asia: uncertainty and implications for above- and below-ground carbon estimates. *Forest Ecology and Management* 360: 323–340.

Zhang, H., Luo, Y., Zhou, G., Jiang, Y., Shah, S. T., Meng, S., and Liu, Q. 2019. Exploring the sensitivity of subtropical stand aboveground productivity to local and regional climate signals in south China. *Forests* 10: 71. https://doi.org/10.3390/f10010071

Zhang, W., Wu, B., Ren, Y., and Yang, G. 2023. Regionally compatible individual tree growth model under the combined influence of environment and competition. *Plants* 12: 2697. https://doi.org/10.3390/plants12142697

Zhang, C., Zhu, R., Sui, X., Li, X. and Chen, Y. (2021). Understanding patterns of taxonomic diversity, functional diversity, and ecological drivers of fish fauna in the Mekong River. *Global Ecology and Conservation*, 28: e01711.

Zhu, S. D., He, P. C., Li, R. H., Fu, S. L., Lin, Y. B., Zhou, L. X., Cao, K. F., and Ye, Q. 2018. Drought tolerance traits predict survival ratio of native tree species planted in a subtropical degraded hilly area in South China. *Forest Ecology and Management* 41: 1–46.

9 Significance of Plant Functional Traits in Soil Carbon and Nutrient Dynamics

Amit Kumar, Sarita Bisht, Narendra Kumar, and Rajesh Kaushal

9.1 INTRODUCTION

Burning fossil fuels and deforestation have increased carbon dioxide concentrations by 30% in the last three centuries. Concentrations of methane and other gases have tripled, contributing to global warming (Le Quere et al., 2009). Plant efficiency is one of the most important factors that will improve/decrease the carbon storage performance of plants. Climate change is altering the functioning of forest ecosystems, especially their carbon sequestration rates. Carbon sequestration rates are the result of many climate, soil, and vegetation factors and characteristics. Plants actively control many processes in the soil, including the storage of soil carbon, which is a significant part of the world's carbon. Cultivation conserves soil water by controlling carbon assimilation, carbon transfer, and storage in soil biomass, and its release from the soil through respiration and filtration (De Deyn et al., 2008). Functional traits are morphological, biochemical, physiological, structural, phenological, or behavioral characteristics that are phenotypically expressed in human organisms and are thought to be related to the response of these organisms to the environment and/or their impact on ecosystem strength (Violle et al., 2007). The importance of plant functional traits (PFTs) and their interaction with the environment leads to ecosystem function and services. This is important for both evolutionary biologists and functional ecologists. Basically, plants that perform well are divided into two main groups: The effective traits of a species underlie its effects on ecosystem dynamics and human services (Lavorel and Garnier, 2002), for example, leaf nitrogen content (LNC) in vascular plants (accelerating nutrient cycling rates). Responses trait affects the species' ability to thrive in the area and to survive environmental changes (such as bark thickness, which determines the fire length and leaf size, which in turn causes a difference in thermal balance).

The composition of the plant community is largely dependent on the availability of nutrients in the soil. Plants also have species specific to the fertility of the soil through feedbacks from the nutrient cycle. Although planting strategies vary greatly

DOI: 10.1201/9781003545231-9

from place to place (Verma et al., 2023), this approach can be generally understood as a plant growing in a rich environment producing more nutritious food, thus releasing more nutrients and thus maintaining a high soil fertility level. Some functions of the plant, including leaves, such as specific leaf area (SLA), carbon (C), nitrogen (N) ratio, and root quality, can have different effects on the availability of nutrients (long roots, carboxylate), resource acquisition, carbon storage, and bacterial communities in the soil (Roumet et al., 2016). However, the influence of plant activity and diversity on soil activity is not yet understood (Laliberte et al., 2015). The main influencing factors for soil carbon and nutrient dynamics are leaf quality (leaf material, leaf mass, leaf dry content, LNC, leaf phosphorus content (LPC), litter decomposition, etc.) and roots; e.g., root exudates, nitrogen-fixing bacteria, and arbuscular mycorrhizal (AM) fungi, play a role according to root length. Leaf nitrogen concentration, one of the main parameters for plant growth, is closely related to many ecosystem processes such as photosynthesis rate, plant growth, and litter decomposition (LeBauer and Treseder, 2008). Phosphorus, an essential element in genetic material, energy storage, and cell structure, is limited in most environments. Therefore, plants have developed many strategies to capture nutrients and respond to changes in their habitat and time (Lambers et al., 2008). The main factors related to soil carbon sequestration and nutrient dynamics include root structure, such as root diameter and root length, which determine the overall root system morphology of individual roots, or root size, such as root depth and root length, which determine the overall root system morphology. Physiological characteristics, such as root respiration, nutrient kinetics, and tissue content, are also affected by the release of root exudates. Finally, roots can interact directly with soil biota, including mycorrhizal fungi and rhizobia, and play an important role in the retention of nutrients from the soil. These principles vary greatly among species and genotypes; however, they are also highly plastic, allowing plants to respond to environmental changes, especially nutrient and water availability. Here we focus on understanding the link between plant performance and main ecosystem services, i.e., soil carbon and nutrient dynamics, which are important for ecological processes and intensive growth. The proposed approach will help develop strategies to maintain and promote nutrient dynamics and cycling.

9.2 PLANT FUNCTIONAL TRAITS AND SOIL CARBON DYNAMICS

Functional diversity affects C storage in the standing vegetation, litter, and soil via the effects of the dominant (locally most abundant) functional trait values on C dynamics. Plant communities dominated by plants with a higher investment in structure (taller and with a denser stem) and showing a more conservative resource use syndrome (lower values of SLA and LNC, higher values of leaf dry matter content (LDMC) and leaf thickness (LT)) should accumulate more C in standing biomass; this should also lead to a larger C stock in the soil, because of a higher C input and a longer C residence time. It also affects C accumulation in plant and soil stocks via the effects of the presence of diverging functional trait values on C dynamics. The coexistence of plants with contrasting functional trait values should lead to a fuller use of resources by the community as a whole and thus to higher biomass production and higher C input into the soil, increasing both aboveground and soil C stocks. They also affect carbon in plants and soils through the effect of different activities on carbon dynamics. The integration of plants that differ in productivity should lead to a higher use of resources in the community, resulting in higher biomass production and higher carbon input to the soil, thus increasing soil carbon stocks. The ability of many plant species to grow rapidly increases photosynthetic capacity according to species, but this leads to shorter lifespans, dry matter content, and lower carbon concentration in the aboveground and belowground plant (Kumar et al., 2022a).

BOX 9.1: KEY DEFINITIONS IN THE FUNCTIONAL TRAIT APPROACH

Plant Functional Traits: PFTs are the characteristics of plant species that determine the ecological performance in an ecosystem. These are morphological, biochemical, physiological, structural, and phenological characteristics that are expressed in phenotypes of individual plants (Violle et al., 2007).

PFTs are of two types:

1. **Effect Traits**: Effect traits of a species underlie its impacts on ecosystem properties and the services or disservices that human societies derive from them (Lavorel and Garnier, 2002), whether or not such traits represent an adaptive advantage to the individual itself. Examples of effect traits include water-retention capacity in bryophytes (regulating ecosystem hydrology), LNC in vascular plants (accelerating nutrient cycling rate), and burrowing behavior (altering soil structure).
2. **Response Traits**: Response traits influence the abilities of species to colonize or thrive in a habitat and to persist in the face of environmental changes. Plant examples include seed size (related to recruitment capacity under different disturbance regimes), bark thickness (conferring fire tolerance), and leaf size (leading to different heat balances).

Plasticity: Plasticity of any given trait, which has a genetic basis and which may or may not be adaptive, can intensify or attenuate evolved responses and can itself evolve in response to selection depending on the scale of spatial or temporal heterogeneity (Singh, 2024).

Ecosystem Service: An ecosystem process confers either direct or indirect benefits to humans. Especially focus on the goods that are directly used by humans (e.g., food, energy, and fibers) and the ecological processes affecting the provision of these goods (e.g., pollination and soil fertility) that are related to human health.

Functional Diversity: The value, range, and relative abundance of the functional traits of the organisms in communities that respond to the environment and influence ecosystem functioning. Functional diversity also includes the richness of functional types, which refers to a non-phylogenetic group of species functioning in a similar way with respect to environmental variation or presenting similar effects on ecosystem properties. Functional diversity measures could bridge the gap between ecosystem functioning and community ecology.

The source of soil carbon is the root exudates of rapidly growing plants in the soil, which are the main source of gaseous carbon input (approximately 5%–33%), as well as the decay of tissues in the above and below ground (Bardgett et al., 2013). Also, slow-growing plants have only low productivity and other plant materials. The carbon input of plant species (slow-growing and fast-growing) into the soil depends on environmental conditions such as light and nutrients. Therefore, soil carbon input in biomes with less nutrients and a short season will be due to insufficient litter, whereas biomes with higher productivity will lead to carbon sequestration in the soil. Many environmental factors as well as physiographic conditions of plants can affect carbon dynamics (Tewari et al., 2024). Under current climate conditions, forest soil carbon is vulnerable to loss, with negative consequences for soil biodiversity, productivity, and climate feedbacks (De Deyn, 2013). Soil carbon dynamics in soil and plants is an important ecosystem service for mitigating climate change; therefore, a clear understanding of soil carbon processes and mechanisms is required. In this context, elucidating the impact of plant performance (i.e., litter characteristics and root physiological characteristics) on soil carbon sequestration and carbon stabilization has become a major environmental challenge (De Deyn et al., 2008).

Plant traits affect aboveground litter accumulation and decomposition (Cortez et al., 2007), which in turn affects soil carbon balance. The results show that the SOC content above the emission nitrogen concentration is very similar. Effluents with high nitrogen content are mineralized, thus producing more microbial residues, which then persist in the mineral soil matrix, leading to an increase in the stable SOC content (Lehmann and Kleber, 2015). In turn, the SOC content is affected not only by weather and soil conditions (in maritime climates, by mean annual temperature, soil pH, and soil moisture content) but also by the community abundance-weighted mean leaf nitrogen concentration (in nitrogen-rich conditions).

The factors that determine the response of plants to symbionts are important for soil carbon inputs, since such symbionts can benefit plants by providing less access to resources. The most widespread plants on Earth are nitrogen-fixing bacteria and mycorrhizal fungi (mycorrhizae). Symbionts that absorb nitrogen from the air can live in symbiosis with legumes, actinomycetes, *Parasponia* species, cycads (Vessey et al., 2005), and many bryophytes and lichens (Cornelissen et al., 2007). Mycorrhizal fungi enable plants to absorb nutrients (especially nitrogen and phosphorus) from the soil (Kumar et al., 2021d). Additionally, mycorrhizal organisms can reduce carbon emissions through carbon sequestration in mycelium, extend root lifespan, and increase carbon sequestration in soil aggregates (Langley et al., 2006). Mycorrhizal fungi are associated with most terrestrial plants and are considered most beneficial for both plant and soil communities. However, their effects on soil carbon sequestration may depend on the combination of plant and symbiont species and soil fertility (Allen et al., 2003). The following features require further elucidation, but it is clear that many mycorrhizal fungal species dominate ecosystems and are associated with specific plants and growth-limiting nutrients (Verma et al., 2023).

9.3 FUNCTIONAL TRAITS DIVERSITY AND NUTRIENT CYCLING

The availability of nutrients in the soil is an important factor in determining the types of plant communities. At the same time, plants exert a specific influence on the value of the soil through feedback from nutrient cycling. There are many studies on the relationship between soil and plant interactions (Kumar et al., 2020a), but the combination of positive responses at different scales is still limited (Kumar et al., 2021a). One way to address these issues is to determine the quality of permanent vegetation for quantitative measurements of soil fertility. Recent international evidence suggests the control of food supply, the integration of plant activities, the climate control of plant quality, and the interaction of leaf nitrogen and phosphorus patterns with climate (Reich and Oleksyn, 2004). However, there are some studies on the relationship between soil plant and nutrient quality. A study conducted by Ordoñez et al. (2009) reported that there was a consistent and significant shift from species with low SLA, LNC, and LPC at low soil fertility toward species with high SLA, LNC, and LPC at high soil fertility, although variability within sites was typically large. In line with their expectations, soil total N was not a good predictor of leaf trait variation: the relations of soil total nitrogen (N) with SLA, LPC (mass and area basis), and leaf N:P were not significant, and those with LNC mass and LNC area were in the opposite direction (with respect to high versus low fertility) to the other soil nutrient measures, including C:N. In general, SLA and LNC (mass and area basis) were related to most soil nutrient measures, positively to N mineralization and soil total P, and negatively to soil C:N (reflecting the increase in nutrient availability with decreasing soil C:N). Relationships of leaf N on an area basis were weaker than those on a mass basis. In contrast, leaf P (mass and area basis) and leaf N:P were only related to soil total P, and as expected, leaf P increased and N:P decreased with increasing soil P. According to Kumar et al. (2021b), lower LNC and LPC are characteristics of higher nutrient plant use. In barren environments, plants have been found to increase leaf carbon content by accumulating more carbon, including lignin and tannins (Kumar et al., 2021c), and these species have been found to have high carbon content of these compounds. Other studies have shown that nutrient content, such as nitrogen, phosphorus, sodium, potassium, and calcium in leaves, controls the turnover of nitrogen and phosphorus in mature leaves (Verma et al., 2023). Plants with high levels of nutrients in their leaves have a greater nitrogen and phosphorus balance than plants with low levels of nutrients; all of the above traits indicate plasticity. Plasticity is particularly important across moisture gradients, being smallest in dry forests and greatest in humid forest types (Kumar et al., 2020b).

9.4 INTERACTION BETWEEN LEAF TRAITS, SOIL CARBON, AND NUTRIENT DYNAMICS

Decomposition of plant litter is a major driver of carbon and nutrient cycling in terrestrial and forest ecosystems, controlling the provision of important ecosystem services, such as soil formation, nutrient availability, and atmospheric composition, and providing advice on plant composition. One of the major goals in ecology is to model how plant activity of different species in different biomes feeds back into soil carbon turnover and hence atmospheric chemistry and climate (Tewari et al., 2021).

Although the rate of decomposition of litter depends on climate and soil microbial composition, litter quality (species identity) is an important factor across biomes (Cornwell et al., 2008). The properties and chemistry of leaves have 'afterlives' that affect the ability of the litter to decompose (Cornelissen et al., 2006). In fact, the differences between fresh leaves and leaf litter are often correlated. Therefore, lignin content, body strength, polyphenol content, or some leaves and dry matter content (Kazakou et al., 2006) may affect the decomposition rate of leaves. In addition, the high nutritional value of decomposers creates unlimited nutrients for the decomposition process (Verma et al., 2023). Therefore, nitrogen (N), phosphorus (P), and calcium (Ca) content and pH (as an indicator of alkali cation content and antimicrobial organic acids) are commonly used as critical estimates of decomposition rate (Kumar et al., 2020a). Therefore, leaf dry matter content is responsible for increasing soil organic carbon in the soil and thus contributes to soil carbon sequestration. Leaf nitrogen and phosphorus concentrations set limits for plant growth, and leaf N:P ratios vary between nitrogen limits and phosphorus limits.

It is known that leaf mass in the analysis is generally related to soil nutrients: leaf area (SLA, m kg^{-1}), leaf nitrogen concentration per unit mass (LNC mass mg g^{-1}), leaf phosphorus concentration (LPC) per unit mass mass mg g^{-1}), leaf carbon content (LCC mass mg g^{-1}), and leaf N:P. Leaf parameters such as SLA, LNC, LCC, and LPC have a positive relationship with the relative growth of the plant, leaf carbon assimilation rate, and energy supply (Lavorel and Garnier, 2002). The analysis also includes area-based leaf nitrogen and leaf phosphorus concentrations (in gm^{-2} for both LNC area and LC area). Leaf N:P is representative of restricted species (Gusewell, 2004). SLA of all materials ranged from 2 to 41 m^2kg^{-1}, LNC weight from 5 to 54 mg g^{-1}, LNC area from 0.3 to 6.6 gm^{-2}, LPC group ranged from medium to 0.1 to 6 mg g^{-1}, LPC ranged from 0.02 to 4.74 gm^{-2}, and leaf N:P ranged from 2.6 to 89. The effect of soil nutrients was different. The response pattern of plants to nutrients is represented in terms of the linkage between growth, structure, and conservation of nutrients (Kumar et al., 2021d), which may be accompanied by the contract model that describes the plant's functioning according to a specific axis (Westoby et al., 2002). It is proven by studies that the axes of specialization emerged from the analysis of the quality of leaves in relation to the nutrients that have existed worldwide in the last few years (Wright et al., 2004).

Leaf N, P, and N:P vary among species and locations, and changes in P are larger than changes in N and N:P ratios (Kumar et al., 2021d). However, drivers of greater variability in leaf P than in leaf N and N:P have not been systematically studied in part because of differences in the effects of environmental changes and competition among strategies to improve leaf chemical variability (Table 9.1) (Elser et al., 2010). Many theories have been proposed to explain the structure of leaf chemistry. Among these, climate, soil, or plant growth morphology is thought to be an important factor affecting leaf N and P. Similarly, several theories have been proposed to explain patterns across regions. For example, the plant physiology hypothesis suggests that all plant metabolic processes are temperature sensitive and that high nitrogen and phosphorus concentrations will help reduce metabolic rates at low temperatures (Woods et al., 2003). However, little evidence and simulations suggest that the temperature of biological leaves varies across species and ecosystems, but our understanding of the temperature of medicinal leaves is still limited (Lukac et al., 2010).

9.5 INTERACTION BETWEEN ROOT TRAITS, SOIL CARBON, AND NUTRIENT DYNAMICS

The carbon cycle is the process of supply and release of carbon to the soil and is an important process in the ecosystem (De Deyn et al., 2008). Therefore, root structural traits such as root length and coarse and fine root intensity in soil determine the distribution of different roots (Tewari et al., 2021). However, the carbon added to the soil from dead roots also depends on root length, which is related to many aspects of the stem and roots (Kumar et al., 2022b). Many studies have shown that root vitality is negatively correlated with root length, root nitrogen/carbon ratio, and soluble compound concentration (Kumar et al., 2021b); there is a good correlation with microorganisms diversity, which is also associated with high lignin and fiber (Kumar et al., 2021d). Some root systems also affect root production and root diseases caused by fungi can shorten lifespan and cause root rot; mycorrhizal fungi themselves act as persistent soil carbon sinks (Langley et al., 2006). Recent studies have also suggested that AM and ectomycorrhizal (ECM) fungi play different roles in the soil carbon cycle: the former promote decomposition by providing an unstable carbon source (Herman et al., 2012), while the latter interfere with decomposition by competing for nitrogen (Averill et al., 2014; Bardgett et al., 2013). These conclusions are generally based on the knowledge that the microbial population has a different metabolism: not only do fungi take up less carbon per unit of fungal biomass, but this leads to higher carbon utilization in bacteria (Six et al., 2006). Thus, fungal mycelium contains poorly formed, non-nutrient substances such as chitin and melanin, whereas bacterial mycelium usually contains phospholipids that are rapidly mineralized upon death (Kumar et al., 2021c). This new method clearly demonstrates that root metabolism, especially root exudates, has a significant effect on soil fertility.

Plants can produce phosphorus by removing organic acid-rich root exudates (Lambers et al., 2008) and can cause desorption and dissolution of soil minerals (George et al., 2011). Promoting microbial mineralization of organic matter by releasing root exudates is another strategy that allows plants to take up more nutrients from the soil. This priming process is particularly effective in nitrogen-deficient soils and involves the induction of microbial extracellular enzymes involved in the decomposition of organic nitrogen, thereby enhancing the nitrogen cycle and providing nitrogen to plants. However, as bacteria grow via exudation, microbial fixation of nitrogen

TABLE 9.1
Association of Plant Functional Traits with Biological Filters, Biogeochemical Cycles, and Disturbance Regimes (Lavorel and Garnier, 2002)

	Biological Filters		Biogeochemical and Disturbance Regimes	
Plant Functional Traits	**Competitive Strength**	**Plant Defense/ Protection**	**Effect on Biogeochemical Cycles**	**Effect on Disturbance Regimes**
		Whole Plant Traits		
Growth form	*	*	*	*
Life form	*		*	*
Plant height	*	*	*	*
Clonality	*			?
Spinescence	*	*		?
Flammability	*	?	*	*
		Leaf Traits		
Specific leaf area	*	*	*	
Leaf size	*	*	*	
Leaf dry matter content		*	*	*
Leaf N and P content	*	*	*	
Physical strength of leaves		*	*	
Leaf lifespan	*	*	*	*
Leaf phenology	*		*	*
Photosynthetic pathway	*			
Leaf frost resistance	*	*		
		Stem Traits		
Stem-specific density		*	*	*
Twig dry matter content		*	*	*
Twig drying time			?	*
Bark thickness		*		?
Specific root length	*	*		?
Diameter of fine root				
Distribution of rooting depth	*		*	*
95% rooting depth	*			*
Nutrient uptake strategy	*		*	
		Regenerative Traits		
Dispersal mode				
Dispersule shape and size				
Seed mass	*	*		
Resprouting capacity after major disturbance			*	

* Traits that are critical to the ecological challenges.
? Traits that need to be studied.

may occur, reducing nutrient availability (Bengtson et al., 2012). These comparative responses suggest a positive effect of root exudates on microbial mineralization–immobilization kinetics. Changes in the quality of root exudates may be related to microbial nutrient limitation (Dijkstra et al., 2013) and the response to certain disease populations associated with changes in nutrition. Recently, Phillips et al. (2013) reported that the estimation of the economic value of forest ecosystems is based on the management of trees by AM or ECM fungi, which have previously become an inorganic nutrient industry with a rapid rate of decomposition and nutrient mineralization, and second, an organic industry with a slower turnover. The theory is based on two pieces of information: first, tree species are often associated with a type of mycorrhizal fungi, namely, AM or ECM fungi in addition to the absorption of inorganic nitrogen present in complex organic nitrogen forms. This process is not consistent with the process that links microbial communities and their functions to the economy; for example, the quality of the plant, the benefits associated with bacteria in the microbial community, high nitrogen mineralization rates, and nitrogen losses. Well-protected plants are associated with fungal-dominated microbial communities and the cost of slow nitrogen cycling (Kumar et al., 2020b).

9.6 PLANT FUNCTIONAL TRAITS INTERACTION WITH SOIL FOR NUTRIENT CYCLING

PFT plays a major role in ecosystem processes, effectively influencing nutrient supply and soil interactions. Root structure, including the depth and distribution of roots, increases nutrient availability through contact with the soil, while mycorrhizal organization enhances phosphorus uptake. Traits such as leaf size, thickness, and nutrient content affect photosynthetic efficiency and nutrient reabsorption before leaf age can conserve the soil nutrient pool (Kumar et al., 2020a,b; Kaushal et al., 2021). Leaf quality is determined by chemical composition, which affects the rate of decomposition and release of nutrients, thereby affecting the soil microbial community and mineralization. Roots also contribute to soil structure and stability by binding material and creating pore space, improving aeration, infiltration, and reducing erosion (Tewari et al., 2024; Kumar et al., 2021a,b, 2022a). Plants take up organic matter to improve soil fertility, water retention, and carbon sequestration, while root exudates promote microbial diversity, which is important for healthy cycling. By selecting and managing plants that perform well, we can improve soil health and enhance ecosystem resilience by better understanding soil management for sustainability and ecological development. Nutrient cycling and soil interactions affect ecosystem processes and functions (Tewari et al., 2022; Kumar et al., 2018, 2021c,d). These characteristics include morphological, physiological, and phenological characteristics that determine how plants obtain, use, and transform nutrients, thus affecting soil and overall ecosystem health. Deep-rooted species can uptake nutrients from deeper layers of the soil, while deep-rooted species can utilize nutrients from deeper soil layer. Plants with high nutrient value use more biomass per unit of nutrient absorbed, thus optimizing nutrient cycling in the ecosystem. High-quality litter decomposes faster and releases nutrients more quickly for plants to absorb (Kumar et al., 2020c; Kumar et al., 2022a; Bisht et al., 2022).

Roots affect soil structure by binding the soil and creating pore space, thus increasing soil aeration, water infiltration, and root penetration. This increases the stability of the soil and reduces erosion. SOM also plays an important role in carbon sequestration and climate change mitigation. The soil microbes are also important for nutrient cycling because they break down organic matter and release nutrients in a form that plants can absorb. By selecting and managing plants with specific characteristics, soil health can be improved, nutrient utilization improved, and ecosystem resilience increased. Understanding these interactions provides valuable information about sustainable land management, agroforestry practices, and ecological restoration efforts to maintain sustainable and healthy ecosystems (Tewari et al., 2021; Kumar et al., 2021f, 2024).

9.7 CONCLUSION

PFTs regulate many ecosystem functions, including soil carbon sequestration and soil nutrient dynamics. Understanding the impact of plant performance on soil carbon and nutrient dynamics by studying the functional ecology of natural ecosystems is the first step in understanding the relationship between PFTs, soil carbon, and nutrients. Leaf traits, including dry leaf content and litter decomposition, contribute to the addition of soil carbon and other nutrients (especially nitrogen and phosphorus). Root traits such as root exudates, long-lived roots, AM fungi, and nitrogen-fixing bacteria also affect soil carbon and nutrient dynamics. Stem traits contribute less to soil carbon and nutrient dynamics than leaves and roots. The challenge in managing soil carbon and soil nutrients based on the concept of PFTs is to determine the variability and flexibility of plant performance. This can be done by examining the response of specific traits in different soils and climates. The aim of this chapter is to understand the relationship between PFTs, soil properties, particularly carbon and nutrients, and soil organisms in different soils and climates.

REFERENCES

Allen, M.F., Swenson, W., Querejeta, J.I., Egerton-Warburton, L.M. and Treseder, K.K. 2003. Ecology of mycorrhizae: a conceptual framework for complex interactions among plants and fungi. *Annual Review of Phytopathology*, 41: 271–303.

Averill, C., Turner, B.L. and Finzi, A.C. 2014. Mycorrhiza-mediated competition between plants and decomposers drives soil carbon storage. *Nature*, 505(7484): 543–545.

Bardgett, R.D., Manning, P., Morriën, E. and De Vries, F.T. 2013. Hierarchical responses of plant–soil interactions to climate change: consequences for the global carbon cycle. *Journal of Ecology*, 101(2): 334–343.

Bengtson, P., Barker, J. and Grayston, S.J. 2012. Evidence of a strong coupling between root exudation, C and N availability, and stimulated SOM decomposition caused by rhizosphere priming effects. *Ecology and Evolution*, 2(8): 1843–1852.

Bisht, S., Kumar, A., Kumar, N., Singh, H. and Kumar, P. 2022. *Biofuel Production by Using Biomass and Its Application*. In *Renewable Energy and Green Technology*, CRC Press.

Cornelissen, J.H.C., Lang, S.I., Soudzilovskaia, N.A. and During, H.J. 2007. Comparative cryptogam ecology: a review of bryophyte and lichen traits that drive biogeochemistry. *Annals of Botany*, 99: 987–1001.

Cornelissen, J.H.C., Quested, H.M., van Logtestijn, R.S.P., Perez-Harguindeguy, N., Gwynn-Jones, D., Dı´az, S., Callaghan, T.V., Press, M.C. and Aerts, R. 2006. Foliar pH as a new plant trait: can it explain variation in foliar chemistry and carbon cycling processes among subarctic plant species and types? *Oecologia*, 147: 315–326.

Cornwell, W.K., Cornelissen, J.H.C., Amatangelo, K., Dorrepaal, E., Eviner, V.T., Godoy, O., Hobbie, S.E., Hoorens, B., Kurokawa, H., Perez-Harguindeguy, N. and Quested, H. M. 2008. Plant species traits are the predominant control on litter decomposition rates within biomes worldwide. *Ecology Letters*, 11: 1065–1071.

Cortez, J., Garnier, E., Pérez-Harguindeguy, N., Debussche, M. and Gillon, D. (2007). Plant traits, litter quality and decomposition in a Mediterranean old-field succession. *Plant and Soil*, 296:19–34.

De Deyn, G.B. (2013). Ecosystem carbon and soil biodiversity. *Ecosystem services and carbon sequestration in the biosphere*, pp. 131–153.

De Deyn, G.B., Cornelissen, J.H. and Bardgett, R.D. 2008. Plant functional traits and soil carbon sequestration in contrasting biomes. *Ecology Letters*, 11(5): 516–531.

Dijkstra, F.A., Carrillo, Y., Pendall, E. and Morgan, J.A. 2013. Rhizosphere priming: a nutrient perspective. *Frontiers in Microbiology*, 4: 216.

Elser, J.J., Fagan, W.F., Kerkhoff, A.J., Swenson, N.G. and Enquist, B.J. 2010. Biological stoichiometry of plant production: metabolism, scaling and ecological response to global change. *New Phytologist*, 186(3): 593–608.

George, T.S., Fransson, A.M., Hammond, J.P. and White, P.J. 2011. Phosphorus nutrition: rhizosphere processes, plant response and adaptations. In *Phosphorus in Action: Biological Processes in Soil Phosphorus Cycling*, pp. 245–271, Berlin, Heidelberg: Springer.

Gusewell, S. 2004. N:P ratios in terrestrial plants: variation and functional significance. *New Phytologist*, 164: 243–266.

Herman, D.J., Firestone, M.K., Nuccio, E. and Hodge, A., 2012. Interactions between an arbuscular mycorrhizal fungus and a soil microbial community mediating litter decomposition. *FEMS Microbiology Ecology*, 80(1): 236–247.

Kaushal, R, Tewari, S., Thapliyal, SD., Kumar, A., Roy, T., Islam, S., Lepcha, S.T.S. and Durai, J. 2021. Build-up of labile, non-labile carbon fractions under fourteen-year-old bamboo plantations in the Himalayan foothills. *Heliyon*, 7(3): e07850.

Kazakou, E., Vile, D., Shipley, B., Gallet, C. and Garnier, E. 2006. Co-variations in litter decomposition, leaf traits and plant growth in species from a Mediterranean old-field succession. *Functional Ecology*, 20: 21–30.

Kumar, A., Dwivedi, G.K., Tewari, S., Sah, V.K., Singh, H., Kumar, P., Kumar, N. and Kaushal, R. 2020a. Soil organic carbon pools under *Terminalia chebula* Retz. based agroforestry system in Himalayan foothills, India. *Current Science*, 118(7): 1098–1103.

Kumar, A., Dwivedi, G.K., Tewari, S., Paul, J., Anand, R., Kumar, N., Kumar, P., Singh, H. and Kaushal, R. 2020b. Carbon mineralization and inorganic nitrogen pools under *Terminalia chebula* Retz.-Based agroforestry system in Himalayan foothills, India. *Forest Science*, 66(5): 634–643.

Kumar, A., Kumar, P., Singh, H. and Kumar, N. 2021a. Impact of plant functional traits on near saturated hydraulic conductivity of soil under different forests of Kempty watershed in Garhwal Himalaya. *Indian Journal of Soil Conservation*, 49(1): 38–44.

Kumar, A., Kumar, P., Singh, H. and Kumar, N. 2021b. Adaptation and mitigation potential of roadside trees with bio-extraction of heavy metals under vehicle emissions and their impact on physiological traits during seasonal regimes. *Urban Forestry & Urban Greening*, 58: 126900.

Kumar, A., Kumar, P., Singh, H. and Kumar, N. 2021c. Modulation of plant functional traits under essential plant nutrients during seasonal regime in natural forests of Garhwal Himalayas. *Plant and Soil*, 465(1): 1–16.

Kumar, A., Kumar, P., Singh, H. and Kumar, N. 2021d. Relationship of physiological plant functional traits with soil carbon stock in the temperate forest of Garhwal Himalaya. *Current Science*, 120(7): 1064–1068.

Kumar, A., Singh, H., Kumari, G., Bisht, S., Malik, A., Kumar, N., Singh, M., Raturi, A., Barthwal, S., Thakur, A. and Kaushal, R. 2022a. Adaptive resilience of roadside trees to vehicular emissions via leaf enzymatic, physiological, and anatomical trait modulations. *Environmental Pollution*, 313: 120191.

Kumar, A., Tewari, S., Singh, I., Panday, R., Kumar, D. and Anand, R. 2018. Effect of nutrient sources on growth, yield and quality of turmeric under harad (*Terminalia chebula*) based agroforestry system. *Indian Journal of Agroforestry*, 20(2): 1–6.

Kumar, N., Jeena, N., Kumar, A., Khwairakpam, R. and Singh, H. 2021f. Comparative response of rice cultivars to elevated air temperature in Bhabar region of Indian Himalaya: status on yield attributes. *Heliyon*, 7(6): e07474.

Kumar, N., Kumar, A., Jeena, N., Singh, R. and Singh, H. 2020c. Factors influencing soil ecosystem and agricultural productivity at higher altitudes. In *Microbiological Advancements for Higher Altitude Agro-Ecosystems & Sustainability*, Singapore: Springer, pp. 55–71.

Kumar, N., Singh, H., Giri, K., Kumar, A., Joshi, A., Yadav, S., Singh, R., Bisht, S., Kumari, R., Jeena, N., Khairakpam, R. and Mishra, G. 2024. Physiological and molecular insights into the allelopathic effects on agroecosystems under changing environmental conditions. *Physiology and Molecular Biology of Plants*, 248: 119255.

Kumar, N., Verma, S., Kumar, A., Dashila, H., Suyal, D.C., Kumari, G., Rawat, S., Jeena, N., Singh, M. and Kumar, A. 2022b. Molecular insights into stress-responsive genes in the mitigation of environmental stresses. Chandra, D., and Bhatt, P. (Eds.1). *Unravelling Plant-Microbe Synergy*. Academic Press, New York, United States.

Laliberté, E., Lambers, H., Burgess, T.I. and Wright, S.J. (2015). Phosphorus limitation, soil-borne pathogens and the coexistence of plant species in hyperdiverse forests and shrublands. *New Phytologist*, 206(2): 507–521.

Lambers, H., Raven, J.A., Shaver, G.R. and Smith, S.E., 2008. Plant nutrient-acquisition strategies change with soil age. *Trends in Ecology & Evolution*, 23(2): 95–103.

Langley, J.A., Chapman, S.K. and Hungate, B.A. 2006. Ectomycorrhizal colonization slows root decomposition: the postmortem fungal legacy. *Ecology Letters*, 9: 955–959.

Lavorel, S. and Garnier, E. (2002). Predicting changes in community composition and ecosystem functioning from plant traits: revisiting the Holy Grail. *Functional Ecology*, 16(5): 545–556.

LeBauer, D.S. and Treseder, K.K. (2008). Nitrogen limitation of net primary productivity in terrestrial ecosystems is globally distributed. *Ecology*, 89(2): 371–379.

Lehmann, J. and Kleber, M. (2015). The contentious nature of soil organic matter. *Nature*, 528(7580): 60–68.

Le Quere, C., Raupach, M. R., Canadell, J. G., Marland, G., Bopp, L., Ciais, P. and Friedlingstein, P. 2009. Trends in the sources and sinks of carbon dioxide. *Nature Geoscience*, 2(12): 831.

Lukac, M., Calfapietra, C., Lagomarsino, A. and Loreto, F., 2010. Global climate change and tree nutrition: effects of elevated CO_2 and temperature. *Tree Physiology*, 30(9): 1209–1220.

Ordoñez, J.C., Van Bodegom, P.M., Witte, J.P.M., Wright, I.J., Reich, P.B. and Aerts, R.. 2009. A global study of relationships between leaf traits, climate, and soil measures of nutrient fertility. *Global Ecology and Biogeography*, 18: 137–149.

Phillips, R.P., Brzostek, E. and Midgley, M.G. 2013. The mycorrhizal-associated nutrient economy: a new framework for predicting carbon–nutrient couplings in temperate forests. *New Phytologist*, 199(1): 41–51.

Reich, P.B. and Oleksyn, J. 2004. Global patterns of plant leaf N and P in relation to temperature and latitude. *Proceedings of the National Academy of Sciences USA*, 101: 11001–11006.

Roumet, C., Birouste, M., Picon-Cochard, C., Ghestem, M., Osman, N., Vrignon-Brenas, S., Cao, K.F. and Stokes, A. 2016. Root structure–function relationships in 74 species: evidence of a root economics spectrum related to carbon economy. *New Phytologist*, 210(3): 815–826.

Singh, H. 2024. Exploring adaptive modulation in plant functional traits and its impact on the productivity of *Acacia auriculiformis* under CO_2-enriched environment. *Industrial Crops and Products*, 210: 118186.

Six, J., Frey, S.D., Thiet, R.K. and Batten, K.M. 2006. Bacterial and fungal contributions to carbon sequestration in agroecosystems. *Soil Science Society of America Journal*, 70(2): 555–569.

Tewari, S., Tewari, L., Dubey, A., Kumar, A., Kumar, N. and Kaushal, R. 2022. Use of the RAPD marker to determine the genetic diversity of various *Dalbergia sissoo* Roxb. (Shisham) genotypes and their evolutionary relationship in Indian subcontinents. *Vegetos*, *35*(3): 850–857.

Tewari, S., Tewari, L., Kumar, A., Bisht, S, Kaushal, R. and Kumar, A. 2024. Germination behaviour of teak genotypes and effect of different treatments on seed germination characteristics in Himalayan Foothills, India. *Indian Forester*, 150(3): 244–254.

Tewari, S., Tewari, L., Kumar, A. and Kaushal, R. 2021. Performance of *Glycine max* L. varieties under *Tectona grandis* L. and *Dalbergia sissoo* Roxb. based agroforestry system. *Indian Journal of Agroforestry*, 20(2): 1–6.

Verma, T., Bhardwaj, D.R., Sharma, U., Sharma, P., Kumar, D., Kumar, A. and Kumar, A. 2023. Aroforestry systems in the mid-hills of the north-western Himalaya: a sustainable pathway to improved soil health and climate resilience. *Journal of Environmental Management*, 348: 119264.

Vessey, J.K., Pawlowski, K. and Bergman, B. 2005. Root-based N-2-fixing symbioses: legumes, actinorhizal plants, *Parasponia* sp. and cycads. *Plant Soil*, 274: 51–78.

Violle, C., Navas, M.L., Vile, D., Kazakou, E., Fortunel, C., Hummel, I. and Garnier, E. 2007. Let the concept of trait be functional. *Oikos*. 116(5): 882–892.

Westoby, M., Falster, D.S., Moles, A.T., Vesk, P.A. and Wright, I.J. 2002. Plant ecological strategies: some leading dimensions of variation between species. *Annual Review of Ecology and Systematics*, 33: 125–159.

Woods, H.A., Makino, W., Cotner, J.B., Hobbie, S.E., Harrison, J.F., Acharya, K. and Elser, J.J. 2003. Temperature and the chemical composition of poikilothermic organisms. *Functional Ecology*, 17(2): 237–245.

Wright, I.J., Reich, P.B., Westoby, M., Ackerly, D.D., Baruch, Z., Bongers, F., Cavender-Bares, J., Chapin, T., Cornelissen, J.H., Diemer, M. and Flexas, J. (2004). The worldwide leaf economics spectrum. *Nature*, 428(6985): 821–827.

10 Carbon Sequestration and Storage

Traits that Influence Forest Carbon Dynamics

Kalyani Supriya

10.1 INTRODUCTION

10.1.1 ROLE OF FORESTS IN THE CARBON CYCLE: AN UNDERSTANDING OF CARBON SEQUESTRATION

Sequestering carbon is crucial to the worldwide endeavour to tackle climate change. The process involves capturing CO_2 from the atmosphere and storing it in reservoirs. This reduces the concentration of greenhouse gases that cause global warming. Effective methods to improve carbon sinks are required since human activities, especially the burning of fossil fuels and deforestation, have increased atmospheric CO_2 levels (Roy et al., 2023; Yadav et al., 2023). Natural ecosystems such as woods, wetlands, and agricultural lands play an important role in carbon sequestration. For example, photosynthesis enables forests to absorb CO_2 and store it in biomass (leaf, branch, and trunk material) and soil. By retaining carbon in their soils, grasses and wetlands also make a contribution (McLaughlin et al., 2023). The technological techniques like carbon capture and storage (CCS), which collect CO_2 emissions from industrial activities and store them underground in geological formations, supplement these natural processes. Direct air capture, one of the emerging technologies, has the potential to remove CO_2 directly from the atmosphere (Kumar et al., 2020a,b; Amirthan and Perera, 2023). The carbon sequestration has advantages for the environment that go beyond reducing global warming. In addition, they promote sustainable land management practices, improve air quality, and improve soil health (Kumar et al., 2021d; Nguyen et al., 2023; Liáng et al., 2023).

10.1.2 CHALLENGES AND STRATEGIES FOR EFFECTIVE CARBON SEQUESTRATION

The mitigation of climate change through carbon sequestration is a promising technique; nevertheless, its success is dependent on the resolution of multiple challenges. It is crucial to ensure the sustainability of carbon storage in order to prevent sequestered CO_2 from returning to the atmosphere. To manage possible leakage problems,

DOI: 10.1201/9781003545231-10

this calls for careful monitoring and upkeep of storage facilities (Kumar et al., 2021a,b; Roy et al., 2023; Nguyen et al., 2023).

To mitigate negative impacts, sequestration projects must also be evaluated in terms of their impact on adjacent ecosystems and communities. Despite their overall effectiveness and differing land needs, reforestation and planting are essential for improving carbon sequestration. Research has measured the ability of forests to sequester CO_2, and the results show that extensive afforestation can play a major role in mitigating the effects of climate change (Houghton et al., 2015; Kumar et al., 2021c; Menard et al., 2023).

10.2 IMPORTANT FEATURES OF TREES AND SOILS THAT IMPACT CARBON STORAGE CAPACITY: CONSEQUENCES FOR THE DYNAMICS OF FOREST ECOSYSTEMS AND CARBON SEQUESTRATION

Climate, soil properties, and atmospheric nitrogen deposition all have a significant influence on soil organic carbon (SOC) stores. SOC data were adjusted to reflect site means in order to focus on the impact of tree species under uniform circumstances. This method made it possible to assess the characteristics of plants and forests. It was found that while leaf dry matter content (LDMC) showed a positive correlation with SOC pool sizes, many functional traits, including wood density, specific leaf area (SLA), specific root length (SRL), and leaf nitrogen content (LNC), exhibited negative correlations (Kumar et al., 2017, 2019).

10.2.1 Procedures and Standardization: Examining SOC Pool Dimensions

The volumes of the SOC pools are essential for comprehending the dynamics of carbon in forest ecosystems. A study that examined these pools in mineral soil layers and the forest floor showed a significant correlation between SOC levels and various functional traits. Below is a thorough analysis of these characteristics and their meanings:

10.2.1.1 Leaf Nitrogen Content (LNC)

The increased nitrogen normally denotes more vigorous photosynthesis and development in leaves, which can result in increased carbon fixation. On the other hand, research indicates that greater LNC may be associated with litter breakdown rates that are quicker, which could eventually result in lower SOC levels. Higher LNC species typically have faster litter breakdown rates, which affects SOC dynamics, as Reich et al. (2004) showed.

10.2.1.2 Specific Leaf Area (SLA)

The leaf area to dry mass ratio is referred to as SLA. Higher SLA species tend to have thinner leaves that are more susceptible to breakdown and herbivory. Because these species may not contribute as much to stable SOC because of their faster turnover rates, this feature frequently has a negative correlation with SOC. Wright et al. (2004) observed that species with high SLA levels produce more litter, which degrades more quickly and affects SOC stability (Kumar et al., 2015, 2016).

10.2.1.3 Specific Root Length (SRL)

The number of roots per unit biomass is indicated by the SRL. Greater root systems are suggested by higher SRL, which may improve nutrient uptake but also increase root turnover and breakdown. This dynamic commonly contributes to lower SOC in quicker cycle systems and can have complex interactions with SOC levels. According to research by Eissenstat et al. (2000), species with high SRL frequently have higher nutrient absorption efficiency, but their quick root turnover can also result in reduced SOC pools.

10.2.1.4 Wood Density

Wood density indicates the compactness of wood and is an important trait affecting carbon storage. The higher wood density species frequently have more resilient, longer-living carbon pools, which helps promote SOC buildup. The higher wood density is linked to slower rates of breakdown, which results in more stable SOC pools, according to Chave et al. (2009).

10.2.1.5 Leaf Dry Matter Content (LDMC)

LDMC and SOC have a positive connection, in contrast to the other features indicated. Denser and more resilient leaves, as indicated by higher LDMC, decompose more slowly and accumulate SOC over time with greater stability. The higher LDMC was associated with better litter quality and slower decomposition, which raised SOC levels (Westoby et al., 2002).

10.2.2 Spectrum of Plant Economics (PES)

The characteristics mentioned above are essential parts of the plant economics spectrum (PES), a theoretical model that explains how plants balance resource acquisition and conservation. The PES is frequently depicted as a continuum ranging from "fast" to "slow" tactics:

10.2.2.1 Fast Strategy

Although they often grow quickly, species with high SLA, high LNC, and low wood density also tend to have high litter turnover and lesser SOC buildup.

10.2.2.2 Slow Strategy

On the other hand, species with high wood density, low SLA, and high LDMC grow more slowly but make a greater contribution to persistent litter, which stores carbon over time.

10.2.3 Influence of Functional Traits on Soil Organic Carbon (SOC)

Augusto and Boča (2022) in their investigation indicated that leaf photosynthetic maximum capacity (Amax) is a significant predictor of SOC, explaining approximately

48.5% of the variance in SOC pool sizes. This relationship is particularly evident in forest floor layers, highlighting the intricate interplay between plant functional traits and soil carbon dynamics. The key functional traits and their impact are mentioned below:

10.2.3.1 Leaf Photosynthetic Greatest Capacity (e.g., Amax)

Under ideal conditions, Amax is the maximum rate of photosynthesis that a leaf can achieve. The lower Amax species frequently grow more slowly and fix carbon at a lower rate, which causes more organic matter to build up in the soil. This could be a result of these species' increased investments in root biomass and high-quality litter, which raise SOC levels. The lower Amax is associated with slower litter breakdown rates and higher SOC stability.

10.2.3.2 Leaf Nitrogen Content (LNC)

The lower LNC is typically associated with slower turnover of litter and roots. The species with reduced nitrogen content may also invest more in structural compounds, which decompose more slowly and enhance SOC accumulation. Reich et al. (2004) found that the species with lower LNC contribute to higher SOC through the production of more recalcitrant litter.

10.2.3.3 Leaf Dry Matter Content (LDMC)

A high LDMC indicates denser leaves that decompose more slowly and are less appetizing to herbivores. Since their litter adds stable organic matter to the soil, species with higher LDMC are frequently associated with greater SOC accumulation. Westoby et al. (2002) emphasized the importance of leaf toughness in litter quality by establishing a link between greater LDMC and increased SOC.

10.2.3.4 Specific Leaf Area (SLA)

The low SLA indicates tougher, thicker leaves that break down more slowly and build up SOC. The lower SLA species tend to be more frugal with their use of resources, which increases the amount of organic matter that is added to the soil. Wright et al. (2004) supported SOC retention by highlighting the connection between delayed litter decomposition and reduced SLA.

10.2.3.5 Specific Root Length (SRL)

A lower SRL frequently denotes a higher biomass per unit length of root, which increases the amount of root mass that the soil receives. As a primary source of organic matter, roots may lead to greater SOC levels. According to Eissenstat et al. (2000), higher root turnover and biomass intake are linked to lower SRL and enhanced capacity for SOC storage.

10.2.3.6 Phosphorus Content

Less phosphorus can have a favourable effect on SOC levels by promoting slower development rates and longer litter lifespans, much like nitrogen. The supply of phosphorus influences plant growth and litter breakdown, which in turn influences SOC dynamics, according to the research by Hawkes, (2005).

10.2.4 Stability of SOC and Tree Species

The stability of SOC is largely dependent on the type of tree. Functional characteristics linked to various tree species have been found to have a major impact on SOC dynamics in a few studies.

10.2.4.1 Mycorrhizalarcticus (AM) Species

Many tree species have symbiotic partnerships with AM fungus, which improve nutrient uptake, especially phosphorus, which can encourage tree growth. Studies have revealed that SOC pools are typically smaller under AM species, despite the fact that these species frequently produce nutrient-rich litter. This contradiction might arise from the nutrient-rich litter breaking down quickly, which, although initially encouraging development, might not be a significant factor in the formation of SOC. According to Willis et al. (2013), there may be less stable SOC pools as a result of the quick turnover of litter from AM species, which affects nitrogen cycling and litter dynamics.

10.2.4.2 Nitrogen-Fixing Species

These species promote soil fertility by the fixation of atmospheric nitrogen, which can enhance plant growth and litter formation. Nevertheless, they frequently produce litter that breaks down quickly, just like AM species, which results in smaller SOC pools. While highlighting the significance of nitrogen-fixing species in the nutrient cycle.

10.2.4.3 Preserved Animals

Tree species that produce more resistant organic matter, or litter that breaks down more slowly, on the other hand, typically accumulate SOC more efficiently. These species frequently have characteristics like increased lignin contents, which support SOC stability. Angst et al. (2019) found that tree species with high lignin content contributed to more stable SOC pools, suggesting a strategy for enhancing SOC accumulation in forests.

10.2.5 Comparative Impacts on the Storage of Carbon

The capacity of forests to store carbon is considerably modified by environmental context, which means that a one-size-fits-all strategy to carbon management is inadequate.

10.2.5.1 Tropical Regions

The maintenance of natural forests in these areas is critical for carbon sequestration. Land use changes and deforestation have the potential to release enormous amounts of carbon stored in the atmosphere. It has been observed that tropical forests are among the world's most significant carbon sinks. Houghton et al. (2009) underlined the need to safeguard tropical forests.

10.2.5.2 Boreal and Temperate Areas

The planting of particular tree species that are known to improve SOC storage may be advantageous in these areas. The selection of the right species is essential because some are better at sequestering carbon than others. The optimization of SOC storage in temperate and boreal forests can be achieved through particular species selection, hence augmenting the forests' function in sequestering carbon.

10.2.6 Carbon Pools in Forests

Forests are essential carbon sinks, storing carbon in various pools that each contribute differently to the carbon cycle:

10.2.6.1 Aboveground Biomass

This refers to any type of living vegetation, such as shrubs and trees, that uses photosynthesis to store carbon. When trees grow, die, or are harvested, the carbon stored here is susceptible to turnover. A substantial amount of the carbon stocks in forests can be attributed to aboveground biomass.

10.2.6.2 Belowground Biomass

Tree roots make a substantial contribution to the organic matter in the soil. By introducing organic materials into the soil, root turnover and exudation improve SOC.

10.2.6.3 Litter and Deadwood

These act as short-term carbon sinks. Unless it is fixed in the soil, carbon is progressively released back into the atmosphere when litter breaks down. The dynamics of deadwood and litter are covered by Wijas et al. (2024), along with their significance for forest carbon budgets.

10.2.6.4 Soils

Soils store around twice as much carbon as the atmosphere does, making them long-term carbon sinks. A number of mechanisms, such as interactions with minerals and the synthesis of stable organic molecules, stabilize SOC in soils. Jobbagy and Jackson (2000) examine the vital role that soils play as carbon reservoirs, highlighting the significance of soil management in the process of sequestering carbon.

10.2.7 Dynamics of Carbon Sequestration

Due to photosynthesis, forests absorb about 2.4 petagrams (Pg) of carbon annually, which is a significant contribution to the global carbon cycle (Augusto & Boča, 2022). Because of this process, forests are crucial for reducing CO_2 emissions from human activity and addressing climate change. Analysing the equilibrium between carbon uptake and release is necessary to comprehend the dynamics of carbon sequestration.

10.2.7.1 Comparing Respiration and Photosynthesis

Trees use photosynthesis to transform atmospheric CO_2 into organic stuff, mostly biomass. Trees and soil microbes employ respiration to counteract this process by releasing CO_2 back into the atmosphere when they break down organic matter to produce energy. Whether a forest functions as a carbon sink—absorbing more carbon than it releases or a carbon source releasing more carbon than it absorbs—depends on how these two processes differ from one another. The carbon dynamics of photosynthesis and respiration in forests are described by Gifford (2000), who highlights the significance of these activities in carbon budgeting.

10.2.7.2 Decomposition

Another essential component of the carbon cycle is decomposition. Decomposition of leaves, branches, and other organic things releases carbon that has been trapped in the atmosphere. The pace of decomposition is regulated by environmental conditions such as temperature, moisture, and the chemical content of the organic matter. The effects of organic matter quality and environmental factors on decomposition rates and, in turn, carbon fluxes in forest ecosystems are examined by Davidson and Janssens (2006).

10.2.8 Impact of Disturbances on Carbon Flux

The disturbances can significantly alter the carbon dynamics of forests, often causing them to switch from carbon sinks to carbon sources:

10.2.8.1 Forestry Activities

The logging and land conversion for agriculture release substantial amounts of stored carbon. When trees are harvested, the carbon sequestered in their biomass is emitted back into the atmosphere, contributing to greenhouse gas emissions. The conversion of forested areas to agricultural land further exacerbates this issue, as the natural carbon storage capacity of these ecosystems is lost. Peters et al. (2007) illustrated the impact of logging and land-use change on carbon emissions, highlighting the need for sustainable forestry practices to mitigate these effects.

10.2.8.2 Natural Disturbances

The events, such as wildfires, storms, and insect infestations, can destroy significant amounts of biomass, leading to immediate and large emissions of carbon. For example, wildfires not only burn trees but also release carbon stored in the soil, further contributing to atmospheric CO_2 levels.

10.2.9 Role of Tropical and Subtropical Forests

Tropical and subtropical forests are particularly crucial for carbon sequestration, harbouring a considerable amount of the world's terrestrial carbon storage. However, land conversion, deforestation, and climate change pose serious challenges to these woods.

10.2.9.1 Carbon Emissions and Deforestation

Tropical deforestation produces large amounts of carbon emissions, which greatly exacerbate global climate change. The greenhouse effect is exacerbated when trees are felled because the carbon they contained in their biomass is released. Houghton (2005) calculated the amount of carbon emissions caused by deforestation and emphasized how it affects the world's climate.

10.2.9.2 Biodiversity and Carbon Storage

The tremendous variety of tropical forests boosts their potential to store carbon through complex ecological interactions. The more resistance to perturbations and higher amounts of carbon sequestration can be maintained in healthy ecosystems with a diversity of tree species.

The global forest biomass has lately changed from being a net sink of carbon to a source of carbon due to human activities, specifically deforestation, forest degradation, and climate change, despite its capacity to absorb substantial amounts of carbon. Human activities have upset the natural carbon balance in forests, such as altered land use and more frequent disturbances.

The capacity of forests to store carbon is at risk due to this change, which has significant implications for global efforts to combat climate change. The evidence of the world forests' net carbon source status was presented by Pan et al. (2011), who also connected this status to human activity. A systematic hold of the biological background, forest management techniques, and tree species composition is required to effectively mitigate climate change. The forests can be made more effective carbon sinks by implementing strategies that support their conservation and sustainable management. To maximize forest carbon storage, Friedlingstein et al. (2014) stress the significance of incorporating ecological information into carbon management strategies.

10.3 THE ROLE OF FOREST MANAGEMENT IN ENHANCING CARBON STORAGE

The capacity of forests to store carbon is under threat due to this transformation, which has significant implications for global efforts to combat climate change. Pan et al. (2011) provided evidence of the global forests' position as a net carbon source and linked it to human activity. To effectively prevent climate change, one must understand the ecological backdrop, forest management practices, and tree species composition. By putting conservation and sustainable management principles into practice, forests can become more efficient carbon sinks. Friedlingstein et al. (2014) emphasize the importance of integrating ecological information into carbon management systems to maximize forest carbon storage.

10.3.1 Afforestation of Abandoned or Marginal Agricultural Land

Planting trees in places that haven't had any forest cover for a long period of time, typically more than 50 years, is referred to as afforestation. Afforestation provides a means of repurposing degraded regions that are unsuitable for intensive farming when it is implemented on abandoned or marginal agricultural holdings. By encouraging the growth of trees, these lands, which are frequently marked by low soil quality or unfavourable circumstances for crop production, can be made into carbon sinks.

10.3.1.1 Environmental Benefits

Trees planted on abandoned agricultural land help restore soil quality, reduce erosion, and increase biodiversity. By providing alternative options for biomass and lumber that may otherwise come from deforested regions, reforestation also lessens the strain on the remaining natural forests.

10.3.1.2 Carbon Sequestration

During photosynthesis, trees grown in these areas take in CO_2 from the atmosphere and store it in their biomass and soil. The conversion of these fields into forests effectively captures atmospheric carbon, so it offsets greenhouse gas emissions. Afforestation targets these underutilized lands and is a very economical way to mitigate climate change in addition to being a method for land restoration.

10.3.1.3 Managing Forests to Increase Carbon Density

Effective forest management can significantly enhance the carbon storage capacity of existing forests. Key strategies include:

10.3.1.3.1 Extending Rotation Lengths

By extending the intervals between harvesting cycles, trees are given more opportunity to grow and accumulate more carbon in their biomass. The mature, older trees are better at storing carbon than younger ones. The longer rotation times also provide the soil and deadwood pools more time to store carbon.

10.3.1.3.2 Minimizing Soil Carbon Losses

A considerable quantity of carbon is stored in forest soils, and bad management techniques like heavy logging or clear-cutting can cause soil deterioration and release of carbon. The sustainable forestry techniques, such as reduced-impact harvesting and selective logging, support the preservation of subsurface carbon storage and soil structure. Soil carbon retention can be further improved by actions such as logging and then leaving organic materials (leaves, branches) in the forest.

10.3.1.3.3 Enhancing Growth

The growth of the surviving trees can be stimulated by thinning congested forests and managing invasive species, which will increase the overall uptake of carbon. To maximize carbon sequestration, forest managers might also plant species that grow more quickly or are more suited to the climate where they are planted.

10.3.2 Utilizing Harvested Wood Products to Store Carbon Off-Site

When utilized in durable goods like furniture and building materials, harvested wood can be an effective carbon-storing strategy. Wood products have the ability to store carbon for decades or even centuries, as opposed to fast disintegrating or burning, which releases stored carbon.

10.3.2.1 Carbon Substitution

By substituting carbon-intensive commodities like steel and concrete, wood products help lower the construction industry's overall carbon footprint. The overall carbon sink potential of the forest is maintained or increased when trees are harvested sustainably because new trees are planted to replace the ones that are taken down.

10.3.2.2 Wood as Renewable Energy

Wood can be a renewable alternative to fossil fuels when utilized as bioenergy. When managed effectively, burning wood is thought to be a part of a carbon-neutral cycle even though it releases CO_2 since carbon is reabsorbed by regrowing trees.

10.3.3 Initiatives for Global Reforestation and Afforestation

Global recognition has been accorded to afforestation, which is the planting of new forests in regions that were previously devoid of trees, and reforestation, which is the restoration of degraded or cleared forested areas. Both procedures support ecosystem restoration and atmospheric carbon sequestration.

10.3.3.1 Cost-Efficiency

By utilizing organic carbon sequestration mechanisms, afforestation and reforestation are seen as cost-effective options. When it comes to investment, forests are less expensive than designed carbon capture technology.

10.3.3.2 The REDD+ Program

This project promotes sustainable forest management, conservation, and the development of forest carbon stores while pressuring nations, particularly those in poor nations, to cut back on emissions from deforestation and degradation. REDD+ seeks to provide additional resources for local communities and make forest conservation economically feasible by providing financial incentives (Grassi et al., 2017).

10.3.4 Factors Related to the Implementation of Afforestation/Reforestation Actions

10.3.4.1 Soil Carbon Losses from Ground Preparation

Soil disturbance during preparation for afforestation or reforestation frequently results in significant losses of soil carbon. Deep tillage and ploughing are two examples of intense ground preparation procedures that disrupt soil structure and release organic carbon into the atmosphere, increasing the likelihood of carbon oxidation. A number of variables, such as the mineral content, texture, and pre-existing soil carbon stocks, affect how much carbon is lost from the soil. According to studies, disturbed soils alter and experience increased rates of mineralization, which might result in carbon losses for several years after planting (Riley et al., 2018; Kumar et al., 2022).

10.3.4.2 Climatic and Soil Factors Affecting Carbon Decomposition

The weather and soil conditions have a big impact on how organic carbon breaks down in soils. Higher moisture content and higher temperatures usually accelerate the decomposition process, increasing the amount of carbon mineralization. This is especially true in areas with high soil organic matter content since higher temperatures can increase microbial activity, which raises greenhouse gas emissions (Davidson & Janssens, 2006). Furthermore, the stability of carbon pools and the rate of decomposition are also determined by the physical and chemical characteristics of the soil, such as pH and texture.

10.3.4.3 Impact of Young Trees on CO_2 Sequestration

In general, younger trees sequester CO_2 at a lesser rate than older ones. This is because they are unable to absorb as much CO_2 from the atmosphere due to their reduced biomass and slower growth rates. This indicates that, for the first ten years after planting, newly planted trees may not be efficient carbon sinks due to initial soil carbon losses during ground preparation. According to Zhou et al. (2019), the overall carbon sequestration capability of young forests is still impaired during this period because soil carbon stocks may decrease.

10.3.4.4 Influence of Previous Land Use on Soil Carbon Stocks

The carbon storage potential of afforested or reforested areas is significantly affected by their previous land use. The following examples illustrate this relationship:

Croplands: In general, planting trees on formerly farmed land increases carbon stocks in the soil. Lal (2004) states that the development of trees aids in the restoration of organic carbon in the soil, which is commonly depleted in agricultural soils.

Grasslands/Peatlands: The carbon reserves in the soil may become less or stable as a result of afforestation in grasslands or peatlands. According to Frolling et al. (2011), planting trees can disrupt preexisting organic carbon pools, resulting in a reduction in the soil's original carbon concentration.

Forest and Agricultural Lands: After planting trees, areas that were historically agricultural or forests usually have higher soil carbon stocks than severely degraded areas, like those affected by mining activities (Guo & Gifford, 2002).

10.3.4.5 Tree Species and Soil Characteristics

Carbon storage in afforestation and reforestation projects is significantly influenced by the kind of trees used and the features of the surrounding soils.

Coniferous Trees: Certain species, like pines, prefer to sequester more organic carbon from the soil when it is sandy or low in pH. The growth and capacity for carbon storage of coniferous species are supported by this environment (Harris et al., 2015).

Trees with Broad Leaves: In areas where mining has historically had an impact, broadleaf plants have the capacity to retain more carbon than coniferous trees.

According to Chazdon (2008), this is typically brought about by their ability to establish themselves more successfully in disturbed soils and to enhance soil structure, which boosts overall carbon retention.

10.4 INTERACTION OF FOREST CARBON DYNAMICS AND CLIMATE CHANGE

10.4.1 Species Impact and Regional Variability

Forest carbon dynamics are significantly impacted by climate change. The disruption of vital physiological functions, like respiration and photosynthesis, is caused by extreme weather events, rising temperatures, and increased vapour pressure deficits (VPDs). The functions that forests play as carbon sinks, which are essential for reducing the effects of climate change, may be compromised by these changes (Li, 2018; Dusenge et al., 2019). The difficulties that forests encounter will also likely worsen due to an expected increase in the frequency and severity of disturbances including deforestation, forest degradation, and natural disasters like fires, pest infestations, and disease outbreaks (Yang et al., 2018). The feasibility of extensive forest restoration as a mitigation strategy has been extensively discussed. Although studies suggest that restoring forest ecosystems might trap up to 205 gigatons of carbon (Bastin et al., 2019), some argue that these estimates may be unduly optimistic as a result of methodological simplifications. However, in order to effectively combat climate change and meet the goals outlined in the Paris Agreement, forest restoration and sustainable management are still essential. Regrettably, between 1990 and 2020, deforestation worldwide led to the loss of over 420 million hectares of forests, greatly lowering the potential for sequestering carbon and increasing CO_2 emissions (Tubiello et al., 2021).

1. **Impacts on Sequestration of Carbon and Forest Growth**: The impacts of climate change on forest ecosystems are complex, especially when combined with rising temperatures and increased concentrations of carbon dioxide (CO_2). These variables have the potential to result in a longer growth season and higher productivity of forests, which could raise the rates of carbon sequestration. However, different tree species and regions respond significantly to these changes. For example, warmer weather may encourage the growth of certain forests, but pests, drought, or changes in precipitation patterns may stress other forests, reducing their capacity to sequester carbon (Gough et al., 2016; Allen et al., 2010).
2. **Variability by Region**: It is anticipated that forests in the Northern Hemisphere's mid-to-high latitudes, including those in Northeast and Central Asia, Northern Europe, and Northwestern America, will benefit from favourable conditions for sequestering carbon. According to the studies, these areas might gain from higher CO_2 and temperature levels, which would improve carbon uptake and forest productivity (McCarthy et al., 2010). On the other hand, it is anticipated that rising temperatures

and altered precipitation patterns will negatively impact tropical forests in places like South America and Central Africa, perhaps causing more drought stress and slower growth (Malhi et al., 2009; Wu et al., 2016).

3. **Changes in Adequacy of Habitat**: The suitability of the local ecosystem for different tree species is expected to change due to rapid climate change, which could be advantageous for some while endangering others. Understanding these changes in habitat appropriateness requires long-term evaluations since they will have a substantial impact on species growth and distribution, which will ultimately alter the capacity of forests to store carbon (Araújo et al., 2006). Iverson et al. (2008), for instance, discovered that some tree species are probably going to migrate northward in reaction to climate change, which could cause changes in the composition of forests and the dynamics of carbon.

10.4.2 Equitable Methods of Reforestation and Afforestation

To ensure that afforestation and reforestation significantly contribute to a resilient and sustainable future, a balanced approach is needed. This strategy should combine urgent steps to address the difficulties of climate change adaptation and mitigation with long-term planning that informs decision-making (Davis et al., 2019). The main difficulty is striking a balance between the necessity for thorough long-term assessments and the pressing requirement for quick action. The study conducted by Baggio-Compagnucci and colleagues (2022) elucidates the intricacies of these methodologies in the Scottish setting. According to their research, various species experience varying effects of climate variability on carbon reserves in tree biomass. According to the research, lowland areas in Scotland may become less suitable for cold-adapted species like Betula spp. by 2050 and more favourable for temperate species like Pedunculate Oak (*Quercusrobur*), Ash (*Fraxinus excelsior*), and Wych Elm (*Ulmusglabra*) (Baggio-Compagnucci et al., 2022). This case illustrated the importance of designing afforestation and reforestation initiatives that consider the adaptability of local species as well as anticipated changes in habitat suitability.

10.4.3 Increasing Disturbances Due to Climate Change

It is anticipated that the intensity and frequency of disturbances endangering forest ecosystems would rise unevenly as a result of climate change. The ability of forests to store CO_2 and reduce carbon stocks can be greatly hampered by biotic agents such as insects, heatwaves, wildfires, droughts, and other events (Cameron et al., 2020; IPCC, 2021). For example, wildfires might result in short- and long-term losses in forest biomass and carbon storage capacity since they are predicted to become more frequent and intense as a result of climate change (Abatzoglou & Williams, 2016). According to Peters et al. (2011), forests may become grasslands or savannas as a result of the combined effects of climate change, air pollution, and invasive species pushing them past their limits of resilience.

REFERENCES

Abatzoglou, J. T., and Williams, A. P. 2016. "Impact of Anthropogenic Climate Change on Wildfire across Western US Forests." *Proceedings of the National Academy of Sciences* 113 (42): 11770–11775.

Allen, C. D., et al. 2010. "A Global Overview of Drought and Heat-Induced Tree Mortality Reveals Emerging Climate Change Risks for Forests." *Forest Ecology and Management* 259 (4): 660–684.

Amirthan, V., and Perera, P. 2023. "Direct Air Capture: Potential and Challenges in Carbon Dioxide Removal." *Nature Climate Change* 13: 112–119.

Angst, G., Mueller, K. E., Eissenstat, D. M., Trumbore, S., Freeman, K. H., Hobbie, S. E., Chorover, J., Oleksyn, J., Reich, P. B. and Mueller, C. W. 2019. "Soil organic carbon stability in forests: Distinct effects of tree species identity and traits." *Global Change Biology*, 25 (4):1529–1546.

Araújo, M. B., et al. 2006. "Climate Change and the Design of Protected Areas in a Biodiversity Hotspot." *Global Ecology and Biogeography* 15 (1): 73–81.

Augusto, L., and Boča, A. 2022. "Tree Functional Traits and Forest Carbon Dynamics: The Role of Site Properties." *Forest Ecology and Management* 510: 120138. https://doi.org/10.1016/j.foreco.2022.120138

Baggio-Compagnucci, J., et al. 2022. "Climate Variability Impacts Carbon Stocks in Tree Biomass Differently Across Species in Scotland." *Forest Ecology and Management* 503: 119710.

Bastin, J.-F., et al. 2019. "The Global Tree Restoration Potential." *Science* 366 (6470): 76–79. https://doi.org/10.1126/science.aax0848

Cameron, D. R., et al. 2020. "Climate Change and Biodiversity: A Climate Change Risk Assessment for the Management of Forests." *Forest Policy and Economics* 113: 102210

Chave, J., Coomes, D., Jansen, S., Lewis, S. L., Swenson, N. G. & Zanne, A. E. 2009. "Towards a worldwide wood economics spectrum." *Ecology Letters* 12: 351–366.

Chazdon, R. L. 2008. "Beyond Defeating Degradation: A New Vision for Forest Restoration." *The International Forestry Review* 10 (4): 492–502.

Davidson, E. A., and I. A. Janssens. 2006. "Temperature Sensitivity of Soil Carbon Decomposition and Its Global Implications." *Nature* 440 (7081): 165–173.

Davis, M. A., et al. 2019. "Afforestation and Reforestation: Solutions to Climate Change?" *Nature Sustainability* 2 (3): 133–135.

Dusenge, M. R., et al. 2019. "Climate Change Impacts on Forest Carbon Dynamics: Integrating "Physiological and Ecological Responses." *Ecological Applications* 29 (4): e01987. https://doi.org/10.1002/eap.1987

Eissenstat, D. M., Wells, C. E., Yanai, R. D., Whitbeck, J. L. 2000. "Building fine roots in a changing environment: implications for root longevity." *New Phytologist* 147: 33–42.

Friedlingstein, P., Andrew, R. M., Rogelj, J., Peters, G. P., Canadell, J. G., Knutti, R., Luderer, G., Raupach, M. R., Schaeffer, M., van Vuuren, D. P. and Le Quéré, C. 2014. "Persistent growth of CO_2 emissions and implications for reaching climate targets." *Nature Geoscience* 7 (10): 709–715.

Gifford, R. M., Barrett, D. J., Lutze, J. L. and Samarakoon, A. B. 2000. *The CO_2 fertilizing effect: relevance to the global carbon cycle. The carbon cycle*. Cambridge: Cambridge University Press, pp. 77–92.

Gough, C. M., et al. 2016. "Climate Change Impacts on Forest Carbon Dynamics in the Northern Hemisphere: Insights from a Meta-Analysis of the Available Literature." *Global Change Biology* 22 (3): 1232–1246.

Grassi, G., House, J. I., and den Elzen, M. 2017. "The Role of Forests in Climate Change Mitigation: Opportunities and Challenges." *Global Change Biology* 23 (10): 3484–3496. https://doi.org/10.1111/gcb.13647

Guo, L. B., and Gifford, R. M. 2002. "Soil Carbon Stocks and Land Use Change: A Meta Analysis." *Global Change Biology* 8 (3): 345–360. https://doi.org/10.1046/j.1354-1013.2002.00486.x

Harmon, M. E., et al. 2011. "The Role of Deadwood in the Carbon Cycle." *Forest Ecology and* Management 345: 115–127. https://doi.org/10.1016/j.foreco.2015.02.007

Hawke, D. J. 2005. "Soil P in a forested seabird colony: inventories, parent material contributions, and N: P stoichiometry." *Soil Research* 43 (8):957–962.

Houghton, 2005. "Tropical deforestation as a source of greenhouse gas emissions." *Tropical Forests and Climate Change*, 13–21.

Houghton, R. A., Hall, F. and Goetz, S. J., 2009. "Importance of biomass in the global carbon cycle." *Journal of Geophysical Research: Biogeosciences* 114(G2): 1–13.

Houghton, R. A., Goodall, J., and Goodall, J. 2015. "Global Carbon Emissions from Land Use and Land Cover Change: A Historical Perspective." *Global Change Biology* 21 (11): 3763–3776. https://doi.org/10.1111/gcb.12959

IPCC. 2021. *Climate Change 2021: The Physical Science Basis. Contribution of Working Group I to the Sixth Assessment Report of the Intergovernmental Panel on Climate Change.* Cambridge University, England.

Iverson, L. R., et al. 2008. "The Effect of Climate Change on Tree Species Distribution in the Eastern United States." *Global Change Biology* 14 (12): 2712–2727.

Jobbagy, E. G., and Jackson, R. B. 2000. "The Vertical Distribution of Soil Organic Carbon and Its Relation to Climate and Vegetation." *Ecological Applications* 10(2): 423–436.

Kumar, A., Dwivedi, G. K., Tewari, S., Paul, J., Anand, R., Kumar, N., Kumar, P., Singh, H., and Kaushal, R. 2020a. "Carbon Mineralization and Inorganic Nitrogen Pools under *Terminalia chebula* Retz.-Based Agroforestry System in Himalayan Foothills, India." *Forest Science* 66 (5): 634–643.

Kumar, A., Dwivedi, G. K., Tewari, S., Paul, J., Sah, V. K., Singh, H., Kumar, P., Kumar, N., and Kaushal, R. 2020b. "Soil Organic Carbon Pools under *Terminalia chebula* Retz. Based Agroforestry System in Himalayan Foothills, India." *Current Science* 118 (7): 1098–1103.

Kumar, A., Kumar, P., Singh, H., Bisht, S., and Kumar, N. 2021a. "Relationship of Physiological Plant Functional Traits with Soil Carbon Stock in Temperate Forest of Garhwal Himalaya." *Current Science* 120 (8): 1368–1373.

Kumar, A., Kumar, P., Singh, H., and Kumar, N. 2021b. "Modulation of Plant Functional Traits under Essential Plant Nutrients during Seasonal Regime in Natural Forests of Garhwal Himalayas." *Plant and Soil* 465: 197–212.

Kumar, A., Kumar, P., Singh, H., and Kumar, N. 2021c. "Impact of Plant Functional Traits on Near Saturated hydraulic Conductivity of Soil under Different Forests of Kempty Watershed in Garhwal Himalaya." *Indian Journal of Soil Conservation* 49 (1), 38–44.

Kumar, A., Singh, H., Kumari, G., Bisht, S., Malik, A., Kumar, N., Singh, M., Raturi, A., Barthwal, S., Thakur, A., and Kaushal, R. 2022. "Adaptive Resilience of Roadside Trees to Vehicular Emissions Via Leaf Enzymatic, Physiological, and Anatomical Trait Modulations." *Environmental Pollution* 313: 120191.

Kumar, N., Jeena, N., and Singh, H. 2019. "Elevated Temperature Modulates Rice Pollen Structure: A Study from Foothill Himalayan Agro-Ecosystem in India." *3 Biotech* 9: 175.

Kumar, N., Jeena, N., Kumar, A., Khairakpam, R., and Singh, H. 2021d. "Comparative Response of Rice Cultivars to Elevated Air Temperature in Bhabar Region of Indian Himalaya: Status on Yield Attributes." *Heliyon* 7: e07474.

Kumar, N., Kumar, N., Shukla, A., Shankhdhar, S. C., and Shankhdhar, D. 2015. "Impact of Terminal Heat Stress on Pollen Viability and Yield Attributes of Rice (*Oryza sativa* L.)." *Cereal Research Communications* 43 (4): 616–626.

Kumar, N., Shankhdhar, S. C., and Shankhdhar, D. 2016. "Impact of Elevated Temperature on Antioxidant Activity and Membrane Stability in Different Genotypes of Rice (*Oryza sativa* L.)." *Indian Journal of Plant Physiology (Now Plant Physiology Report)* 21 (1): 37–43.

Kumar, N., Suyal, D. C., Sharma, I. P., Verma, A., and Singh, H. 2017. "Elucidating stress proteins in rice (*Oryza sativa* L.) genotype under elevated temperature: a proteomic approach to understand heat stress response." *3 Biotech* 7: 205.

Lal, R. 2004. "Carbon Sequestration in Dryland Soils." *Environmental Management* 33 (4): 501–508

Li, Y. 2018. "Climate Change and Forest Carbon Dynamics: The Effects of Rising Temperatures and Extreme Weather Events on Photosynthesis and Respiration." *Forest Ecology and Management* 429: 131–141. https://doi.org/10.1016/j.foreco.2018.06.012

Liáng, Y., Wang, Y., and Zhang, X. 2023. "Sustainable Land Management Practices for Enhancing Carbon Sequestration." *Sustainability* 15 (2): 800.

Malhi, Y., et al. 2009. "Tropical Forests in the Changing Climate." *Nature* 462 (7271): 56–61.

McCarthy, J. J., et al. 2010. *Climate Change Impacts in the United States: The Third National Climate Assessment*. U.S. Global Change Research Program.

McLaughlin, O., Smith, P., and Brown, T. 2023. "Carbon Sequestration in Wetlands and Grasslands: A Review of Strategies and Opportunities." *Journal of Environmental Management* 325: 116–130.

Menard, R., Aubert, D., and Dufresne, A. 2023. "Large-Scale Afforestation: Impacts on Carbon Sequestration and Local Ecologies." *Ecological Applications* 33 (4): e2617. https://doi.org/10.1002/eap.2617

Nguyen, H., Liu, J., and Thompson, M. 2023. "Evaluating the Benefits of Carbon Sequestration: Air Quality, Soil Health, and Climate Change Mitigation." *Environmental Research Letters* 18 (3): 034025.

Pan, Y., et al. 2011. "A large and persistent carbon sink in the world's forests." *Science* 333 (6045): 988–993.

Peters, W., Jacobson, A. R., Sweeney, C., Andrews, A. E., Conway, T. J., Masarie, K., Miller, J. B., Bruhwiler, L. M., Pétron, G., Hirsch, A. I. and Worthy, D. E. 2007. An atmospheric perspective on North American carbon dioxide exchange: Carbon Tracker. *Proceedings of the National Academy of Sciences* 104(48): 18925–18930.

Peters, G. P., et al. 2011. "The Role of Forests in Climate Change Mitigation." *Nature Climate Change* 1 (8): 511–516.

Reich, P. B., Tilman, D., Naeem, S., Ellsworth, D. S., Knops, J., Craine, J., Wedin, D. and Trost, J., 2004. Species and functional group diversity independently influence biomass accumulation and its response to CO_2 and N. *Proceedings of the National Academy of Sciences* 101(27): 10101–10106.

Riley, H., Smith, P., and Goudie, A. 2018. "Soil Carbon Losses during Ground Preparation for Afforestation: A Review." *Agriculture, Ecosystems & Environment* 262: 116–128. https://doi.org/10.1016/j.agee.2018.04.001

Roy, S., Yadav, R., and Chatterjee, S. 2023. "Carbon Sequestration: A Comprehensive Review of Techniques, Opportunities, and Challenges." *Environmental Science and Pollution Research* 30 (1): 567–587.

Tubiello, F. N., et al. 2021. "Global and Regional Estimates of Land Use Change, 1990–2020." *Environmental Research Letters* 16 (12): 124044.

Westoby, M., Falster, D. S., Moles, A. T., Vesk, P. A. and Wright, I. J. 2002. "Plant ecological strategies: some leading dimensions of variation between species." *Annual Review of Ecology and Systematics* 33 (1): 125–159.

Wijas, B. J., Allison, S. D., Austin, A. T., Cornwell, W. K., Cornelissen, J. H. C., Eggleton, P., Fraver, S., Ooi, M. K., Powell, J. R., Woodall, C. W. and Zanne, A. E. 2024. "The role of deadwood in the carbon cycle: Implications for models, forest management, and future climates." *Annual Review of Ecology, Evolution, and Systematics* 55: 131–155.

Willis, A., Rodrigues, B. F. and Harris, P. J. 2013. "The ecology of arbuscular mycorrhizal fungi." *Critical Reviews in Plant Sciences* 32 (1): 1–20.

Wright, I. J., Reich, P. B., Westoby, M., Ackerly, D. D., Baruch, Z., Bongers, F., Cavender-Bares, J., Chapin, T., Cornelissen, J. H., Diemer, M. and Flexas, J., 2004. "The worldwide leaf economics spectrum." *Nature* 428 (6985): 821–827.

Wu, J., et al. 2016. "Global Climate Change and Forests: A Synthesis of Research Findings and Implications for Management." *Forest Ecology and Management* 378: 9–15.

Yadav, R., Roy, S., and Chatterjee, S. 2023. "The Role of Land Use Changes in Carbon Sequestration: A Global Perspective." *Global Change Biology* 29 (5): 1234–1250. https://doi.org/10.1111/gcb.16354

Yang, Y., et al. 2018. "The Impact of Climate Change on Forest Disturbance Dynamics." *Global Change Biology* 24 (8): 3843–3860. https://doi.org/10.1111/gcb.14398

Yu, Z., Beilman, D. W., Frolking, S., MacDonald, G. M., Roulet, N. T., Camill, P., and Charman, D. J. (2011). "Peatlands and Their Role in the Global Carbon Cycle. Eos." *Transactions American Geophysical Union* 92 (12): 97. DOI: 10.1029/2011EO120001

Zhou, G., Wang, Z., and Liu, S. 2019. "Assessing the Initial Carbon Sequestration Potential of Young Forests: Implications for Climate Mitigation." *Forest Ecology and Management* 438: 207–217. https://doi.org/10.1016/j.foreco.2019.01.007

11 Drought Tolerance and Water-Use Efficiency
Adapting to Climate Change

Tuhina Ghosh, Subrata Debnath, Aditya Pratap Singh, Victor Thingujam, Saptarshi Mondal, and Kousik Atta

11.1 INTRODUCTION

A long-term, substantial shift in climate-related variables, such as rainfall, temperature, wind, or snowfall patterns, is referred to as climate change (Pope et al., 2022). Global warming and greenhouse gas emissions are thought to be the main causes of the unfavorable acceleration of change in climate. The global average temperature has risen by 0.9°C throughout the 19th century as a result of continuously rising anthropogenic activity, and by 2050, it is anticipated that it will rise to 1.5°C (Sovacool et al., 2021; Kumar et al., 2020a; Kumar et al., 2021a,b). It is anticipated that climate change would have a major impact on agriculture, both directly and indirectly, through effects on crops, soils, cattle, and pests. Changes in soil moisture content and subsequent rises in soil temperature and CO_2 levels are anticipated to be the primary impacts of climate change on soils. The soil properties and activities that are essential for restoring soil fertility and productivity are expected to be affected differently by global climate change. Plant productivity and carbon sequestration are adversely affected by drought (Choat et al., 2018). The ecosystems of forests can be impacted by climate change. Deforestation can result in desertification and thus fuel climate change. There are detrimental effects of rising temperatures and drought on ecosystem products and services that benefit humankind, in addition to decreasing species diversity.

Through the combined effects of several climatic conditions on bio-geochemical processes connected to water consumption and carbon sequestration, climate change will have a significant impact on forest water-use efficiency (WUE). Through affecting the Rubisco activity, stomatal conductance, photosystem II, and mitochondrial respiration, temperature changes impact the overall photosynthetic productivity by reducing soil water availability, speeding up soil nutrient cycles, extending the growing season, and altering stand structure (Teskey et al., 2015; Kumar et al., 2015, 2021c). Rapid temperature increases have more detrimental effects on ecosystem carbon and water processes than gradual temperature increases. Heatwaves can alter stomatal aperture and metabolic rate, which can affect photosynthesis in

DOI: 10.1201/9781003545231-11

vegetation and cause water stress (Bastos et al., 2020). Consequently, the severity and length of the hydraulic tolerance, heatwaves, and water-use strategies linked to the crucial adaptation strategy of various forests to extreme heat all influenced the forest WUE, which displayed varying responses to heatwaves (Teskey et al., 2015; Kumar et al., 2021d).

11.2 DROUGHT-TOLERANCE MECHANISMS IN FOREST PLANTS

The drought tolerance of plants is influenced not only by external circumstances such as geographical location, competition, drought severity, and surrounding environment but also by internal characteristics like their phenotype and genetic makeup. Research on internal factors has explored the species' tolerance mechanisms at the molecular, physiological, biochemical, morphological, and anatomical levels.

11.2.1 Morphological Adaptations

The different morphological adaptations of different tree species is shown in Table 11.1

11.2.2 Physiological Adaptations

Osmotic adjustment has been regarded as an indirect selection criterion in numerous breeding programs aimed at developing plants with improved drought tolerance. Organic solutes, known as compatible osmolytes or compatible solutes, aid in osmoregulation and can also stabilize the structure of various biomolecules

TABLE 11.1
Different Morphological Adaptation of Different Tree Species

Traits	Plants	Adaptations	Reference
Root collar diameter	*Pinus koraiensis*	The species that have higher root collar diameter values are more resistant to drought.	Bushal et al. (2021)
Leaf size	*Pinus koraiensis*	Smaller leaves are generally more effective during drought stress than larger ones because they allow better air flow, which helps prevent heat buildup and reduces the temperature on the leaf surface.	Bushal et al. (2021)
Leaf mass per area	The species belong to 72 genera and 35 families	A higher leaf mass per area means more resources are invested in the leaf's structure and a thicker cuticle layer, which improves drought tolerance.	Arinawa et al. (2022)

(Continued)

TABLE 11.1 (*Continued*)
Different Morphological Adaptation of Different Tree Species

Traits	Plants	Adaptations	Reference
Leaf thickness	*Larix gmelinii*, *Pinus koraiensis*, *Fraxinus mandshurica* and *Tilia amurensis*	Increased leaf thickness may result in more guard cells within a given leaf area.	Khan et al. (2022)
Stomatal pore length & stomatal density	*Fagus sylvatica* L.	Severe drought significantly reduced the length of stomatal pores, but it also caused a noticeable increase in stomatal density. This higher stomatal density might help improve the plant's adaptability to drought conditions.	Petrík et al. (2020)
Stomatal conductance	*Picea abies* L.	In response to drought conditions, most tree species reduce their stomatal conductance to use less water for every unit of carbon assimilated, increasing their water-use efficiency in the process.	Pan et al. (2024)
Root-to-shoot ratio	Eucalyptus species	Drought tolerance is achieved by increasing the below-ground biomass accumulation or by dropping the aboveground biomass accumulation, which aids in lowering the shoot evaporative demand and optimizing water absorption. This adaptive approach can improve leaf gas exchange, boosting seedling productivity and enhancing survival chances in regions prone to drought.	Gonçalves et al. (2017)
Leaf shedding (tropical dry forests)	*Bursera simaruba* (L.), *Cojoba rufescens*.	This process is controlled by a complex crosstalk of hormones such as ethylene, abscisic acid (ABA), and auxin. By manipulating this mechanism, it is possible to improve the drought tolerance of trees.	Wolfe et al. (2017)
The conduit diameter (CD) and vascular bundle (VBD)	*Larix gmelinii*, *Pinus koraiensis*, *Fraxinus mandshurica*, and *Tilia amurensis*	A decrease in conduit diameter leads to reduced water transport efficiency, but this may be partially offset by an increase in conduit density. However, this change reduces the xylem area, which in turn limits water absorption.	Khan et al. (2022)
Leaf vein density	*Cercocarpus betuloides*	The smaller leaf size mainly enhances vein density. Moreover, high vein density is associated with greater tolerance to leaf hydraulic failure. These strategies protect trees against mechanical damage and embolism-induced blockages by offering alternative water transport routes through a more redundant vascular system.	Scofoni et al. (2011)

and membranes. Additionally, they may serve as free-radical scavengers, shielding DNA from the detrimental effects of reactive oxygen species (ROS). Some examples of osmolytes are as follows: glycine betaine (GB) is an effective osmoprotectant against stresses like drought due to its small size, water solubility, and non-toxic nature (Bhowal et al., 2021; Kumar et al., 2020b; Kumar et al., 2021d). It is synthesized in the chloroplast through choline oxidation and helps stabilize enzymes, proteins, and membranes under osmotic stress. GB also aids water retention in cells, preventing dehydration. Proline, an amino acid, is a key osmoprotectant involved in osmotic adjustment. It accumulates in the cytoplasm of many plant species under environmental stress and aids in stabilizing membranes and proteins, scavenges ROS, and regulates cytoplasmic pH during stress conditions. Moreover, in response to drought stress, plants accumulate soluble sugars, which serve as accessible organic osmolytes. The most prevalent sugars include sucrose, glucose, and fructose (Afzal et al., 2021).

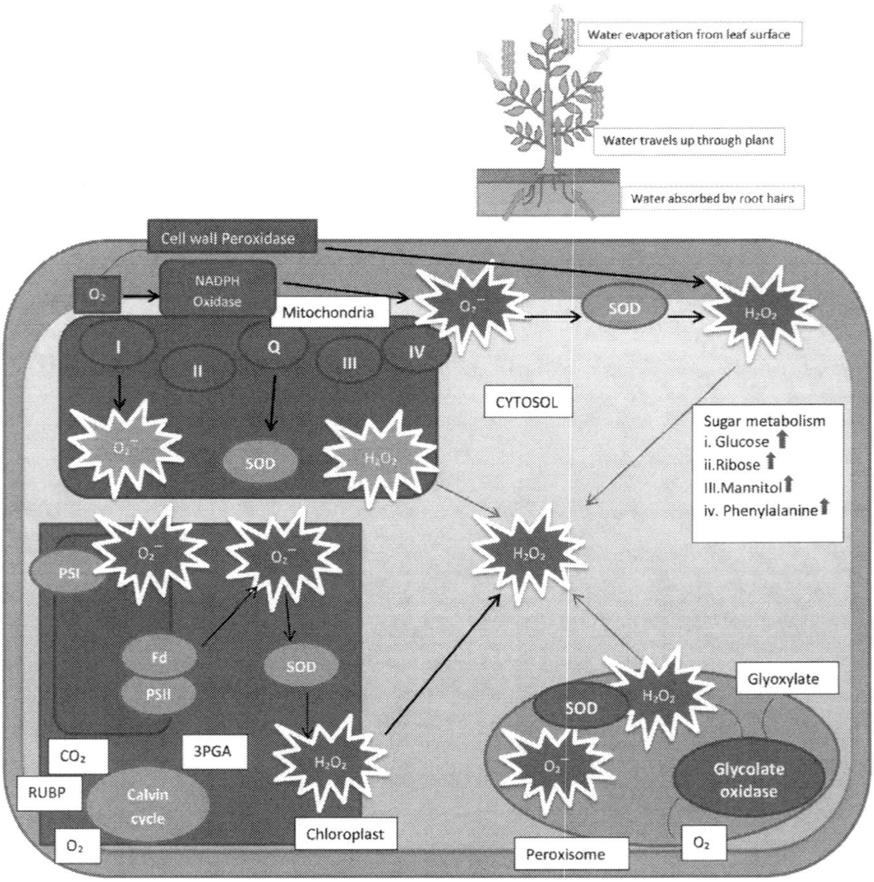

FIGURE 11.1 Production of ROS and other osmolytes in cell under drought stress.

11.2.3 Biochemical Responses

Water shortages due to drought create oxidative stress in plants, resulting in the production of various ROS such as hydroxyl radicals (OH•), hydrogen peroxide (H_2O_2), singlet oxygen ($^1O_2$), and superoxide (O_2–). Within plants, O_2– and H_2O_2 influence various cellular reactions, including the Fenton reaction (an iron-catalyzed process) and reactions involving enzymes such as peroxidases (PODs), xanthine oxidase, NADPH oxidase, and lipoxygenases. The primary cellular components most susceptible to damage from these radicals include nucleic acids, proteins, carbohydrates, and lipids, particularly through the peroxidation of unsaturated fatty acids in membranes. Enzymes are also vulnerable, often undergoing denaturation. Under water deficit conditions, about 70% of total hydrogen peroxide (H_2O_2) is generated through photorespiration, which is a significant contributor to ROS production (Kumar et al., 2017, 2019; Sachdev et al., 2021).

Drought stress leads to stomatal closure, resulting in the accumulation of ROS. H_2O_2 and other ROS can cause cellular damage, toxicity, and inhibit photosynthesis during stressful times. Conversely, under normal conditions, these molecules play a role in signaling pathways within plant cells (Kumar et al., 2016; Mansoor et al., 2022). However, significant increases in ROS levels and their scavenging are crucial processes that enable plants to effectively manage ROS as a primary defense mechanism against various stressors, both biotic and abiotic. To maintain a balance of ROS levels during drought stress, plant tissues generate various enzymatic and non-enzymatic antioxidants (Figure 11.2) that act as ROS scavengers. Enhanced tolerance to abiotic stress in plants is believed to be linked to the quick detoxification of high ROS levels (Martinez et al., 2018).

11.3 WATER-USE EFFICIENCY (WUE) IN FOREST PLANTS

WUE is the amount of carbon assimilated as biomass or grain produced per unit of water used by the crop. WUE is related to a plant's ability to control stomatal closure to stop water loss and absorb carbon amounts (Tomás et al., 2012). WUE is

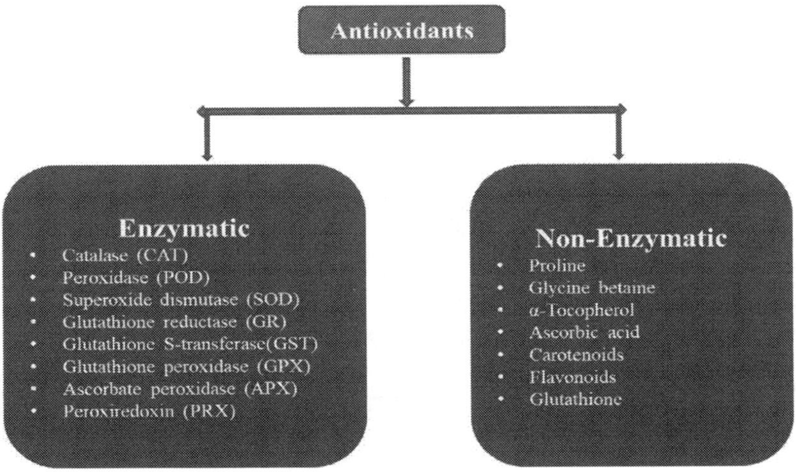

FIGURE 11.2 Different types of antioxidants produced during drought stress.

impacted by soil water content and is strongly correlated with the transpiration rate and net photosynthetic rate. Because stomatal movements control the intake of CO_2 for photosynthesis and the loss of water through transpiration, they are crucial to plant productivity and WUE. The water balance of plants and cells is influenced by water taken from the soil and water lost to the atmosphere through evapo-transpiration (Kaiser, 1987).

11.3.1 Types of WUE

11.3.1.1 Instantaneous Water-Use Efficiency (WUEs)

WUEs is a measure of how efficiently a plant uses water for photosynthesis at a specific moment. It is the ratio of the rate of water loss through transpiration to the rate of carbon assimilation (photosynthesis). This measure is taken within a certain time frame and reflects the plant's immediate response to environmental conditions. Instantaneous WUE provides insight into how efficiently a plant is using water at a specific moment, which is particularly important in water-scarce environments or during periods of drought. Plants with higher instantaneous WUE can conserve water while maintaining carbon uptake, making them better suited to dry climates or conditions with limited water availability (Zhang et al., 2019).

Instantaneous WUE can fluctuate due to several environmental and physiological factors:

- **Stomatal Conductance**: As opening and closing of stomata regulate gas exchange, the amount of water lost through transpiration can vary. Lower stomatal conductance generally improves WUE but can limit photosynthesis.
- **Light Intensity**: Higher light intensity typically increases photosynthesis, improving WUE up to a point.
- **Humidity**: Under low humidity, transpiration rates tend to increase, reducing WUE.
- **Temperature**: Higher temperatures can increase transpiration rates, often leading to lower WUE.
- **Soil Moisture**: Plants under drought stress may close their stomata, reducing transpiration and photosynthesis, which can increase or decrease WUE depending on the severity of the stress.

11.3.1.2 Intrinsic Water-Use Efficiency (iWUE)

iWUE is a physiological metric that reflects the efficiency with which plants use water while balancing carbon dioxide (CO_2) assimilation with stomatal conductance. Unlike instantaneous water-use efficiency (WUE), which is based on the actual rate of transpiration, intrinsic WUE focuses more on the physiological potential for efficient water use and is commonly calculated based on the relationship between photosynthesis and stomatal conductance. Intrinsic WUE is particularly useful for understanding a plant's ability to regulate water use without being directly influenced by environmental fluctuations in transpiration (Linares & Camarero, 2012).

iWUE is influenced by various physiological and environmental factors:

- **Stomatal Conductance (gs):** A decrease in stomatal conductance (closing stomata) generally increases iWUE because less water is lost through transpiration, but this can also limit CO_2 uptake and thus photosynthesis.
- **Photosynthesis:** An increase in the rate of photosynthesis, with little change in stomatal conductance, will improve iWUE (Kumar et al., 2022).
- **Environmental Conditions:** Higher light intensity and moderate temperatures typically improve photosynthesis, potentially increasing iWUE. Drought conditions may increase iWUE if plants reduce stomatal opening to conserve water.

11.4 FUNCTIONAL TRAITS LINKED TO DROUGHT TOLERANCE AND WUE

Drought tolerance and WUE are important traits that help plants survive in environments with limited water. These traits are shaped by different physical, structural, and chemical features in plants that help them take in water efficiently, reduce water loss, and manage the trade-off between growth and survival.

11.4.1 Hydraulic Conductivity and Cavitation Resistance

Hydraulic conductivity is the plant's ability to move water through its vascular system, mainly through the xylem. Proper water movement is essential for the plant to carry out its functions, especially during drought. The structure of the xylem is key to how well water is transported, with things like the size of the vessels, thickness of the walls, and how the xylem elements are arranged all affecting water flow. Plants that can survive in dry conditions often have xylem structures that strike a balance between effective water transport and the ability to resist cavitation. Cavitation happens when air bubbles form in the xylem, blocking water flow. This is more likely in drought when water pressure in the plant increases, raising the risk of these blockages. Plants that resist cavitation usually have narrower xylem vessels, which may not move water as efficiently in normal conditions but provide better protection against air bubbles. By keeping water moving and avoiding cavitation, drought-tolerant plants can maintain their water flow and essential processes even when water is scarce (Kumar et al., 2021d).

11.4.2 Stomatal Conductance and Regulation

Stomatal openings on leaves control both gas exchange and water loss through a process called transpiration. Stomatal conductance is the rate at which gases, including water vapor, move through these openings. In drought conditions, it's especially important for plants to manage their stomatal activity to reduce water loss while still allowing enough carbon dioxide (CO_2) to enter for photosynthesis. Different plant species have varying stomatal densities and functions, and these differences play a big role in how well a plant conserves water. Plants with fewer stomata or better control over their stomatal openings are more capable of limiting water loss during dry periods. They can quickly close their stomata when soil moisture drops, which helps reduce transpiration and save water. However, closing stomata also restricts CO_2 entry, which can slow down photosynthesis. Some drought-tolerant plants have developed the ability to partially close their stomata, maintaining a balance between conserving water and still taking in enough CO_2. These plants often show better WUE, as they can continue photosynthesis even with less water. The ability to adjust stomatal conductance in response to environmental factors like humidity and soil moisture is a key survival strategy in dry environments.

11.4.3 Carbon Assimilation and Allocation

During drought, carbon assimilation can be heavily impacted due to reduced stomatal conductance, limited water availability, and potential damage to the plant's photosynthetic system. Drought-tolerant plants often have adaptations that help them keep higher levels of carbon assimilation even with less water. One approach is the use of specialized photosynthetic processes like crassulacean acid metabolism (CAM) or C4 photosynthesis. CAM plants open their stomata at night, when it's cooler, allowing them to capture CO_2 while losing less water. C4 plants have a more efficient carbon-fixing process that uses less water while keeping photosynthesis rates high, even under dry conditions.

11.4.4 Deep Rooting and Resource Allocation

Root structure plays a major role in a plant's ability to access water during drought. Plants with deep roots are generally better at handling drought because their roots can reach deeper soil layers where moisture is still available. These deep roots allow for a steady water supply even when the topsoil dries out, which is especially helpful in dry areas. Developing deep roots often requires plants to shift resources away from shoot growth and focus on root development. This strategy helps the plant cover more soil area to find water and nutrients, improving WUE. Balancing resources between above-ground and below-ground growth is crucial for plants in water-limited environments. Additionally, plants with deep roots are more resistant to prolonged drought because they can tap into water sources that plants with shallow roots cannot reach. While developing a deep-root system demands a lot of energy and resources, which can slow down growth in the short term, it ultimately helps the plant survive better in the long run-in drought-prone areas.

11.5 CLIMATE ADAPTATION STRATEGIES IN FOREST CONSERVATION

Assisted migration is the deliberate relocation of species or genotypes to locations where they do not presently exist but are projected to thrive under future climate conditions (Hällfors et al., 2014). Types of assisted migration include intra-range assisted migration, which involves moving species or genotypes within their existing distribution to higher elevations or microclimates that are more suitable for future circumstances. Range expansion assisted migration entails moving species to locations just outside their present native range where climatic conditions are becoming more favorable. Ex situ assisted migration involves introducing species to completely new environments where they have no natural history but where future conditions are expected to support their survival (Park and Talbot, 2018).

Reforestation with drought-adapted species involves several key strategies. The choice of species is critical; choosing native species, where possible, that are better adapted to local ecosystems is essential. Selecting species and genotypes from similar climate conditions, even if they are geographically separated, assures successful establishment under future climate circumstances. Resilience through biodiversity is important, as diverse forests are likely to tolerate numerous stresses such as pests, diseases, or harsh weather. Water conservation techniques, including water-efficient irrigation methods, mulching, soil moisture conservation techniques, contour planting, and other land management practices that reduce runoff and improve water infiltration into the soil, are essential. Agroforestry approaches integrate drought-tolerant species into agroforestry systems to benefit both the environment and local communities. Genetic selection for future climates is another crucial aspect; trees adapt to drought stress via a wide range of cellular and physiological acclimation methods, which serve as the foundation for genetic gains in drought tolerance. Tree breeding programs aimed at creating drought-resistant characteristics involve choosing individuals or genotypes within a species that have traits such as deep roots, efficient water usage, and high tolerance to heat and drought.

11.6 CONCLUSION

Drought tolerance and WUE are critical for forests to adapt to climate change, as they play a vital role in biodiversity conservation, carbon storage, and water cycle regulation, all of which are increasingly threatened by severe droughts. Combining traditional breeding with genetic engineering can enhance drought resistance in forest species, while forest management practices like promoting biodiversity, improving soil health, and encouraging mixed-species planting are essential for maintaining productivity. Strengthening genetic diversity through selective breeding, natural regeneration, and assisted gene flow further supports adaptation. Practical strategies such as agroforestry, reforestation with drought-tolerant species, and assisted migration can enhance resilience, while functional diversity ensures that various species contribute to ecosystem stability. Creating resilient forests is urgent, and integrating scientific advancements with traditional knowledge will help sustain them for future generations.

REFERENCES

Afzal, S., Chaudhary, N., & Singh, N. K. 2021. Role of soluble sugars in metabolism and sensing under abiotic stress. In *Plant Growth Regulators: Signalling under Stress Conditions*, T. Aftab and K. R. Hakeem (eds.), pp. 305–334. Singapore: Springer Nature.

Arinawa, L. F., Adam, K., Jan, A., Michael, P. N., Alan, C., Fritz, S., & Jiří, D. 2022. Comparative anatomy of leaf petioles in temperate trees and shrubs: the role of plant size, environment and phylogeny. *Annals of Botany*, *129*(5): 567–582. https://doi.org/10.1093/aob/mcac014

Bastos, A., Ciais, P., Friedlingstein, P., Sitch, S., Pongratz, J., Fan, L., & Wigneron, J. P. 2020. Direct and seasonal legacy effects of the 2018 heat wave and drought on European ecosystem productivity. *Science Advances*, *6*(1): 1–14.

Bhowal, B., Chandra, P., & Saxena, S. C. 2021. Engineering glycine betaine biosynthesis in alleviating abiotic stress effects in plants. In *Compatible Solutes Engineering for Crop Plants Facing Climate Change*, pp. 97–127, Springer Nature Switzerland AG.

Bhusal, N., Lee, M., Lee, H., Adhikari, A., Han, A. R., Han, A., & Kim, H. S. 2021. Evaluation of morphological, physiological, and biochemical traits for assessing drought resistance in eleven tree species. *Science of the Total Environment*, *779*: 146466.

Choat, B., Brodribb, T. J., Brodersen, C. R., Duursma, R. A., López, R. and Medlyn, B. E., 2018. Triggers of tree mortality under drought. *Nature*, 558(7711): 531–539.

Gonçalves, J. L., Alvares, C. A., Rocha, J. H., Brandani, C. B., & Hakamada, R. 2017. Eucalypt plantation management in regions with water stress. *Southern Forests: A Journal of Forest Science 79*(3): 169–183. https://doi.org/10.2989/20702620.2016.1255415

Hällfors, M. H., Vaara, E. M., Hyvärinen, M., Oksanen, M., Schulman, L. E., Siipi, H., & Lehvävirta, S. 2014. Coming to terms with the concept of moving species threatened by climate change – a systematic review of the terminology and definitions. *PLoS One 9*(7), e102979.

Kaiser, W. M. 1987. Effect of water deficit on photosynthetic capacity. *Physiologia Plantarum 71*(1): 142–149.

Khan, A., Shen, F., Yang, L., Xing, W., & Clothier, B. 2022. Limited acclimation in leaf morphology and anatomy to experimental drought in temperate forest species. *Biology 11*(8): 1186.

Kumar, A., Dwivedi, G. K., Tewari, S., Paul, J., Anand, R., Kumar, N., Kumar, P., Singh, H., & Kaushal, R. 2020a. Carbon mineralization and inorganic nitrogen pools under *Terminalia chebula* Retz.-based agroforestry system in Himalayan foothills, India. *Forest Science*, *66*(5): 634–643.

Kumar, A., Dwivedi, G. K., Tewari, S., Paul, J., Sah, V. K., Singh, H., Kumar, P., Kumar, N., & Kaushal, R. 2020b. Soil organic carbon pools under *Terminalia chebula* Retz. based agroforestry system in Himalayan foothills, India. *Current Science*, *118*(7), 1098–1103.

Kumar, A., Kumar, P., Singh, H., Bisht, S., & Kumar, N. 2021a. Relationship of physiological plant functional traits with soil carbon stock in temperate forest of Garhwal Himalaya. *Current Science*, *120*(8): 1368–1373.

Kumar, A., Kumar, P., Singh, H., & Kumar, N. 2021b. Modulation of plant functional traits under essential plant nutrients during seasonal regime in natural forests of Garhwal Himalayas. *Plant and Soil*, *465*: 197–212.

Kumar, A., Kumar, P., Singh, H., & Kumar, N. 2021c. Impact of plant functional traits on near saturated hydraulic conductivity of soil under different forests of Kempty watershed in Garhwal Himalaya. *Indian Journal of Soil Conservation*, *49*(1), 38–44.

Kumar, A., Singh, H., Kumari, G., Bisht, S., Malik, A., Kumar, N., Singh, M., Raturi, A., Barthwal, S., Thakur, A., & Kaushal, R. 2022. Adaptive resilience of roadside trees to vehicular emissions via leaf enzymatic, physiological, and anatomical trait modulations. *Environmental Pollution*, *313*: 120191.

Kumar, N., Jeena, N., & Singh, H. 2019. Elevated temperature modulates rice pollen structure: A study from foothill Himalayan agro-ecosystem in India. *3 Biotech*, *9*: 175.

Kumar, N., Jeena, N., Kumar, A., Khairakpam, R., & Singh, H. 2021d. Comparative response of rice cultivars to elevated air temperature in Bhabar region of Indian Himalaya: status on yield attributes. *Heliyon*, *7*: e07474.

Kumar, N., Kumar, N., Shukla, A., Shankhdhar, S. C., & Shankhdhar, D. 2015. Impact of terminal heat stress on pollen viability and yield attributes of rice (*Oryza sativa* L.). *Cereal Research Communications*, *43*(4): 616–626.

Kumar, N., Shankhdhar, S. C., and Shankhdhar, D. 2016. Impact of elevated temperature on antioxidant activity and membrane stability in different genotypes of rice (*Oryza sativa* L.). *Indian Journal of Plant Physiology (Now Plant Physiology Report)*, *21*(1): 37–43.

Kumar, N., Suyal, D. C., Sharma, I. P., Verma, A., & Singh, H. 2017. Elucidating stress proteins in rice (*Oryza sativa* L.) genotype under elevated temperature: a proteomic approach to understand heat stress response. *3 Biotech*, 7: 205.

Linares, J. C., & Camarero, J. J. 2012. From pattern to process: linking intrinsic water-use efficiency to drought-induced forest decline. *Global Change Biology*, *18*(3): 1000–1015.

Mansoor, S., Ali Wani, O., Lone, J. K., Manhas, S., Kour, N., Alam, P., & Ahmad, P. 2022. Reactive oxygen species in plants: from source to sink. Antioxidants, *11*(2): 225.

Martinez, V., Nieves-Cordones, M., Lopez-Delacalle, M., Rodenas, R., Mestre, T. C., Garcia-Sanchez, F., Rubio, F., Nortes, P.A., Mittler, R. and Rivero, R.M., 2018. Tolerance to stress combination in tomato plants: new insights in the protective role of melatonin. *Molecules*, 23(3): 535.

Pan, S., Wang, X., Yan, Z., Wu, J., Guo, L., Peng, Z., & Liu, L. 2024. Leaf stomatal configuration and photosynthetic traits jointly affect leaf water use efficiency in forests along climate gradients. *New Phytologist*, 244(4):1250–1262.

Park, A., & Talbot, C., 2018. Information underload: ecological complexity, incomplete knowledge, and data deficits create challenges for the assisted migration of forest trees. *BioScience*, *68*(4): 251–263.

Petrík, P., Petek, A., Konôpková, A., Bosela, M., Fleischer, P., Frýdl, J., & Kurjak, D. 2020. Stomatal and leaf morphology response of European Beech (*Fagus sylvatica* L.) provenances transferred to contrasting climatic conditions. *Forests*, *11*(12): 1359.

Pope, J. O., Brown, K., Fung, F., Hanlon, H. M., Neal, R., Palin, E. J., & Reid, A. 2022. Investigation of future climate change over the British Isles using weather patterns. *Climate Dynamics*, *58*(5–6): 2405–2419.

Sachdev, S., Ansari, S. A., Ansari, M. I., Fujita, M., & Hasanuzzaman, M. 2021. Abiotic stress and reactive oxygen species: generation, signaling, and defense mechanisms. *Antioxidants*, *10*: 277.

Scoffoni, C., Rawls, M., McKown, A., Cochard, H. and Sack, L., 2011. Decline of leaf hydraulic conductance with dehydration: relationship to leaf size and venation architecture. *Plant Physiology*, *156*(2): 832–843.

Sovacool, B. K., Griffiths, S., Kim, J., & Bazilian, M. 2021. Climate change and industrial F-gases: a critical and systematic review of developments, sociotechnical systems and policy options for reducing synthetic greenhouse gas emissions. *Renewable and Sustainable Energy Reviews*, *141*: 110759.

Teskey, R., Wertin, T., Bauweraerts, I., Ameye, M., McGuire, M. A., & Steppe, K. 2015. Responses of tree species to heat waves and extreme heat events. *Plant, Cell & Environment*, *38*(9): 1699–1712.

Tomás, M., Medrano, H., Pou, A., Escalona, J. M., Martorell, S., Ribas-Carbó, M. and Flexas, J., 2012. Water-use efficiency in grapevine cultivars grown under controlled conditions: effects of water stress at the leaf and whole-plant level. *Australian Journal of Grape and Wine Research*, *18*(2): 164–172.

Wolfe, B. T., 2017. Retention of stored water enables tropical tree saplings to survive extreme drought conditions. *Tree Physiology*, *37*(4): 469–480.

Zhang, Y., Yu, X., Chen, L., & Jia, G. 2019. Whole-plant instantaneous and short-term water-use efficiency in response to soil water content and CO_2 concentration. *Plant and Soil*, *444*(1–2): 281–298.

12 Functional Traits as Indicators of Ecosystem Health and Biodiversity

Amit Joshi and Jharna Maiti

12.1 INTRODUCTION

Functional traits have emerged as a valuable tool for investigating and assessing ecosystem health and biodiversity. They describe how plants' features influence their role and performance in ecosystems. Functional features offer information about how species contribute to ecosystem processes and react to environmental changes, in contrast to traditional biodiversity measurements that concentrate on species richness or abundance (Kumar et al., 2020a,b; Borja et al., 2024). These traits are critical in understanding how ecosystems function and how robust they are to disruptions. They include physical characteristics such as leaf size or body size, physiological attributes such as metabolic rates or drought tolerance, and behavioral patterns such as foraging techniques or migrating behaviors. A more thorough understanding of ecosystem stability, productivity, and resilience can be attained by researchers by examining the functional diversity found within ecosystems as well as the range and distribution of these attributes (Kumar et al., 2021d; Triviño et al., 2023). Because it improves an ecosystem's capacity to sustain itself and bounce back from shocks, functional variety is essential for protecting ecosystems from stresses like pollution, habitat loss, and climate change. Monitoring functional features can also serve as a precursor to traditional measurements, like species decrease, by indicating possible degradation of the environment (Wudu et al., 2023).

This chapter investigates the categorization of functional traits, their significance in assessing ecosystem health, and their utility in biodiversity conservation, ecosystem restoration, and climate change adaptation. In addition, it discusses issues with trait-based evaluations, such as missing data and the requirement for consistent measurement methods, and it looks into new technologies that could improve monitoring programs. In the end, the chapter emphasizes how important functional features are to preserving ecological balance and guaranteeing the survival of ecosystems over the long run.

12.2 FUNCTIONAL TRAITS

Forest trees' functional characteristics are essential for maintaining biodiversity and influencing ecological processes. These features stand in for the qualities that dictate how different tree species interact with their surroundings, adjust to adversity,

 DOI: 10.1201/9781003545231-12

and support ecosystem services. By focusing on these traits, scientists and environmentalists may acquire a better understanding of forest ecosystems' robustness, stability, and health, especially when faced with challenges such as habitat loss, deforestation, and climate change. Morphological traits, physiological traits, reproductive qualities, and phenological traits are the four main categories into which forest tree functional traits can be divided (Kumar et al., 2021a,b; Yan et al., 2023). The physical characteristics of trees, such as leaf size and structure, tree height and canopy structure, and root architecture that affect their survival, growth, and ecological interactions, are referred to as morphological features. The internal processes of trees that control their capacity to develop, procreate, and endure in various environmental circumstances are described in physiological traits. Important physiological traits include photosynthetic rate, nutrient-uptake efficiency, and water-use efficiency. Reproductive factors govern the phenology of a tree's blooming phenology, seed size and quantity, and dispersal techniques; phenological traits control the tree's cold hardiness, dormancy, reaction to climatic fluctuation, bud burst, and leaf senescence (Kumar et al., 2015, 2016; Dowtin et al., 2023). These functional traits are crucial for evaluating the health of forest ecosystems because they provide a clearer picture of how different tree species contribute to ecosystem services including carbon storage, water control, and habitat provision. High functional diversity forests, or those that exhibit a wide variety of characteristics, are thought to be more adaptable to various disturbances, including changes in land use, pest outbreaks, and extreme weather. Keeping an eye on the functional characteristics of forest trees might help spot signs of stress or degradation in the ecosystem early on. A forest may be stressed as a result of climate change if, for instance, water usage efficiency varies or phenological patterns alter (Baskent, 2023). For the purpose of ecosystem restoration, climate adaptation plans, and conservation planning, these features provide useful information.

12.3 MATHEMATICAL MODELS FOR ACCESSING FUNCTIONAL TRAITS OF ECOSYSTEM AND BIODIVERSITY

A thorough evaluation of the functional characteristics of forest trees is necessary to comprehend their place in ecosystems and biodiversity. Functional traits are essential features that affect how trees interact with their surroundings and support ecosystem functions like carbon sequestration, nutrient cycling, and habitat development (Hou et al., 2023). Examples of these traits are tree height, leaf area, wood density, and nutrient uptake. The survival, procreation, and resource usage of species under various ecological situations are determined by these qualities. By offering a strong framework for measuring and evaluating these characteristics, mathematical models (Table 12.1) enable ecologists to investigate the ways in which species interactions, the environment, and climate change affect forest ecosystems. These models cover a wide range, from simple linear correlations between traits and environmental factors to complex simulations that account for ecological interactions throughout time. Through the use of these models, scientists are able to forecast the distribution of species, comprehend the resilience of ecosystems, and pinpoint patterns of

TABLE 12.1
Mathematical Models

Model Category	Model Name	Description	Example
Functional diversity indices	Functional richness (FRic)	Measures the range of functional traits, quantifying the niche space occupied by species in the community.	**Example**: In a tropical forest, calculate the trait range of leaf size, wood density, and seed mass for ten tree species to assess the diversity of functional traits (Jeldu et al., 2024).
	Functional evenness (FEve)	Measures the evenness of distribution of species traits in the multidimensional trait space.	**Example**: Analyzing whether tree species in a temperate forest have a more even distribution of traits like specific leaf area (SLA) and wood density (Pantaleão et al., 2024).
	Functional divergence (FDiv)	Measures how much species differ from the centroid of trait space, focusing on functional dissimilarity.	**Example**: Measuring divergence in leaf area and maximum tree height to determine if some species dominate certain niches in a boreal forest (Tang et al., 2023).
Trait–environment relationship models	Generalized linear models (GLMs)	Extends linear regression to model trait–environment relationships with non-normal error distributions.	**Example**: Investigating how altitude (environmental variable) affects tree height and leaf area (functional traits) in mountain ecosystems (Janizadeh et al., 2023).
	Structural equation models (SEMs)	Models complex relationships between traits and environmental factors, including indirect effects.	**Example**: Modeling the indirect effects of soil pH on tree biomass through leaf nitrogen content in a subtropical forest (Bakshi et al., 2024).
Community assembly models	Niche models	Describes how species coexist by partitioning resources through differences in functional traits.	**Example**: Modeling resource use based on root depth and leaf area index to see how tree species share water and light in a savannah ecosystem (Carrillo-García et al., 2023).
	Trait-filtering models	Analyzes how environmental filters select species with particular functional traits.	**Example**: Investigating how a wetland's flooding regime selects tree species with specific traits like high wood density and rapid root growth (Hoenle et al., 2023).

(Continued)

TABLE 12.1 (*Continued*)
Mathematical Models

Model Category	Model Name	Description	Example
Species distribution models (SDMs)	Ecological niche models (ENMs)	Predicts the geographical distribution of species based on their functional traits and environmental variables.	**Example**: Predicting where tree species with high drought-tolerance traits will be distributed under future climate change scenarios in Mediterranean regions (Gomes et al., 2024).
	Maximum entropy model (MaxEnt)	Estimates species distribution by finding the probability distribution that maximizes entropy given trait and environmental data.	**Example**: Using MaxEnt to predict the distribution of tree species based on traits like leaf longevity and temperature tolerance in a rainforest ecosystem (Zhang et al., 2023).
Allometric scaling models	Allometric models	Relates tree size (e.g., height and diameter) to functional traits like biomass and nutrient cycling.	**Example**: Using an allometric model to estimate tree biomass based on trunk diameter in a temperate forest, allowing carbon stock calculations (Lin et al., 2023).
Vegetation dynamics models	Dynamic global vegetation models (DGVMs)	Simulates changes in forest ecosystems over time, linking species functional traits with environmental and ecological processes.	**Example**: Using DGVMs to simulate how tree species with different drought tolerance traits respond to climate change in a tropical forest over decades (Zou et al., 2023).
Neutral models of biodiversity	Unified Neutral Theory of Biodiversity (UNTB)	Assumes that biodiversity is driven by random processes, with traits distributed neutrally across species.	**Example**: Using UNTB to model how tree species with similar traits spread across a tropical forest after a disturbance like a hurricane (Overcast et al., 2023).
Demographic models	Trait-based demographic models	Links tree functional traits with population dynamics like growth, reproduction, and mortality.	**Example**: Analyzing how specific leaf area (SLA) and seed size influence tree mortality rates in a temperate forest (Li et al., 2024).

functional diversity that are essential for conservation planning. The subsequent segments delve into diverse mathematical models that are frequently employed in the examination of forest tree functional traits. These models include diversity indices, species distribution models (SDMs), and trait–environment relationship models (Kumar et al., 2017; Bourhis et al., 2023). Additionally, instances of their utilization in research pertaining to forest ecosystems and biodiversity are provided. These models provide important insights for sustainable forest management and biodiversity conservation in the face of global environmental concerns by bridging the gap between tree-level features and large-scale ecosystem dynamics (Pour et al., 2023).

12.4 FUNCTIONAL DIVERSITY AND ECOSYSTEM HEALTH

The range of functional characteristics seen in an ecosystem, which represents the diversity of ways in which species contribute to ecological processes, is referred to as functional diversity. Functional diversity highlights the ecological roles that species play and how their features affect ecosystem functioning (Kumar et al., 2023), as opposed to species diversity, which is primarily concerned with the number of distinct species that are present (Table 12.2). Understanding ecosystem health requires an emphasis on trait diversity since it reveals how well an ecosystem can carry out essential tasks like carbon storage, nutrient cycle, and disturbance tolerance. A great range of features enable organisms in highly functionally diverse environments to inhabit distinct ecological niches. The stability and productivity of the ecosystem are increased by this diversity of functions. For example, in a forest, different tree species might make the best use of resources like light, water, and nutrients due to their varied root depths, water-use techniques, and leaf shapes. Because other species with similar activities can compensate for the loss of some species, diversity ensures that the ecosystem can continue to function even in the event of a change in the environment or the extinction of some species (Kumar et al., 2019; Rao et al., 2024). By increasing ecosystems' resilience, this "insurance effect" makes it possible for them to recover from shocks like fires, droughts, or disease outbreaks more successfully. Because functional diversity allows for the simultaneous performance of several ecological roles, it is also essential for preserving the health of ecosystems. For instance, several species may be involved in pollination, nitrogen cycling, and soil formation processes that are critical to maintaining ecosystem services that promote human well-being. Conversely, a decrease in functional diversity can impair ecosystem functions and increase an ecosystem's susceptibility to external stresses like pollution, habitat loss, and climate change. Species diversity alone cannot provide a more accurate evaluation of ecosystem health than functional diversity monitoring (Kumar et al., 2021c,d; Sharma & Birman, 2024). Early signs of ecosystem degradation might include changes in functional features, such as a decline in species with qualities supporting nutrient intake or water retention. For example, an ecosystem may become more vulnerable to erosion and drought if species with deep roots that encourage water infiltration disappear. Conservationists can identify possible ecosystem imbalances before they worsen by monitoring changes in functional features, which enables proactive management measures (Faria et al., 2023).

TABLE 12.2
Functional Traits Contribute to the Health and Stability of Their Respective Ecosystems

Ecosystem	Functional Traits	Functional Diversity Example	Trees that show Functional Diversity	Impact on Ecosystem Health	Reference
Forest	Leaf size, root depth, drought tolerance	Trees with varying root depths (shallow and deep) optimize water and nutrient use	Maple (*Acer* spp.) (large leaves), Pine (*Pinus* spp.) (deep roots)	Increases resilience to drought and nutrient cycling efficiency	Coyle (2023)
Grassland	Grazing tolerance, seed dispersal mechanisms	Presence of both grazing-tolerant grasses and wind-dispersed species	Big Bluestem (*Andropogon gerardii*) (grazing-tolerant), Prairie Dropseed (*Sporobolus heterolepis*) (wind-dispersed)	Maintains plant diversity and enhances seed dispersal and regeneration	Boerman et al.(2023)
Wetland	Water-use efficiency, flood tolerance, root types	Plants with high flood tolerance coexist with species that promote water filtration	Cattail (*Typha latifolia*) (high flood tolerance), Bulrush (*Schoenoplectus* spp.) (promotes water filtration)	Enhances water quality, reduces soil erosion, and supports flood control	Antonijević et al.(2023)
Coral Reef	Feeding strategies, reproductive modes	Coral species with different feeding modes (filter feeders, predators)	Elkhorn Coral (*Acropora palmata*) (filter feeder), Brain Coral (*Diploria* spp.) (predator)	Supports biodiversity and productivity, improving reef resilience	Chapron et al.(2023)
Desert	Water retention, heat tolerance, seed dormancy	Species with high heat and drought tolerance dominate, while others lie dormant	Saguaro Cactus (*Carnegiea gigantea*) (high water retention), Desert Marigold (*Baileya multiradiata*) (heat tolerant)	Maintains ecosystem stability during extreme temperatures and droughts	Emming (2023)
Agricultural	Nitrogen fixation, pest resistance	Crop rotation with nitrogen-fixing plants (e.g., legumes) and pest-resistant species	Soybean (Glycine max) (nitrogen-fixing), Tomato (*Solanum lycopersicum*) (pest-resistant varieties)	Improves soil fertility and reduces pest outbreaks, supporting sustainability	Zhang et al. (2024)

12.5 USING FUNCTIONAL TRAITS AS ECOSYSTEM HEALTH INDICATORS

The importance of functional traits as indicators for assessing ecosystem health and biodiversity is increasingly recognized. Compared to traditional metrics, which simply reflect species richness or abundance, functional characteristics offer a deeper understanding of how species contribute to ecosystem functions and respond to environmental changes. These characteristics are crucial for keeping an eye on and controlling ecological systems because they offer a more complex understanding of ecosystem dynamics and resilience (Li et al., 2023). The following are some advantages of assessing ecosystem health using functional traits:

- **Holistic View**: Functional traits illustrate how species contribute to critical activities such as primary production, nutrient cycling, and habitat structure, providing a complete picture of ecosystem health. This method takes into account how species functions and interactions impact the resilience and stability of ecosystems, going beyond simple species counts (Liu et al., 2024). For instance, the big leaves of mahogany (*Swietenia macrophylla*) optimize photosynthesis and support high primary output. The canopy structure, which is essential for preserving the habitat for many species and controlling the microclimate inside the forest, is supported by the enormous leaf size. Another illustration is the deep root systems of eucalyptus (*Eucalyptus* spp.) trees, which have access to groundwater and support stable forests and the cycling of nutrients (Yuniati et al., 2024). By lowering erosion and preserving soil structure, this characteristic promotes the general health of the forest ecosystem.
- **Early Warning Indicators**: Changes in functional traits can serve as early warning signs of ecological stress before more traditional indicators, such as species extinction, become apparent. For instance, a decrease in species that can withstand prolonged droughts may indicate that an ecosystem's water stress is getting worse. For instance, cacti, such as *Carnegiea gigantea*, the saguaro cactus, have unique characteristics that enable them to endure protracted droughts, such as thick, waxy skins. Before more significant effects on other plant species become evident, a decrease in cactus populations can indicate the emergence of water stress in arid environments (Peng et al., 2024). Another illustration is the Creosote Bush (*Larrea tridentata*), which has deep root systems that store water and small, thick leaves that help it adapt to dry conditions. A decline in the number of creosote bushes could be a sign of increased water stress or habitat degradation in desert environments (Gerstner et al., 2024).
- **Resilience and Adaptation**: Enhancing ecological resilience is largely dependent on the variety of functional features that exist within an ecosystem. This is known as functional diversity. An ecosystem with a high functional diversity can withstand shocks and encourage recovery by maintaining critical functions in the face of extinction or stress to specific species. (Müller et al., 2023). For instance, legumes (such as soybean and

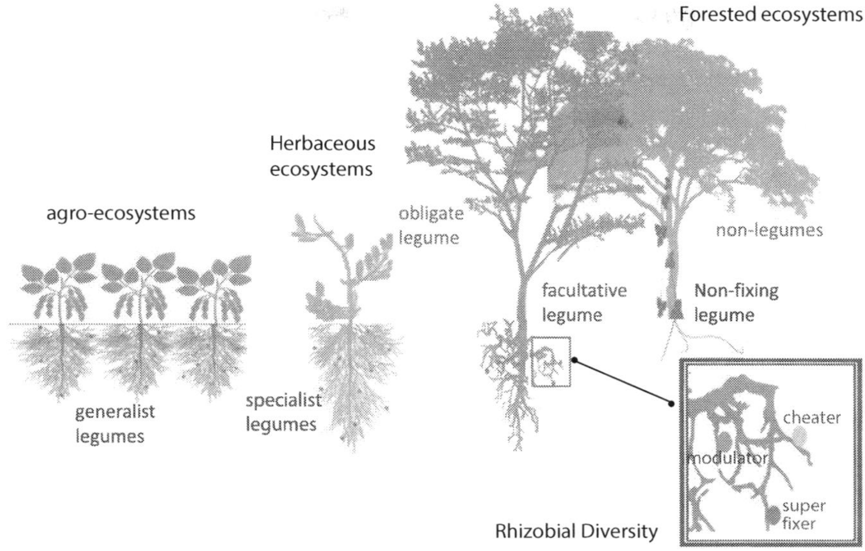

FIGURE 12.1 Legumes work in symbiosis with rhizobial bacteria.

Glycine max) work in symbiosis with rhizobial bacteria to fix atmospheric nitrogen into the soil (Figure 12.1). This characteristic promotes ecosystem stability and recovery following disturbances by improving soil fertility and enabling plant growth even in soils deficient in nitrogen. Another illustration is the fact that lupins, such as *Lupinus* spp., can yield seeds that can lie dormant in the ground for long periods of time. Because of this characteristic, they can withstand adverse environments and recover from shocks like fires or droughts, enhancing the resilience of ecosystems and recovery (Siger et al., 2023).

12.6 EXAMPLES OF FUNCTIONAL TRAITS IN ECOSYSTEM HEALTH ASSESSMENT

Because they shed light on how organisms interact with their surroundings and contribute to ecosystem functions, functional features are essential for the conservation of biodiversity. By emphasizing the functional roles of species rather than just their presence or abundance, an understanding of these features might improve conservation measures (Kumar et al., 2022; Chacón-Labella et al., 2023). This strategy aids in maintaining both the variety of species and the vital ecological functions that they sustain.

12.6.1 Role of Functional Traits in Biodiversity Conservation

Understanding Ecological Roles: Functional traits can help identify the specific roles that species play within an ecosystem. For example, characteristics such as leaf size, root depth, and water-use efficiency affect how plants interact with their surroundings and support activities like soil stabilization, nutrient cycling, and primary productivity (Table 12.3) (Griffith et al., 2023). Conservationists can guarantee that ecosystems maintain the functional variety required for their resilience and stability by emphasizing these characteristics.

Maintaining Ecosystem Functions: The goal of functional trait-focused conservation is to protect the ecological systems that support biodiversity. For instance, maintaining species like legumes that have characteristics linked to nitrogen fixation guarantees soil fertility and promotes plant diversity. By maintaining ecosystem functions, this strategy contributes to the preservation of biodiversity and the provision of ecosystem services.

TABLE 12.3
Functional Traits of Species Contribute to the Implications for Biodiversity Conservation

Functional Trait	Example Species	Conservation Implication
Leaf size	Rubber tree (*Hevea brasiliensis*)	Ensuring the presence of species with optimal leaf traits supports forest productivity and habitat structure (Joseph, 2024).
Root depth	Pine (*Pinus* spp.)	Protecting species with deep root systems helps prevent soil erosion and maintain soil health (Bose et al., 2023).
Drought tolerance	Agave (*Agave* spp.)	Conserving drought-tolerant species ensures ecosystem stability in arid regions and helps adapt to climate change (Witzel & Matros, 2023).
Nitrogen fixation	Soybean (Glycine max)	Incorporating nitrogen-fixing plants in restoration projects improves soil fertility and supports diverse plant communities (Gannett et al., 2024).
Seed dispersal mechanism	Maple (*Acer* spp.)	Protecting species with effective seed dispersal mechanisms aids in forest regeneration and species distribution (Myers et al., 2023).
Water-use efficiency	Eucalyptus (*Eucalyptus* spp.)	Maintaining species with high water-use efficiency supports forest health and adaptability in fluctuating climates (Costa, 2023).
Phenological timing	Cherry Blossom (*Prunus* spp.)	Monitoring phenological traits helps detect shifts due to climate change and supports conservation planning (Talukder et al., 2023).
Pollination mechanisms	Bee Orchid (*Ophrys* spp.)	Protecting species with specialized pollination traits ensures the survival of both plants and their pollinators (Bateman& Rudall, 2023).
Heat tolerance	Cactus (*Carnegiea gigantea*)	Conserving heat-tolerant species helps maintain ecosystem stability in desert environments and supports biodiversity (Conver et al., 2023).
Root symbiosis	Douglas Fir (*Pseudotsuga menziesii*)	Supporting species with root symbiosis enhances nutrient uptake and soil health (Riesco Muñoz et al., 2024)

12.6.2 Applications of Functional Trait Analysis in Conservation

Species Selection for Restoration: Selecting species based on their functional traits can improve the success of ecosystem restoration initiatives. For example, selecting plant species that support nitrogen cycling and soil stabilization can enhance soil health and ecosystem recovery. This approach guarantees that a wide variety of species are supported by the restored environment and that it functions as intended (Loureiro et al., 2023).

Identifying Key Functional Species: Because of their particular functional characteristics, certain species have special functions within their ecosystems. Apex predators, for instance, that exhibit characteristics associated with hunting and territorial behavior are essential for controlling prey populations and preserving ecological equilibrium. Ecosystem dynamics and biodiversity are preserved when these important functional species are recognized and protected (Merchant et al., 2023).

Monitoring Ecosystem Health: Functional characteristics can be used to track alterations in the health of ecosystems and identify early indicators of deterioration or stress. Changes in the functional features of plant species, for example, diminished resistance to drought or altered timing of flowering, might signal alterations in the environment or effects of human activity. Conservationists can respond quickly to developing problems and protect the health of ecosystems by monitoring these qualities (Pinna et al., 2023).

12.7 CHALLENGES AND FUTURE DIRECTIONS

There are several benefits to using functional features as markers of biodiversity and ecosystem health, including a better comprehension of ecological functions and resilience. This strategy does, however, also highlight a number of issues and potential directions for improvement. By tackling these issues and looking in new directions, functional trait-based assessments can become more successful and help develop conservation and management policies that work better (Nizamani et al., 2024)

12.7.1 Challenges

Data Availability and Quality: The availability and caliber of functional characteristic data represent a major obstacle. For many species, there are frequently insufficient comprehensive data on functional features, particularly in places and ecosystems that are poorly investigated. Trait-based assessments may be less accurate and applicable if the data are inconsistent or incomplete. Additionally, it may be challenging to compare data for this challenge due to the variability in trait measurements across different studies. To address this, efforts should be made to standardize functional trait measurement techniques and create centralized databases that compile trait data from various sources (Stewart et al., 2023). Collaborative research projects and citizen science can also assist in gathering larger amounts of data.

- **Trait Complexity and Interactions**: Functional qualities interact with other traits and the context in which they exist. They do not function in a vacuum. Because of this complexity, interpreting how modifications to functional features impact ecosystem health may be difficult. For example, the context (soil, climate) might affect the effects of a particular feature differently. Advanced modeling and analytical methods, such as trait-based models and multitrait approaches, will be used to assist in accounting for these relationships in this problem (Monzel et al., 2023). Trait–environment connections will be better understood by integrating functional trait data with ecological and environmental models.

- **Standardization of Measurements**: Data inconsistencies may result from the absence of defined procedures for assessing and reporting functional features. Comparing results across studies and geographical areas can be challenging due to the use of different definitions or methods by different researchers. It is imperative to create and execute standardized standards for trait measuring and reporting in order to address this issue (Blanco & Lo, 2023). By offering standardized trait data and methodology, initiatives such as the Functional Trait Database and the TRY Plant Trait Database seek to address this problem (McMillan et al., 2023).
- **Data Integration and Analysis**: It is difficult to integrate data on functional traits with other ecological and environmental data, like species distribution and ecosystem functions. Effectively combining these data sources calls for advanced analytical techniques and tools. We'll make use of developments in data science and technology that provide chances for large-scale dataset integration and analysis, like big data analytics and machine learning (Anderegg, 2023). More thorough studies can also be facilitated by collaborative platforms that unite academics from different domains.

12.7.2 Future Directions

- **Improving Trait Databases**: To encompass a wider variety of species and habitats, future efforts should concentrate on updating and growing functional trait databases. This entails incorporating information on lesser-known species and taking into account characteristics pertinent to recently developing environmental issues, such as climate change (Calbi et al., 2024).
- **Exploring Multitrait Approaches**: Multitrait techniques, which investigate how combinations of traits impact resilience and ecological processes, should be further explored in research. Studying how characteristics interact and support the general health of an ecosystem is one aspect of this (Dias et al., 2023).
- **Advancing Remote Sensing and Technology**: Technological developments like high-throughput phenotyping and remote sensing have the potential to improve trait monitoring and measurement. The ability to evaluate and manage ecosystems can be improved by these technologies, which can offer extensive, precise data on characteristics and environmental conditions (Sheikh et al., 2024).
- **Incorporating Functional Traits into Policy and Management**: For practical implementation, functional characteristic information must be

incorporated into conservation strategies and ecosystem management plans. This involves directing habitat restoration, species protection, and climate adaptation plans with trait-based evaluations (Urbina-Cardona et al., 2023).

- **Addressing Global and Regional Variability**: To better understand how ecosystems react to local and global environmental changes, future research should focus on regional and worldwide variability in functional features. Studying the variations in functional attributes under various soil, climate, and disturbance regimes is part of this (Gerhard et al., 2023).
- **Enhancing Collaborative Research**: It is imperative to foster cooperation between researchers, conservationists, and policymakers in order to progress the application of functional features in ecosystem health evaluation (Pu et al., 2023). The exchange of data, standardization, and the creation of novel solutions to shared problems can all be facilitated by collaborative research.

12.8 CONCLUSION

Functional traits, which highlight the roles that species play in ecological processes and their responses to environmental change, provide an effective framework for assessing ecosystem health and biodiversity. This method offers a more sophisticated knowledge of how ecosystems function and adapt than traditional measures. Even with obstacles like trait complexity, measurement uniformity, and data availability, improvements in technology and methodology could lead to better trait-based assessments. By accepting these developments and tackling the issues at hand, conservation methods will be more successful in keeping ecosystems robust and functional. In the end, we can better perpetuate biodiversity and the essential ecological processes that support life on Earth by incorporating functional features into ecosystem management and conservation efforts.

REFERENCES

Anderegg, L. D. (2023). Why can't we predict traits from the environment? *New Phytologist*, *237*(6), 1998–2004.

Antonijević, D., Hoffmann, M., Prochnow, A., Krabbe, K., Weituschat, M., Couwenberg, J., … & Augustin, J. (2023). The unexpected long period of elevated CH_4 emissions from an inundated fen meadow ended only with the occurrence of cattail (*Typha latifolia*). *Global Change Biology*, *29*(13), 3678–3691.

Bakshi, B., Polasky, S., & Frelich, L. E. (2024). Using structural equation models (SEM) to link climate change, forest composition, deer, and outdoor recreation. *Ecological Modelling*, *493*, 110731.

Baskent, E. Z. (2023). Characterizing and assessing key ecosystem services in a representative forest ecosystem in Turkey. *Ecological Informatics*, *74*, 101993.

Bateman, R. M., & Rudall, P. J. (2023). Morphological continua make poor species: Genus-wide morphometric survey of the European Bee Orchids (*Ophrys* L.). *Biology*, *12*(1), 136.

Blanco, J. A., & Lo, Y. H. (2023). Latest trends in modelling forest ecosystems: New approaches or just new methods? *Current Forestry Reports*, *9*(4), 219–229.

Boerman, N. A., Gillen, R. L., & Masters, R. A. (2023). Establishment of sand bluestem with the aid of pre-emergence-applied herbicides. *Crop, Forage & Turfgrass Management*, *9*(1), e20198.

Borja, A., Berg, T., Gundersen, H., Hagen, A. G., Hancke, K., Korpinen, S., …& Uyarra, M. C. (2024). Innovative and practical tools for monitoring and assessing biodiversity status and impacts of multiple human pressures in marine systems. *Environmental Monitoring and Assessment*, *196*(8), 694.

Bose, T., Hammerbacher, A., Slippers, B., Roux, J., & Wingfield, M. J. (2023). Continuous replanting could degrade soil health in short-rotation plantation forestry. *Current Forestry Reports*, *9*(4), 230–250.

Bourhis, Y., Bell, J. R., Shortall, C. R., Kunin, W. E., & Milne, A. E. (2023). Explainable neural networks for trait-based multispecies distribution modelling—A case study with butterflies and moths. *Methods in Ecology and Evolution*, *14*(6), 1531–1542.

Calbi, M., Boenisch, G., Boulangeat, I., Bunker, D., Catford, J. A., Changenet, A., …& Roccotiello, E. (2024). A novel framework to generate plant functional groups for ecological modelling. *Ecological Indicators*, *166*, 112370.

Carrillo-García, C., Girola-Iglesias, L., Guijarro, M., Hernando, C., Madrigal, J., & Mateo, R. G. (2023). Ecological niche models applied to post-megafire vegetation restoration in the context of climate change. *Science of the Total Environment*, *855*, 158858.

Chacón-Labella, J., Hinojo-Hinojo, C., Bohner, T., Castorena, M., Violle, C., Vandvik, V., & Enquist, B. J. (2023). How to improve scaling from traits to ecosystem processes. *Trends in Ecology & Evolution*, *38*(3), 228–237.

Chapron, L., Kuffner, I. B., Kemp, D. W., Hulver, A. M., Keister, E. F., Stathakopoulos, A., …& Grottoli, A. G. (2023). Heterotrophy, microbiome, and location effects on restoration efficacy of the threatened coral *Acropora palmata*. *Communications Earth & Environment*, *4*(1), 233.

Conver, J. L., Raleigh, K. N., & Swann, D. E. (2023). Temporal and habitat-based growth rate variability in the Saguaro Cactus (*Carnegiea gigantea*). *Natural Areas Journal*, *43*(4), 235–242.

Costa, J. R. D. S. (2023). Comparing functional traits of planted Eucalyptus sp. genotypes managed as high tree and coppice, pp. 21–86.

Coyle, D. R. (2023). Tip, Shoot, Root, and Regeneration Pests. In: D. Allison, J., Paine, T. D., Slippers, B., Wingfield, M. J. (eds) *Forest Entomology and Pathology*. Springer, Cham. https://doi.org/10.1007/978-3-031-11553-0_15.

Dias, H. Q., Sukumaran, S., Neetu, S., Mulik, J., Janakiram, P., & Kumar, D. S. (2023). Deciphering benthic ecosystem functioning and resilience in a major port and marine protected area via the multi-trait approach. *Environmental Research*, *237*, 116988.

Dowtin, A. L., Cregg, B. C., Nowak, D. J., & Levia, D. F. (2023). Towards optimized runoff reduction by urban tree cover: A review of key physical tree traits, site conditions, and management strategies. *Landscape and Urban Planning*, *239*, 104849.

Emming, J. (2023). Landscaping with large succulents: A watering guide for ocotillos, cacti, and yuccas part one: General information, and tips regarding ocotillos. *Cactus and Succulent Journal*, *95*(1), 67–77.

Faria, D., Morante-Filho, J. C., Baumgarten, J., Bovendorp, R. S., Cazetta, E., Gaiotto, F. A., …& Benchimol, M. (2023). The breakdown of ecosystem functionality driven by deforestation in a global biodiversity hotspot. *Biological Conservation*, *283*, 110126.

Gannett, M., DiTommaso, A., Sparks, J. P., & Kao-Kniffin, J. (2024). Microbial nitrogen immobilization reduces competitive advantage of nitrophilous plants with soybean. *Plant and Soil*, 16, 1–20.

Gerhard, M., Koussoroplis, A. M., Raatz, M., Pansch, C., Fey, S. B., Vajedsamiei, J., …& Striebel, M. (2023). Environmental variability in aquatic ecosystems: Avenues for future multifactorial experiments. *Limnology and Oceanography Letters*, *8*(2), 247–266.

Gerstner, B. P., Laport, R. G., Rudgers, J. A., & Whitney, K. D. (2024). Plant–soil microbe feedbacks depend on distance and ploidy in a mixed cytotype population of *Larrea tridentata*. *American Journal of Botany*, *111*(8), e16298.

Gomes, I., Chozas, S., Barradas, M. C. D., Alves, F. L., & Ascensão, F. (2024). Predicting the climatic suitability of non-native drought resistant trees in the face of desertification–A case study with *Argania spinosa* in the Iberian Peninsula. *Agriculture, Ecosystems & Environment, 376*, 109232.

Griffith, P., Lang, J. W., Turvey, S. T., & Gumbs, R. (2023). Using functional traits to identify conservation priorities for the world's crocodylians. *Functional Ecology, 37*(1), 112–124.

Hoenle, P. O., Staab, M., Donoso, D. A., Argoti, A., & Blüthgen, N. (2023). Stratification and recovery time jointly shape ant functional reassembly in a neotropical forest. *Journal of Animal Ecology, 92*(7), 1372–1387.

Hou, J., Fanin, N., & Ni, Z. (2023). A network prediction model to quantify relationship between biodiversity and ecosystem functioning (BEF). *Methods in Ecology and Evolution, 14*(12), 2907–2916.

Janizadeh, S., Bateni, S. M., Jun, C., Im, J., Pai, H. T., Band, S. S., & Mosavi, A. (2023). Combination four different ensemble algorithms with the generalized linear model (GLM) for predicting forest fire susceptibility. *Geomatics, Natural Hazards and Risk, 14*(1), 2206512.

Jeldu, F. A., Tolera, M., & Woldeamanuel, T. (2024). Predictive role of species diversity on aboveground carbon via functional diversity and dominance in the moist afromontane forest: Implications for conservation policies. *International Journal of Forestry Research, 2024*(1), 9040431.

Joseph, M. (2024). Natural Rubber (*Hevea brasiliensis* Muell Arg.). In *Soil Health Management for Plantation Crops: Recent Advances and New Paradigms* (pp. 281–308). Singapore: Springer Nature Singapore.

Kumar, A., Dwivedi, G. K., Tewari, S., Paul, J., Anand, R., Kumar, N., ...& Kaushal, R. (2020a). Carbon mineralization and inorganic nitrogen pools under *Terminalia chebula* Retz.-Based agroforestry system in Himalayan foothills, India. *Forest Science, 66*(5), 634–643.

Kumar, A., Dwivedi, G. K., Tewari, S., Paul, J., Sah, V. K., Singh, H., ...& Kaushal, R. (2020b) Soil organic carbon pools under *Terminalia chebula* Retz. based agroforestry system in Himalayan foothills, India. *Current Science, 118*(7), 1098–1103.

Kumar, A., Kumar, P., Singh, H., Bisht, S., & Kumar, N. (2021a). Relationship of physiological plant functional traits with soil carbon stock in temperate forest of Garhwal Himalaya. *Current Science, 120*(8), 1368–1373

Kumar, A., Kumar, P., Singh, H., & Kumar, N. (2021b). Modulation of plant functional traits under essential plant nutrients during seasonal regime in natural forests of Garhwal Himalayas. *Plant and Soil, 465*, 197–212.

Kumar, A., Kumar, P., Singh, H., & Kumar, N. (2021c). Impact of plant functional traits on near saturated hydraulic conductivity of soil under different forests of Kempty watershed in Garhwal Himalaya. *Indian Journal of Soil Conservation, 49*(1), 38–44.

Kumar, A., Singh, H., Kumari, G., Bisht, S., Malik, A., Kumar, N., ...& Kaushal, R. (2022). Adaptive resilience of roadside trees to vehicular emissions via leaf enzymatic, physiological, and anatomical trait modulations. *Environmental Pollution, 313*, 120191.

Kumar, N., Jeena, N., & Singh, H. (2019). Elevated temperature modulates rice pollen structure: A study from foothill Himalayan agro-ecosystem in India. *3Biotech, 9*, 175.

Kumar, N., Jeena, N., Kumar, A., Khairakpam, R., & Singh, H. (2021d). Comparative response of rice cultivars to elevated air temperature in Bhabar region of Indian Himalaya: Status on yield attributes. *Heliyon, 7*, e07474.

Kumar, N., Kumar, N., Shukla, A. Shankhdhar, S. C., & Shankhdhar, D. (2015). Impact of terminal heat stress on pollen viability and yield attributes of rice (*Oryza sativa* L.). *Cereal Research Communications, 43*(4), 616–626.

Kumar, N., Shankhdhar, S. C., & Shankhdhar, D. (2016). Impact of elevated temperature on antioxidant activity and membrane stability in different genotypes of rice (*Oryza sativa* L.). *Indian Journal of Plant Physiology (Now Plant Physiology Report)*, *21*(1), 37–43.

Kumar, N., Suyal, D. C., Sharma, I. P., Verma, A., & Singh, H. (2017). Elucidating stress proteins in rice (*Oryza sativa* L.) genotype under elevated temperature: A proteomic approach to understand heat stress response. *3 Biotech*, 7, 205.

Kumar, R., Bhardwaj, A. K., & Chandra, K. K. (2023). Levels of natural and anthropogenic disturbances and assessment of their impact on plant community functional diversity. *Forestist*, *73*(1), 108–116.

Li, C., Wu, Y., Gao, B., Zheng, K., Wu, Y., & Wang, M. (2023). Construction of ecological security pattern of national ecological barriers for ecosystem health maintenance. *Ecological Indicators*, *146*, 109801.

Li, T., Wang, Q., Tissue, D. T., Wright, I. J., Luo, P., Lai, C., …& Wang, X. (2024). Forest demography depends on stand structure, functional traits, and climate in the eastern region of the Qinghai-Tibet Plateau. *Agricultural and Forest Meteorology*, *355*, 110143.

Lin, J., Gamarra, J. G., Drake, J. E., Cuchietti, A., & Yanai, R. D. (2023). Scaling up uncertainties in allometric models: How to see the forest, not the trees. *Forest Ecology and Management*, *537*, 120943.

Liu, L., Gao, Y., Yang, W., Liu, J., & Wang, Z. (2024). Community metagenomics reveals the processes of nutrient cycling regulated by microbial functions in soils with P fertilizer input. *Plant and Soil*, *499*(1), 139–154.

Loureiro, N., Mantuano, D., Manhães, A., & Sansevero, J. (2023). Use of the trait-based approach in ecological restoration studies: A global review. *Trees*, *37*(5), 1287–1297.

McMillan, J. R., Morrison, B., Chambers, N., Ruggerone, G., Bernatchez, L., Stanford, J., & Neville, H. (2023). A global synthesis of peer-reviewed research on the effects of hatchery salmonids on wild salmonids. *Fisheries Management and Ecology*, *30*(5), 446–463.

Merchant, T. K., Henn, J. J., de Silva, I., Van Cleemput, E., & Suding, K. N. (2023). Four reasons why functional traits are not being used in restoration practice. *Restoration Ecology*, *31*(3), e13788.

Monzel, A. S., Enríquez, J. A., & Picard, M. (2023). Multifaceted mitochondria: Moving mitochondrial science beyond function and dysfunction. *Nature Metabolism*, *5*(4), 546–562.

Müller, J., Mitesser, O., Cadotte, M. W., van der Plas, F., Mori, A. S., Ammer, C., …& Eisenhauer, N. (2023). Enhancing the structural diversity between forest patches—A concept and real-world experiment to study biodiversity, multifunctionality and forest resilience across spatial scales. *Global Change Biology*, *29*(6), 1437–1450.

Myers, A. L., Storer, A. J., Dickinson, Y. L., & Bal, T. L. (2023). A review of propagation and restoration techniques for American beech and their current and future application in mitigation of beech bark disease. *Sustainability*, *15*(9), 7490.

Nizamani, M. M., Hughes, A. C., Qureshi, S., Zhang, Q., Tarafder, E., Das, D., …& Zhang, Z. G. (2024). Microbial biodiversity and plant functional trait interactions in multifunctional ecosystems. *Applied Soil Ecology*, *201*, 105515.

Overcast, I., Achaz, G., Aguilée, R., Andújar, C., Arribas, P., Creedy, T. J., …& Morlon, H. (2023). Towards a genetic theory of island biogeography: Inferring processes from multidimensional community-scale data. *Global Ecology and Biogeography*, *32*(1), 4–23.

Pantaleão, L. C., de Moraes, L. F. D., Cesário, F. V., Moser, P., Dias, A. T. C., Amorim, T. A., & Sansevero, J. B. B. (2024). Linking plant functional traits to soil properties in tropical forest restoration. *Forest Ecology and Management*, *563*, 121976.

Peng, Y., Zhang, W., Yang, X., Zhang, Z., Zhu, G., & Zhou, S. (2024). Current status and prospects of algal bloom early warning technologies: A Review. *Journal of Environmental Management*, *349*, 119510.

Pinna, M., Zangaro, F., Saccomanno, B., Scalone, C., Bozzeda, F., Fanini, L., & Specchia, V. (2023). An overview of ecological indicators of fish to evaluate the anthropogenic pressures in aquatic ecosystems: From traditional to innovative DNA-based approaches. *Water, 15*(5), 949.

Pour, M. D., Barati, A. A., Azadi, H., Scheffran, J., & Shirkhani, M. (2023). Analyzing forest residents' perception and knowledge of forest ecosystem services to guide forest management and biodiversity conservation. *Forest Policy and Economics, 146*, 102866.

Pu, X., Ding, W., Ye, W., Nan, X., & Lu, R. (2023). Ecosystem service research in protected areas: A systematic review of the literature on current practices and future prospects. *Ecological Indicators, 154*, 110817.

Rao, A. S., Chhawri, R., Chauhan, A., Yadav, S. S., Meena, K. C., & Bansal, P. (2024). Plant functional traits: A key framework for understanding and managing ecosystem responses to global environmental challenges. In *Plant Functional Traits for Improving Productivity* (pp. 287–299). Singapore: Springer Nature Singapore.

Riesco Muñoz, G., Rodríguez da Cámara, A. T., & Campos Fernández, B. (2024). Variation in physical properties and estimation of the density of thinned wood in Douglas fir (*Pseudotsuga menziesii*). *European Journal of Wood and Wood Products*, 82(5): 1685–1696.

Sharma, I., & Birman, S. (2024). Biodiversity loss, ecosystem services, and their role in promoting sustainable health. In *The Climate-Health-Sustainability Nexus: Understanding the Interconnected Impact on Populations and the Environment* (pp. 163–188). Cham: Springer Nature Switzerland.

Sheikh, M., Iqra, F., Ambreen, H., Pravin, K. A., Ikra, M., & Chung, Y. S. (2024). Integrating artificial intelligence and high-throughput phenotyping for crop improvement. *Journal of Integrative Agriculture, 23*(6), 1787–1802.

Siger, A., Grygier, A., & Czubinski, J. (2023). Comprehensive characteristic of lipid fraction as a distinguishing factor of three lupin seed species. *Journal of Food Composition and Analysis, 115*, 104945.

Stewart, K., Carmona, C. P., Clements, C., Venditti, C., Tobias, J. A., & González-Suárez, M. (2023). Functional diversity metrics can perform well with highly incomplete data sets. *Methods in Ecology and Evolution, 14*(11), 2856–2872.

Talukder, P., Mishra, M., Sinha, B., Roy, S., Roy, D. D., Barua, S., …& Dutta, D. (2023). Chemical interaction between Cherry Blossom and its pest. *American Journal of Applied Bio-Technology Research, 4*(2), 1–12.

Tang, R., Li, S., Lang, X., Huang, X., & Su, J. (2023). Rare species contribute greater to ecosystem multifunctionality in a subtropical forest than common species due to their functional diversity. *Forest Ecology and Management, 538*, 120981.

Triviño, M., Potterf, M., Tijerín, J., Ruiz-Benito, P., Burgas, D., Eyvindson, K., …& Duflot, R. (2023). Enhancing resilience of boreal forests through management under global change: A review. *Current Landscape Ecology Reports, 8*(3), 103–118.

Urbina-Cardona, N., Cardona, V. O., & Cuellar, S. (2023). Uncovering thematic biases in ecosystem services mapping: Knowledge shortfalls and challenges for use in conservation. *Biological Conservation, 283*, 110086.

Witzel, K., & Matros, A. (2023). Fructan accumulators in a changing world climate: Chances for new functional plants. In *The Book of Fructans* (pp. 311–331). Academic Press.

Wudu, K., Abegaz, A., Ayele, L., & Ybabe, M. (2023). The impacts of climate change on biodiversity loss and its remedial measures using nature based conservation approach: A global perspective. *Biodiversity and Conservation, 32*(12), 3681–3701.

Yan, P., Fernández-Martínez, M., Van Meerbeek, K., Yu, G., Migliavacca, M., & He, N. (2023). The essential role of biodiversity in the key axes of ecosystem function. *Global Change Biology, 29*(16), 4569–4585.

Yuniati, R., Handayani, W., Nugrahani, L., Khoerunnisa, I., Putrika, A., & Hemelda, N. M. (2024). The capacity of trees to remove particulate matter and lead and its impact on tree health. *Biodiversitas: Journal of Biological Diversity*, *25*(8): 2541–2551.

Zhang, S., Han, W., Liu, T., Feng, C., Jiang, Q., Zhang, B., …& Zhang, Y. (2024). Tetracycline inhibits the nitrogen fixation ability of soybean (*Glycine max* (L.) Merr.) nodules in black soil by altering the root and rhizosphere bacterial communities. *Science of The Total Environment*, *908*, 168047.

Zhang, Y., Jiang, X., Lei, Y., Wu, Q., Liu, Y., & Shi, X. (2023). Potentially suitable distribution areas of Populus euphratica and Tamarix chinensis by MaxEnt and random forest model in the lower reaches of the Heihe River, China. *Environmental Monitoring and Assessment*, *195*(12), 1519.

Zou, L., Stan, K., Cao, S., & Zhu, Z. (2023). Dynamic global vegetation models may not capture the dynamics of the leaf area index in the tropical rainforests: A data-model intercomparison. *Agricultural and Forest Meteorology*, *339*, 109562.

13 Conservation Efforts
Preserving Rare and Endangered Tree Functional Traits

Rowndel Khwairakpam, Rajneesh Bhardwaj, Charu Bisht, Bokado Kangujam, and Narendra Kumar

13.1 INTRODUCTION

Plant functions are inevitably important for the healthy forest biodiversity, which in turn sustain crucial ecological functions that benefit the human population. In recent years, it has been alarmingly disrupted due to a rapid decline in the tree diversity largely bought about by subsequent events of widespread climate instability spun out of all the contributing factors of climate change. Tree species diversity contributes to an array of ecological functions, which are pivotal for maintenance of ecology. Functional and structural composition is a critical determinant of ecosystem process has well been demonstrated by multiple studies. These determinants have been used to access health, restoration, and influences on both internal and external sources. Plant functional traits provide measurable indicators of ecosystem delivery and ecosystem changes, which allows categorical analysis and formulation of tailored management and recovery strategies to address the challenges posed by changes in climate and land use planning (Kumar et al., 2021d; Miedema Brown & Anand, 2022). Plant traits that affect how plants respond to their environment, other species, and tropic levels have become an affective and accurate way to study the consequences of large-scale land use and climate change, as well as the effective diverse disturbances. In fact, employing plant functional features to explain ecosystem process has been referred to as the holy grail of ecology since they provide both an analytical account and an accurate estimation of various ecosystem processes and function. These functional traits provide precise measurement of many ecosystems process a response. They scale up directly from plants, which are often considered the foundation of an ecosystem, to ecosystem-level functions. This is because changes in plant traits have a direct and often determinative effective effect on the ecosystem function that drive ecosystem process.

Functional traits are characteristics integrally linked to the environment and indicative of species adaptive strategies. These characteristics are typically closely linked to biogeochemical cycles, life histories, interspecific competition, and disease tolerance.

DOI: 10.1201/9781003545231-13

Trait-based approaches have proven to be effective instruments for forecasting the ramification of evolving global climate and implication of biodiversity reduction on ecological functioning (Kumar et al., 2020a,b; Volis, 2023). Functional diversity is a crucial determinant of ecosystem function, as species exhibiting diverse traits are more inclined to fulfil complementary roles. Rare and common species within a community may inhabit distinct niches and utilize distinct resources, resulting in their specific function within a community may inhabit distinct niches and utilize various resources, resulting in their specific contribution to ecosystem functionality (Kearsley et al., 2019). This effective resource utilization through niche differentiation enhances overall productivity. Plant functional traits can influence the ecosystem process by shaping community standards that dictate ecosystem functionality, or conversely, be influenced by these processes, affecting how individual plants or communities respond and filtering which traits are viable for survival (Kumar et al., 2021a,b; Li et al., 2021). The characteristics of the plant community, as the primary producers of an ecosystem processes or function, encompass the nutrient flow or activities within an ecosystem that operates on a broad scale and maintains the overall functionality of the ecosystem. Ecosystem processes encompass the interactions between plant communities and abiotic environmental factors, including soil root dynamics (Faucon et al., 2017). These processes may be complicated or straightforward, and while they transpire on a grand scale, they represent the attributes and characteristics of entire communities within the ecosystem. They exist autonomous from human presence or utility, unlike environmental process that yield direct advantage to human culture. The paraphrase "ecosystem service" especially denotes the benefit that humanity obtains from this natural process, highlighting their inherent focus on human well-being. Rare tree species and their role in ecological function have been subjected to enduring controversies and persistent scholarly debate over and again. Some of these debates pertain to unsurety if these functions are replaced by common tree species against loss of rare species and its function (Kumar et al., 2021c,d; Lyons et al., 2005).

Different studies have established insight into the dynamics of species abundance and scarcity within local communities. Rare species typically inhabit peripheral niches in both environmental and functional domains potentially diminishing competition and promotion traits associated with this niche. Rare species occupying peripheral are more likely to effectively include novel regions (Dee et al., 2019). Although rare species presses a narrower environmental niche width, which correlates within an increased extinction risk, they demonstrate comparable or even greater intraspecific trait variation than common species. This variation enables the utilization of diverse resources, thereby improving long-term population stability and biodiversity (Zhang et al., 2022). Thus, rare species may have a more substantial impact on ecosystem stability and functionality than previously recognized necessitating additional research to evaluate their contributions.

Rare species perform complementary support functions to the other even with the closely related species resulting from the distinctive functional trait unique to each species (Mi et al., 2021). Consequently, those rare species may enrich the function diversity of local communities, sustain ecological functioning, and impart functional traits that enable the primary ecosystem process support in unforeseen environmental conditions. Many rare species have been demonstrated to contribute to along with or as an alternative to dominant species (Kumar et al., 2022; Zhang et al., 2022). Likewise, rare species more likely possess functionally distinctive and characteristics, which could result into unique functional role in ecosystem process and services. Categorizing the contributions of rare species to ecosystem services is essential, as it is commonly presumed that such species have a minimal impact on ecosystems over short time spans and within localized areas. This presumption is grounded on the notion that due to its low abundance rare species exert a minimal effect on the community. A similar point of view may not be accurate as losing such function will lead to harsh impact to the ecosystem process at the local level. Rare species can play a significant role in long-term and large-scale functioning of ecosystems.

Only a few rare species, and more notably a disproportionately high ratio of uncommon species with large-sized individuals, exhibit significant functional specialization and little functional redundancy in features relevant to resource acquisition. All numerous species and most non-abundant/rare species exhibited functional redundancy. Owing of the prevalence, abundant species significantly influence primary ecological services. The majority of rare species has been found to exhibit significant redundancy in their functional feature combinations associated with resource acquisition, suggesting that they may contribute minimally to ecosystem functioning or stability (Kumar et al., 2015; Leitão et al., 2016). Within the framework of biodiversity's role in augmenting productivity and standing biomass, as well as in biodiversity conservation, productivity gains may reach saturation with more diversity due to functional redundancy.

13.2 THE CONSERVATION STRATEGIES OF RET TREE SPECIES BASED ON FUNCTIONAL TRAITS

The essential function of plants as integral elements of Earth's ecosystems and their necessity for existence undergo the urgency of confronting the serious threat to global plant biodiversity. The rising prevalence of endangered plant species, primarily caused by habitat destruction, overexploitation, pollution, and climate change, underscores the severity of the issue. The biodiversity issue is intensified by the decline of genetic variety and the susceptibility of many endemic species, many of which are restricted to isolated natural populations. Such species frequency encounters an increased danger of extinction owing to their distribution and ecological specialization.

The International Union for Conservation of Nature Red List of threatened species is an essential resource for comprehending the global conservation status of plant species. The latest data indicates that 155,774 plant species are threatened among the 38,630 evaluated, highlighting the fragile condition of biodiversity protection. Alarmingly, more than 75% of plant species are endemic to the 35 Global Biodiversity Hot Spot, which are areas of extraordinary ecological significance in maintaining global biodiversity and ecological services (Kumar et al., 2016; Velmurugan et al., 2018).

The convergence of biodiversity reduction statistics and ecological functions of trees presents a strong rationale statistics and ecological functions of trees present a strong rationale for prioritizing the protection of rare, vulnerable, and endangered tree species (Miedema Brown & Anand, 2022). In addition to their evident contribution to biodiversity, tree species are essential for the sustainability of ecological functionality. Their significance is intricately linked to functional features that delineate their ecological identities and roles within an ecosystem. The characteristics, collectively termed plant functional identity, support processes including nutrient cycling, disturbance regulation, and biodiversity preservation, hence guaranteeing the delivery of vital ecological services.

13.3 FUNCTIONAL TRAIT AND ECOSYSTEM VIABILITY

Functional features affect ecosystem dynamics and services by shaping functional diversity, which includes

i. **Functional Evenness**: Guaranteeing equitable ecological contributions.
ii. **Functional Richness**: Encompassing the variety of roles within a system.
iii. **Functional Divergence**: Signifying difference that improves ecological stability.
iv. **Functional Dispersion**: Encompassing trait variety among plant population.

13.4 ESSENTIAL FUNCTIONAL CHARACTERISTICS OF TREES THAT AFFECT THEIR ECOLOGICAL IMPORTANCE

i. **Aboveground Traits**: Plant height aboveground biomass, basal area, and wood density are essential for sustaining canopy structures, facilitating carbon sequestration and providing habitat. Reproduction characteristics like floral dimension, nectar accessibility, pollen tube length, seed weight, and leaf/stem biomass ratios promote biodiversity by facilitating pollination and seed dissemination. Adaptive metrics, including bark thickness, shot mass, tensile strength, and xylem vulnerability, directly affect resilience to fire, pest, and drought.
ii. **Foliar Characteristics**: Attributes such as leaf mass per area, specific leaf area, leaf thickness, stomatal density, and chlorophyll content govern photosynthesis, transpiration, and growth sufficiency. Nutrient-related measurements, including leaf nitrogen content, leaf carbon-to-nitrogen ratio, leaf phosphorus content, and leaf potassium content, facilitate nutrient cycling. Mechanical qualities such as leaf tensile strength, leaf toughness, and leaf turgor loss contribute to stress resilience.
iii. **Characteristics of Root System**: Root-associated characteristics such as depth, thickness, density, biomass, and specific root length affect water absorption, soil stabilization, and nutrient accessibility. Functional elements such as root carbon content, root tensile strength, and root aerenchyma are crucial in governing belowground ecosystem processes.

13.5 FUNCTIONAL ROLE OF TREES IN ECOSYSTEM SERVICES

The functional characteristics of trees facilitate the provision of ecological services, including

i. **Nutrient Cycling**: Trees promote nutrient turnover through characteristics such as wood phosphorus concentration, leaf nitrogen content, and root chemistry.
ii. **Disturbance Mitigation**: Characteristics such as bark thickness and root tensile strength enhance ecosystem stability against natural disasters.
iii. **Climate Regulation**: Canopy density and wood biomass facilitate carbon sequestration and temperature regulation.
iv. **Habitat Provision**: The integration of canopy attributes and floral qualities fosters species diversity.
v. **Soil Fertility and Erosion Mitigation**: Root depth, density, and aerenchyma inhibit soil degradation while augmenting fertility.

13.6 PRESERVATION OF GLOBAL DIVERSITY HOT SPOTS

Considering more than 75% of plant species are isolated as Global Biodiversity Hotspots, it is imperative to protect their distinctive functioning features. Conservation efforts aimed at rare and endangered species must include their functional roles to maintain ecosystem multi-functionality. These initiatives support ecological functions such as gas regulation, food supply, pollination, and resilience to disturbances, which are essential for global ecological and human welfare. The emphasis on functional features connects species conservation with ecological functionality (Kumar et al., 2017, 2019; Velmurugan et al., 2018). By prioritizing rare, threatened, and endangered tree species according to their functional contribution, conservation efforts can concurrently alleviate biodiversity loss, preserve ecosystem stability, and ensure the stability of life-supporting ecological processes.

13.6.1 In Situ Conservation

In situ conservation, which involves the preservation of ecosystem and biodiversity within their natural habitats, is the most suitable method for species conservation, especially endemic species, as it maintains the original genetic and geographical centres of biodiversity. For a comprehensive and effective conservation program, many tactics and methodologies must be employed to enhance and support in situ protection. Ex situ conservation, which involves the preservation of biodiversity beyond its natural environments, is occasionally the sole option for safeguarding the rare and endemic species.

13.6.2 Quasi In Situ Cultivation

Quasi in situ conservation serves as an intermediary strategy between ex situ and in situ conservation it entails the use of ex situ collection within natural or semi-natural environment as a fundamental element of a comprehensive ex situ–in situ conservation strategy (Volis, 2023). This method prioritizes the conservation of plant species by highlighting local adaptation and protecting both neutral and adaptive genetic variation. This strategy is particularly relevant for plant species facing rapid habitat destruction, especially those characterized by a limited number of existing population and lacking protection within nature reserves or national parks. The extinction risks for these RET species are so elevated that preserving genetic diversity and ensuring a minimum viable population size requires relocation. Quasi in situ conservation strategy applies to various plant categories, particularly rare and critically endangered species facing habitat destruction.

Key aspects of its applicability in rare species where reproductive success becomes limited and population size is often reduced due to self-incompatibility (SI) and declining population size, genetic diversity. Due to SI allele becoming scarce or lost, the likelihood of locating suitable partners diminishes, exacerbating reproductive failure. Quasi in situ conservation contributes to the preservation of genetic variety and helps mitigate demographic challenges such as the lack of mates for self-incompatible species. This strategy is also essential for species with limited populations and no protection in reserves, enabling relocation to maintain minimum viable populations and genetic diversity. Living collections facilitate propagule production, site optimization, and population creation.

13.7 THE STRATEGY COMPRISES FIVE KEY STEPS

13.7.1 Survey and Analysis

The starting point of the suggested technique involves species distribution to pinpoint prospects of selectively adapted species or population clusters. This paradigm is essential for comprehending the magnitude and spatial distribution of local adaptation. There is a difference in the intensity of local selection, whether progressive with increasing geographic distance or sudden due to locally adapted populations or population clusters. This approach is essential for understanding the degree of local adaptation and its spatial distribution. The variation in the intensity of local selection, whether it occurs gradually with increasing distance or suddenly due to habitat shift, needs a comprehensive analysis of the local selection regime to evaluate the adaptation of the populations in question (Etterson, 2004). Two methodologies are employed to ascertain local adaptation: Transplantation trials and assessment for outbreeding depression, where in locally adapted genotypes, are hybridized with specimens from different origins. Under local environmental conditions, progeny resulting from crossbreeding with non-adapted genotypes are likely to exhibit reduced fitness due to the disruption of co-adapted gene complexes or differing adaptations to specific environmental variables. When local adaptation plays a critical role, it is essential that the introduced genotypes align closely with the prevailing abiotic and biotic conditions. To achieve this, the genotypes should originate either from regions subject to strong local selective pressures or from habitats with closely analogous environmental conditions that impose similar selection regimes. This ensures compatibility with local adaptive traits and enhances the likelihood of successful integration and survival.

13.7.2 Representative Sampling

When considering collections, multi-species options are favoured in scenarios where multiple coexisting species within an ecosystem face endangerment, assuming adequate space is available. There is a notable exception that encompasses species indicative of various stages in vegetation succession.

13.7.3 Collection Maintenance

The subsequent phase entails the creation of live gene banks, which include the nurturing of sampled germplasm to generate alive collections. In the quasi in situ approach, site selection should emphasize local adaptation, whether confirmed or presumed. This requires a close environmental alignment between the ex situ location and the sites of the sampled natural populations. The environmental compatibility is biologically significant, especially if the ex situ location is inside a natural or at least a semi-natural setting. The chosen site must be protected by national legislation and undergo frequent monitoring to ensure effective upkeep and population viability. An optimal design entails the creation of several populations that accurately represent a particular ecogeographical region or habitat, farmed within a safeguarded area that reflects the same ecogeographical attributes. A significant quantity of genetically diverse individuals from each population should be planted, spaced at intervals that enable the identification of individual genotypes when necessary. This method is essential for assessing survival rates and, if required, executing regulated pollination. The careful organization and upkeep of these live collections are essential for preserving genetic diversity and guaranteeing the enduring effectiveness of quasi in situ conservation efforts (Pavlik 1996; White &Walker 1997; Menges 2008).

13.7.4 Reintroduction and Monitoring

Reintroduction and relocation have been demonstrated to be problematic and frequently uncertain endeavours. The elevated risk of failure in reintroduction depends on the intricate dynamics of biotic and abiotic interactions encountered by introduced plants, which frequently fluctuate over minimal geographic scales, and catastrophic occurrences that typically cluster spatially (Coelho et al., 2020). Releasing propagules among various locations is beneficial compared to introducing them at a single site, as it mitigates the risk of random loss of an individual colony. Conversely, establishing a network of introduced populations necessitates evaluating the success of relocation within a metapopulation framework. The effects of the quantity and spatial factors of a new population, such as size and proximity to the nearest population, are significant in this context. Like selecting a relocation site, testing ought to be a standard process. The final phase in effective conservation is the assessment of relocation success. Consequently, various populations of differing sizes should be established and observed over several years. A relocation is deemed successful if the transplanted population following extensive monitoring exhibits demographic competence comparable to that of its wild counterpart, namely, the reference natural population.

13.7.5 Biotechnological Approaches

Biotechnology functions as an essential instrument in plant conservation, augmenting rather than supplanting conventional techniques. Plant conservation biotechnology encompasses the management, characterization, and sustainable utilization of plant genetic resources to enhance conservation initiatives. In vitro conservation refers to the preservation of plant germplasm within culture collections using tissue culture techniques. This approach enables convenient access for the evaluation, application, and safe exchange of plant material. For short-term preservation, standard tissue culture methods are employed, while long-term storage involves cryopreservation of plant material in liquid nitrogen (LN) (Walters & Pence, 2021). Several reports address the application of biotechnological strategies for the conservation of rare and endangered species, either providing a summary of the topics or concentrating on specific regions/countries or plant groupings. This chapter explores sophisticated conservation methods for species of plants that are characterized by small populations and limited geographical ranges. It emphasizes notable progress in biotechnology approaches and their incorporation with traditional ex situ conservation procedures, particularly with in vitro culture techniques. Focus is placed on different strategies, acknowledged as the approaches capable of guaranteeing the long-term preservation of genetic resources. Notwithstanding its potential, extensive research on the use of different strategies for the conservation of narrow endemic plant species hereafter termed endemic plant species remains scarce in the existing literature. This chapter seeks to fill this gap by analysing the present status of research and investigating prospects in the field.

13.8 SHORT- (IN VITRO PROPAGATION) AND MEDIUM-TERM PRESERVATION (SLOW GROWTH)

13.8.1 In Vitro Propagation

In vitro techniques enhance the large-scale production of desirable plant species more rapidly and economically than conventional methods (Coelho et al., 2020). Nevertheless, the regenerants may exhibit somaclonal variability. Somaclonal differences in regenerants are affecting the tissue culture procedure which needs genetic consistency among regenerants. Conversely, somaclonal variants are sought in plant breeding efforts to cultivate plants with enhanced agronomic features (Chokheli et al., 2020). Highlights of recent advancements in plant conservation biotechnology over the past decade, building on previously assessed progress in the field, are focused here. It explores in vitro techniques, somaclonal variation, the evaluation of genetic uniformity using molecular markers, and potential future developments. A list of endangered tree species with the published methodologies is used in the conservation (Thakur et al., 2021) (Table 13.1).

Tissue culture is an effective method for conserving endangered plant species, facilitating the propagation of species with restricted explant sources, such as those exhibiting seed dormancy, self-incompatibility, and inbreeding depression, which leads to diminished seed production. Therefore, several in vitro techniques utilize to cultivate plants facing the aforementioned hurdles for survival in their natural environment are reviewed here.

TABLE 13.1
Methodologies for Tissue Culture Aimed at Conservation of Endangered Tree Species

S. No.	Species Name	Methodology	Threat Status
1	*Buxus sempervirens*	In vitro conservation and cryopreservation	Endangered
2	*Argania spinosa*	Micrografting	Endangered
3	*Hildegardia populifolia*	Micropropagation	Critically endangered
4	*Zanthoxylum armatum*	Micropropagation and indirect organogenesis	Endangered
5	*Cedrela fissilis*	Organogenesis	Endangered
6	*Fraxinus nigra*	Organogenesis	Endangered
7	*Quercus ilex*	Somatic embryogenesis	Endangered
8	*Shorea tumbuggaia*	In vitro germination	Endangered
9	*Moringa peregrina*	Callus induction and micropropagation	Endangered
10	*Torreya taxifolia*	Somatic embryogenesis and cryopreservation	Endangered
11	*Gomortega keule*	Micropropagation	Endangered
12	*Fraxinus profunda*	Organogenesis	Endangered
13	*Terminalia bellerica*	Micropropagation	Endangered
14	*Arbutus andrachne*	Micropropagation	Endangered

13.8.2 Micropropagation

The generation of plantlets by tissue culture techniques utilizes meristematic plant tissues, such as shoot buds or apical tissues derived from mature herbs, shrubs, trees, or seedlings. The micropropagation process comprises four distinct stages: I, II, III, and IV. Some writers have introduced a stage "o" that necessitates the pre-preparation of donor material cultivated under in situ conditions, involving treatments such as fungicides plant growth regulators (PGRs), etiolation, and pruning.

Stage I: Initiation of the explants—This phase encompasses the selection of explants and their surface sterilization under aseptic conditions.
Stage II: Shoot Multiplication—Standardization of culture conditions enriched with PGR (mostly cytokinins, occasionally auxin in combination with cytokinin).
Stage III: Induction of roots—Shoots are cultured in a culture medium enriched with plant growth regulator (PGR) in both in vitro and ex vitro conditions.
Stage IV: Acclimatization of plantlets—Shoot with roots are later relocated to ex vitro conditions.

Micropropagation offers a crucial strategy from the conservation, propagation, and restoration of endangered plant species, particularly woody species, where traditional propagation techniques fail due to insufficient seed production. Micropropagation is a prevalent technique in biotechnology that entails the rapid multiplication and generation of clones under controlled axenic conditions including suitable culture media, specific PGRs, and optimal environmental factors such as temperature and photoperiod. Recently, the application of micropropagation has extended beyond multiplication and clonal propagation to encompass germplasm conservation, the cultivation of virus-free plants, and the attainment of somatic embryogenesis as well as the production of synthetic seed, genetically modified organisms, pharmaceuticals, and other natural products utilizing cell suspensions and hairy roots. Over the previous decade (2010–2020), numerous research articles have been published utilizing micropropagation techniques to protect vulnerable plant species for large-scale production (Chokheli et al., 2020). Researchers have utilized various plant components as explants, including nodal segment, root segment, axillar buds, seed, hypocotyl, shoot tips, cotyledonary nodes, leaves, zygotic embryo, seedlings, axenically, derived rhizomes, rhizome buds, sprouted rhizomes suckers, petiole, apical segment, and flower receptacle, for effective conservation.

13.8.3 Somatic Embryogenesis

The procedure comprises the transformation of somatic cells into embryoids (somatic embryos) that mimic zygotic embryos. These embryoids display a bipolar architecture, evolving into shoot and root axes, along with separate embryonic structures including cotyledons, the hypocotyl, and the radicle (Thakur et al., 2022). Somatic embryogenesis generally advances through five phases for effective plant regeneration.

Initiation of Embryo Culture: Embryo culture is commenced by positioning explants onto a tissue culture medium supplemented with auxins and cytokinins.

Multiplication of Embryonic Culture: This stage occurs in liquid or semisolid media supplemented with PGRs as outlined in the initial stage.

Prematuration of Somatic Embryo: Cultures are transitioned to a medium devoid of PGRs, facilitating the production and maturation of somatic embryos while inhibiting additional proliferation.

Somatic Embryo Maturation: Maturation of somatic embryos transpires on a medium enriched with abscisic acid and/or modified osmotic potential to facilitate the complete development of the somatic embryos.

Plant Development: Ultimately, cultures are transferred to a medium free of PGRs to facilitate natural plant development.

Embryoids are produced from somatic tissues, usually cultivated in a medium enriched with the auxin 2,4-D (2-dichlorophenoxyacetic acid). Supplementary PGRs, like abscisic acid and cytokinin, are essential in facilitating somatic embryogenesis. The germination, maturation, and development of somatic embryos into plantlets are contingent upon the plant species and the type of explant utilized. In somatic embryogenesis, cells experience epigenetic and biochemical restructuring, dedifferentiating, and re-entering the cell cycle to generate embryoids, akin to the mechanisms of zygotic embryo development. The potential of somatic embryos is significant, prompting an ongoing study to elucidate the molecular mechanisms and metabolic pathways involved in their development. Somatic embryos or embryonic cultures are essential resources for genetic manipulation and can be cryopreserved by various procedures, facilitating the creation of gene banks for future applications. Thus, somatic embryogenesis has emerged as a fundamental method for the long-term conservation of endangered species. A diverse array of plant tissues has been recognized as appropriate explants for somatic embryogenesis, encompassing young leaves, cotyledonary leaves, flower buds, roots, shoot apices, seeds, in vitro-derived hypocotyl segments, protocorm-like entities, corms, petioles, bulbs, and internal segments.

13.8.4 Micropropagation and Bioreactor Advances

Micropropagation is an essential method for the mass propagation of crops, medicinal plants, ornamental species, and trees. It guarantees the swift production of disease-free plants, the propagation of rare species, and the synthesis of phytochemicals. The procedure entails aseptic cultivation of explants, such as meristems, on nutrient media within regulated environments, allowing cells to exhibit totipotency and follow developmental pathways to generate organs or somatic embryos. Micropropagation consists of specific stages: initiation, shoot multiplication, root induction, and acclimatization (Murthy et al., 2023). During organogenesis, explants either directly form shoots or undergo intermediate callus formation, whereas somatic embryogenesis advances through induction, proliferation, maturation, and the development of plantlets. Traditional methods, although efficient, are resource-intensive and expensive due to the necessity for frequent subculturing and solid media utilization. Liquid media and bioreactor systems provide scalable solutions for the efficient production of substantial quantities of plantlets. These systems facilitate management, improve nutrient absorption, and accelerate growth, aided by forced aeration. Nonetheless, liquid cultures present challenges such as asphyxia and physiological disorders, necessitating bioreactor designs specifically adapted for plant propagation. Innovations such as temporary immersion, wave, and bubble bioreactors have resolved these challenges, catering to the distinct requirements of shoots, somatic embryos, and organogenic propagules. Effective circulation, mixing, and aeration are essential for optimal in vitro development. Contemporary bioreactor systems are evolving to accommodate various species' needs, decreasing expenses, enhancing yields, and facilitating the automation of plant propagation procedures (Murthy et al., 2003). These advancements offer potential for enhanced efficiency and sustainability in micropropagation practices.

13.9 CRYOPRESERVATION

The conservation of forest genetic resources is essential for meeting future demands for wood products and services. Genotypes displaying improved traits are presently produced through traditional breeding programs, selection of mature superior trees, and genetic transformation methods, all of which are essential for increasing the productivity of forestry clonal plantations. Current methodologies are well-defined, with cryopreservation serving as a supplementary storage technique, crucial for traits in vegetatively propagated species (Kaviani & Kulus, 2022). Moreover, many hardwood forest trees produce recalcitrant seeds that exhibit restricted viability and vigour under conventional storage conditions, thereby requiring methods such as cryopreservation for long-term preservation. This context will highlight the availability of straightforward, reliable, and cost-effective methods for the conservation of RET hardwood trees in forests with recalcitrant seeds. The quality of germplasm and physiological traits is crucial for successful cryopreservation, and these factors are anticipated to be affected by climate change in diverse ways and extents, contingent upon species and geographic location. Recalcitrant species cultivated in tropical and subtropical environments may encounter substantial developmental challenges due to elevated temperatures and water scarcity. This issue may also affect temperatures of RS species in warm climates that require a specific level of cooling, while potentially promoting their growth in northern regions where current temperatures are inadequate for optimal development. Literature suggests that specific constraints on field collection, particularly nursery methods that impose maladaptive selective pressures on critical traits for coping with abiotic stress, may be intensified by climate change (Streczynski et al., 2019). Nevertheless, we can also utilize these selective factors to our benefit. Advanced nursery and horticultural methods are being devised to improve favourable quality traits associated with plant responses to diverse environmental stresses, facilitating the cultivation of these species in their existing habitats without necessitating relocation or field collection to higher latitudes or altitudes. Emphasis is on critical attributes for addressing abiotic stress. Climate change may not affect the genotypes stored in vitro for prolonged durations, but it may impact newly incorporated genotypes in these collections. However, these outcomes are not universally applicable, as they depend on the species (with tropical and temperate species often responding differently to climate change), genotype, the adaptive genetic potential of each population, the severity of the environmental change, and the ex situ conservation strategy and/or explant utilized.

Emphasis is given towards the following subjects:

i. Medium-term conservation via slow growth storage.
ii. Cryopreservation methodologies.
iii. Selection of explants for cryopreservation, including in vivo collection of embryonic axes and dormant buds, as well as in vitro collection of shoot tips and embryonic cultures.
iv. Genetic stability of the cryopreserved specimens.

The utilization of cryopreservation methods for the conservation of in vitro cultures of arboreal species is a novel advancement, with the first documentation emerging in the early 1990s. Since then, somatic embryos from various woody species have been successfully cryopreserved. Cryopreservation is particularly advantageous for the management of embryonic cultures. Following the establishment of an embryogenic line, regular subculturing is crucial to maintain embryonic viability. These tasks are frequently burdensome due to the labour-intensive nature and inherent risks of losing lines from contamination, technical failures, reduced embryonic viability, or the emergence of soma clones (Bradai et al., 2023).

13.9.1 Cryopreservation Methodologies

13.9.1.1 Conventional Two-Stage Cooling

Cryogenic storage employs ultra-fast, rapid, and slow-cooling techniques. Slow cooling, also known as two-step or controlled rate freezing, is the most common method and employs programmable freezing apparatus to establish precise and uniform thermal conditions. Gradual cooling induces the formation of ice in the extracellular solution, resulting in the efflux of water from the cytoplasm and vacuoles to the external compartments, where it subsequently freezes (Engelmann, 2011). This freeze-induced dehydration attempts to convert cells into amorphous semi-solids, wherein LN vitrifies any remaining intracellular water. The traditional technique entails pre-treating biological specimens with cryoprotective agents, such as dimethyl sulfoxide, and cooling at a controlled rate (0.1–2.0°C/min) to achieve a pre-freezing temperature of approximately −40°C. Following rapid immersion in LN, samples typically cool at a rate of 1,000°C/min. This method has effectively preserved callus and cell suspensions from over 700 cell lines across nearly 600 plant species. Only differentiated explants can cryopreserve shoot tips from species that tolerate low temperatures. An exception is the effective cryopreservation of cassava shoot tips. The high cost of two-step freezing has rendered it less popular as more affordable alternatives emerge.

13.9.1.2 Vitrification-Based Methods

The vitrification technique has been applied to over 100 plant species and is currently the most widely used protocol for plant cryopreservation, particularly for shoot tips from various tropical and subtropical tree species. Often termed "complete vitrification," this method induces vitrification in both intra- and extracellular solutions. The process involves several key steps: treating the samples (loading) with a diluted cryoprotectant mixture (loading solution (LS)) to enhance the explant's tolerance to concentrated and potentially toxic chemicals, dehydrating the samples with a highly concentrated and effective plant vitrification solution (PVS), rapid cooling and rewarming, unloading (removing) the cryoprotectants, and subsequent recovery. Selecting an appropriate LS and determining its exposure time are critical, as these factors significantly influence the tissues' dehydration tolerance and the success of the vitrification process.

13.9.1.3 Contemporary Methods

Vitrification, which turns cellular water into glass, prevents ice formation at liquid nitrogen (LN) temperatures by increasing cellular viscosity and minimizing ice production. This method involves rapid cooling after partial dehydration using concentrated cryoprotectant solutions or desiccation. PVS turns water inside plant cells into "biological glass," effectively preserving sensitive tissues like shoot tips and somatic embryos. Modern cryogenic methods—such as vitrification, droplet-vitrification, encapsulation-dehydration, and cryo-plates—combine standard techniques like preculture, dehydration, and cryoprotectant exposure with rapid immersion in LN, ensuring efficient preservation. These techniques have advanced to suit various plant materials, offering faster cooling and economic efficiency over traditional slow-cooling methods. Encapsulation–vitrification, a hybrid technique, involves embedding explants in an alginate matrix followed by treatments with LS and PVS. While the encapsulation process requires longer PVS incubation, it provides gentler, safer dehydration. Droplet-vitrification combines droplet freezing and vitrification, using aluminium foil to quickly cool samples immersed in PVS, minimizing injury and ensuring high regrowth post-storage. This method is especially effective for preserving tropical and subtropical ornamental plants. Newer methods like V and D cryo-plates further enhance plant cryopreservation by immobilizing explants and increasing cooling rates. V cryo-plates use PVS2 for dehydration and encapsulation-droplet-vitrification, while D cryo-plates utilize air dehydration and alginate encapsulation. These methods streamline handling, reduce chemical stress, and increase regrowth rates. Cryo-mesh techniques, like cryo-plates, use stainless-steel mesh for flexibility and diverse applications. Vacuum infiltration vitrification (VIV) enhances cryoprotectant penetration by applying a vacuum during the loading process (Nadarajan & Pritchard, 2014). This ensures even distribution of cryoprotectants, reduces dehydration damage, and provides more uniform cooling and rewarming. VIV is particularly beneficial for large or delicate plant tissues, improving preservation of a wide range of species and aiding in the long-term conservation of genetic resources.

13.9.1.4 Advances in Cryopreservation of Forest Tree Seeds

Cryopreservation has been effectively utilized for the seeds of various forest tree species, employing diverse techniques according to seed types. Orthodox seeds, capable of enduring desiccation, typically necessitate minimal pretreatment for cryopreservation. Seeds exhibiting diminished moisture levels (e.g., 6%–8%) demonstrate enhanced germination rates, typically accomplished via air-drying. The predominant method utilized is rapid thawing at 37°C–41°C, although slow thawing at ambient temperature has been implemented with inconsistent results. Germination rates for cryopreserved orthodox seeds have demonstrated results like those stored via conventional conservation techniques. Intermediate seeds necessitate meticulous monitoring of maturity and moisture content during collection. Mature seeds generally exhibit superior viability and germination rates compared to immature seeds. Attaining ideal moisture content, typically 7%–20%, via swift desiccation mitigates desiccation injury. Viability is often associated with seed maturity and dry weight, as fully matured seeds typically produce superior results. The cryopreservation of recalcitrant seeds poses a challenge because of their susceptibility to desiccation and freezing. Whole seeds frequently do not endure cryopreservation; conversely, excised zygotic embryos, embryonic axes, and other embryogenic tissues exhibit superior recovery rates. Moisture content levels of approximately 20% in zygotic embryos are optimal for cryostorage (Li et al., 2018). Recovery rates reaching 98% have been documented for certain species, underscoring the feasibility of utilizing embryonic tissues as a substitute for the preservation of recalcitrant seeds.

13.9.1.5 Growth Recovery and Post-Cryopreservation Strategies

The recovery process of cryopreserved materials involves two stages: the initial survival phase and the regrowth phase. During the initial recovery, cells combat damage from cryopreservation, while in the regrowth phase, surviving cells proliferate, forming new tissues such as roots, shoots, or embryos, depending on the material. Different strategies are needed for each phase. Adjusting the medium osmotic potential gradually and using solid media helps mitigate the osmotic shock. Antioxidants reduce oxidative stress, while specific additives absorb toxins from damaged tissues. Sensitive materials benefit from iron-free or ammonium-free media during early recovery, although these ions become crucial for regrowth. PGRs are essential for promoting normal tissue development. Combinations of BA and GA3 with low auxin concentrations are effective for shoot tips, while undifferentiated cultures respond to 2,4-D or NAA. Some tissues, such as somatic embryos or orchid protocorms, recover well without PGRs. Balanced auxin–cytokinin mixtures support embryogenic cultures and adventitious buds. Darkness during initial recovery is often beneficial, though species-specific adjustments and timely exposure to light are critical for proper shoot formation. Red and blue light combinations have shown promise but require further research (Popova et al., 2023).

For stress-sensitive materials, a stepwise recovery approach is recommended. Early recovery involves solid ammonium-free or antioxidant-enriched media under darkness for 5–7 days, followed by full-strength PGR-enriched media with gradual exposure to dim light. Normal light conditions can be introduced after a week, with further transfers to PGR-free media after 2–3 weeks to support complete plant development. Tailoring recovery protocols to specific material needs remains essential. Improved understanding of these processes will aid in developing effective recovery procedures for diverse cryobank materials.

Successful growth recovery is crucial for the long-term viability of seeds following their exposure to ultra-low temperatures. Post-freezing viability and germination are influenced by various factors, including seed genotype, age, cryoprotectants used, and freezing-thawing procedures. Common methods to assess germplasm viability include triphenyltetrazolium chloride and fluorescein diacetate assays, which are often followed by regrowth and maintenance protocols tailored to the germplasm. For post-cryopreservation seed recovery, several approaches have been employed (Gantait et al., 2016). Seeds stored in LN are often thawed quickly by immersion in a 40°C water bath. Dormancy-breaking techniques, such as soaking legume tree seeds in boiling water and cooling overnight, have shown success. These seeds were planted in soil mixes containing composted hardwood sawdust, hardwood fines, and quartz sand, topped with a thin gravel layer to encourage germination.

13.9.2 Ex Situ Conservation

13.9.2.1 Seed Bank

Seedbank remains a potent and effective instrument. Seeds capable of withstanding extreme desiccation and low temperature are expected to endure for a minimum of 100 years under standard circumstances in a typical freezer. The remarkable resilience of seeds across numerous taxa has resulted in the establishment of over 1,750 seed banks across the globe, covering more than 50,000 species. Not all plants generate seeds, nor do all seeds endure freezing conditions, predictive models enable an insight into the number of taxa requiring different techniques, and an initial list of these species is accessible. These "exceptional" species necessitate cryobiotechnologies, which provide efficient, long-term ex situ conservation for a diverse array of tissues beyond seeds. The utilization of cryobiotechnologies enhances the capacity for preserving plant biodiversity necessitating institutional relationships across living collection, seed banks, and cryobanks to facilitate technology transfer, knowledge acquisition and dissemination, and capacity development in biodiversity centres (Breman et al., 2021).

Botanical garden seed banks, primarily located in Europe and North America, have seen significant growth, doubling in number over the past 20 years. Spain pioneered wild flora conservation with its first seed bank in 1966, leading to the formation of the Spanish Network of Genebanks for Wild Plants in 2002. Today, botanical gardens host some of the world's largest and most advanced seed banks, such as China's Germplasm Bank of Wild Species (over 11,000 species) and Australia's PlantBank (over 12,000 seed accessions, including 437 threatened species). Seed bank records in PlantSearch show 67,270 taxa across 100,218 accessions in over 80 institutions, though less than 10% of these taxa are globally threatened. Seed banks are essential for the conservation of plant genetic resources, with 710–1,750 genebanks preserving 5.4–7.4 million accessions of seeds, in vitro cultures, DNA, and cryopreserved materials. Prominent instances comprise Kew's Millennium Seed Bank (MSB) in the UK, the preeminent repository of wild plant diversity with approximately 997,000 accessions, and the Global Seed Vault in Svalbard, Norway, which contains over 1.1 million food crop seed accessions (BGCI, 2021). This review examines seed banks that preserve wild plant populations, emphasizing the MSB and its global collaboration (MSBP), as well as other entities within the BGCI network. More than 350 wild plant seed banks and 710–1,750 genebanks globally preserve 5.4–7.4 million accessions of plant genetic resources via seed, in vitro, DNA, and cryopreservation methods. Prominent instances comprise Kew's MSB in the UK, the preeminent repository of wild plant diversity, and the Svalbard Global Seed Vault in Norway, which contains over 1.1 million food crop seed accessions.

13.9.2.2 The Millennium Seed Bank (MSB)

The MSB preserves wild plant diversity by storing seeds from ~40,000 species in just 300 m^2, offering a cost-effective and space-efficient ex situ conservation method. Since the 1970s, Kew has developed seed banking programs, expanding globally through the Millennium Seed Bank Partnership (MSBP) since 2000. Partners enter access and benefit-sharing agreements and receive support for seed banking, training, and research. MSB collaborates with botanical gardens (27%), forestry and agricultural genebanks (20%), universities (32%), and governmental and other institutions (21%). Facilities are built for 500-year longevity, with seeds stored at −20°C for viability lasting decades to centuries. Unlike short-term storage for restoration (<5 years) or breeding (5–10 years), MSB follows protocols tailored for diverse wild seed types. The MSBP Seed Conservation Standards ensure high-quality collections, assessing seed viability, longevity, and genetic diversity. Seeds are primarily stored in their country of origin, with up to half sent to MSB for duplicate storage (ISTA, 2021). Where national facilities are lacking, entire collections may be stored at MSB until local storage is established. Conservation seed accessions include seeds, a herbarium voucher for identification, and detailed collection data, ensuring proper documentation and long-term conservation.

13.9.2.3 Botanical Garden Database

Botanical gardens have undergone considerable evolution since their inception in 16th-century Italy as physic gardens dedicated to medicinal flora. During the 17th and 18th centuries, gardens such as Kew and Madrid's Real Jardín Botánico transitioned to the cultivation of exotic species derived from tropical explorations, promoting global plant commerce and facilitating the establishment of new gardens in tropical areas. The 19th and 20th centuries witnessed the emergence of municipal pleasure gardens, whereas university gardens prioritized taxonomy and the study of plant diversity. Over the past 50 years, botanical gardens have proliferated significantly, with 60% established since the mid-20th century, especially in China and Indonesia, where they facilitate economic development and conservation efforts. Currently, botanical gardens serve as essential scientific institutions, containing curated plant collections for research, conservation, education, and public engagement. They are instrumental in addressing plant blindness and preserving endangered species, including "exceptional species" that cannot be preserved in seed banks and necessitate genetically diverse living collections.

BGCI assists botanical gardens via databases such as GardenSearch, PlantSearch, and ThreatSearch, which monitor gardens, plant collections, and conservation statuses. Botanical gardens collectively contain over 100,000 plant species, accounting for 30% of global plant diversity and 41% of endangered species. They employ more than 60,000 specialists and attract 750 million visitors annually, rendering them essential for public education and conservation initiatives. Examples encompass collections of non-timber trees such as breadfruit and tropical fruits, which enhance agricultural and forestry conservation efforts. Integrating plant collections from various botanical gardens into metacollections augments their conservation significance by broadening taxonomic representation, diversifying accessibility, minimizing redundancy, and decreasing maintenance expenses. This cooperative method, influenced by zoo species management initiatives, is especially efficacious for taxa with limited individual representation per garden, such as trees. Examples encompass the American Public Gardens Association's Multisite Collections, BGCI's Global Conservation Consortia, and the Center for Plant Conservation's National Collection (BGCI, 2021). Furthermore, BGCI advocates for seed banking as an adjunct to living collections, endorsing initiatives such as the Global Seed Conservation Challenge, training programs, and small grants. The creation of an accession-level module in the PlantSearch database enhances economical and coordinated conservation of endangered species in botanical gardens.

13.10 CONCLUSION

This chapter considers the significance of plant functional traits in sustaining forest biodiversity and ecosystem functions, which are essential for human well-being. Climate change has swiftly diminished tree species diversity, jeopardizing ecological processes. Functional traits, including biogeochemical cycles, life histories, and interspecific competition, can evaluate ecosystem health and forecast reactions to climate change and land use. Functional diversity, characterized by species occupying distinct niches, is crucial for ecosystem stability, with rare species enhancing this stability through their unique attributes. The research highlights the significance of preserving rare and endangered tree species, particularly those possessing functional traits that facilitate ecosystem processes such as nutrient cycling, disturbance regulation, and biodiversity conservation. Here, the role of rare species in enhancing ecosystem services and functionality, highlighting that while certain rare species exhibit functional redundancy with more common species and others provide unique characteristics essential for ecosystem stability, is elaborated. Given that over 75% of endangered plants are endemic to biodiversity hotspots, preserving their functional roles is essential for global ecological stability. The research advocates for prioritizing functional traits in conservation strategies, particularly for rare species, and underscores the importance of in situ conservation while employing ex situ methods when required.

The discussion also addresses quasi-in situ conservation as a transitional approach for safeguarding rare species from habitat degradation. This strategy integrates ex situ collections with natural or semi-natural habitats to conserve genetic diversity and local adaptations. The research emphasizes the significance of biotechnology, specifically tissue culture, cryopreservation, and micropropagation, in the conservation of rare and endangered species. These biotechnological techniques offer efficient and economical means to conserve plant genetic resources, notwithstanding challenges such as somaclonal variability and the necessity for additional refinement, especially for species with limited geographic distributions. Finally, the paper discusses cryopreservation methods for the preservation of forest genetic resources, particularly focusing on recalcitrant seeds and woody species. It evaluates conventional two-stage cooling and contemporary vitrification techniques, highlighting the benefits of vitrification regarding efficiency and cost-effectiveness. The narrative examines the difficulties associated with cryopreserving recalcitrant seeds and explores the viability of excised embryonic tissues as alternatives. Strategies for post-cryopreservation recovery, such as the application of antioxidants and growth regulators, are examined, alongside the necessity to modify cryopreservation techniques to enhance climate change resilience and stress tolerance in species.

REFERENCES

BGCI. ThreatSearch. Available online: https://tools.bgci.org/threat_search.php (accessed on 19 September 2021).

Bradaï, F., Almagro-Bastante, J., & Sánchez-Romero, C. (2023). Effect of sucrose preculture and culture inoculum density on cryopreservation of olive somatic embryos. *Scientia Horticulturae*, *322*, 112385.

Breman, E., Ballesteros, D., Castillo-Lorenzo, E., Cockel, C., Dickie, J., Faruk, A., O'Donnell, K., Offord, C. A., Pironon, S., Sharrock, S., & Ulian, T. (2021). Plant diversity conservation challenges and prospects—The perspective of botanic gardens and the millennium seed bank. *Plants*, *10*(11), 2371. https://doi.org/10.3390/plants10112371

Chokheli, V. A., Dmitriev, P. A., Rajput, V. D., Bakulin, S. D., Azarov, A. S., Varduni, T. V., Stepanenko, V. V., Tarigholizadeh, S., Singh, R. K., Verma, K. K., & Minkina, T. M. (2020). Recent development in micropropagation techniques for rare plant species. *Plants*, *9*(12), Article 12. https://doi.org/10.3390/plants9121733.

Coelho, N., Gonçalves, S., & Romano, A. (2020). Endemic plant species conservation: biotechnological approaches. *Plants*, *9*(3), 345.

Dee, L. E., Cowles, J., Isbell, F., Pau, S., Gaines, S. D., & Reich, P. B. (2019). When do ecosystem services depend on rare species? *Trends in Ecology & Evolution*, *34*(8), 746–758. https://doi.org/10.1016/j.tree.2019.03.010.

Engelmann, F. (2011). Use of biotechnologies for the conservation of plant biodiversity. *In Vitro Cellular & Developmental Biology-Plant*, *47*, 5–16

Faucon, M.-P., Houben, D., & Lambers, H. (2017). Plant functional traits: soil and ecosystem services. *Trends in Plant Science*, *22*(5), 385–394. https://doi.org/10.1016/j.tplants.2017.01.005

Gantait, S., Kundu, S., Wani, S. H., & Das, P. K. (2016). Cryopreservation of forest tree seeds: a mini review. *Journal of Forest and Environmental Science*, *32*(3), 311–322.

ISTA. International Rules for Seed Testing. Available online: https://www.seedtest.org/en/publications/international-rules-seed-testing.html (accessed on 31 August 2021).

Kaviani, B., & Kulus, D. (2022). Cryopreservation of endangered ornamental plants and fruit crops from tropical and subtropical regions. *Biology*, *11*(6), 847. https://doi.org/10.3390/biology11060847.

Kearsley, E., Hufkens, K., Verbeeck, H., Bauters, M., Beeckman, H., Boeckx, P., & Huygens, D. (2019). Large-sized rare tree species contribute disproportionately to functional diversity in resource acquisition in African tropical forest. *Ecology and Evolution*, *9*(8), 4349–4361. https://doi.org/10.1002/ece3.4836.

Kumar, A., Dwivedi, G. K., Tewari, S., Paul, J., Anand, R., Kumar, N., Kumar, P., Singh, H., & Kaushal, R. (2020a). Carbon mineralization and inorganic nitrogen pools under *Terminalia chebula* Retz.-Based agroforestry system in Himalayan foothills, India. *Forest Science*, *66*(5), 634–643.

Kumar, A., Dwivedi, G. K., Tewari, S., Paul, J., Sah, V. K., Singh, H., Kumar, P., Kumar, N., & Kaushal, R. (2020b). Soil organic carbon pools under *Terminalia chebula* Retz. based agroforestry system in Himalayan foothills, India. *Current Science*, *118*(7), 1098–1103.

Kumar, A., Kumar, P., Singh, H., Bisht, S., & Kumar, N. (2021a). Relationship of physiological plant functional traits with soil carbon stock in temperate forest of Garhwal Himalaya. *Current Science*, *120*(8), 1368–1373

Kumar, A., Kumar, P., Singh, H., & Kumar, N. (2021b). Modulation of plant functional traits under essential plant nutrients during seasonal regime in natural forests of Garhwal Himalayas. *Plant and Soil*, *465*, 197–212.

Kumar, A., Kumar, P., Singh, H., & Kumar, N. (2021c). Impact of plant functional traits on near saturated hydraulic conductivity of soil under different forests of Kempty watershed in Garhwal Himalaya. *Indian Journal of Soil Conservation*, 49(1), 38–44.

Kumar, A., Singh, H., Kumari, G., Bisht, S., Malik, A., Kumar, N., Singh, M., Raturi, A., Barthwal, S., Thakur, A., & Kaushal, R. (2022). Adaptive resilience of roadside trees to vehicular emissions via leaf enzymatic, physiological, and anatomical trait modulations. *Environmental Pollution*, *313*, 120191.

Kumar, N., Jeena, N., & Singh, H. (2019). Elevated temperature modulates rice pollen structure: a study from foothill Himalayan agro-ecosystem in India. *3 Biotech*, *9*, 175.

Kumar, N., Jeena, N., Kumar, A., Khairakpam, R., & Singh, H. (2021d). Comparative response of rice cultivars to elevated air temperature in Bhabar region of Indian Himalaya: status on yield attributes. *Heliyon*, *7*, e07474.

Kumar, N., Kumar, N., Shukla, A., Shankhdhar, S. C., & Shankhdhar, D. (2015). Impact of terminal heat stress on pollen viability and yield attributes of rice (*Oryza sativa* L.). *Cereal Research Communications*, *43*(4), 616–626.

Kumar, N., Shankhdhar, S. C., & Shankhdhar, D. (2016). Impact of elevated temperature on antioxidant activity and membrane stability in different genotypes of rice (*Oryza sativa* L.). *Indian Journal of Plant Physiology (Now Plant Physiology Report)*. *21*(1), 37–43.

Kumar, N., Suyal, D. C., Sharma, I. P., Verma, A., & Singh, H. (2017). Elucidating stress proteins in rice (*Oryza sativa* L.) genotype under elevated temperature: a proteomic approach to understand heat stress response. *3 Biotech*, 7, 205.

Leitão, R. P., Zuanon, J., Villéger, S., Williams, S. E., Baraloto, C., Fortunel, C., Mendonça, F. P., & Mouillot, D. (2016). Rare species contribute disproportionately to the functional structure of species assemblages. *Proceedings of the Royal Society B: Biological Sciences*, *283*(1828), 20160084. https://doi.org/10.1098/rspb.2016.0084.

Li, J. W., Ozudogru, E. A., Li, J., Wang, M. R., Bi, W. L., Lambardi, M., & Wang, Q. C. (2018). Cryobiotechnology of forest trees: recent advances and future prospects. *Biodiversity and Conservation*, *27*, 795–814.

Li, R., Zhu, S., Lian, J., Zhang, H., Liu, H., Ye, W., & Ye, Q. (2021). Functional traits are good predictors of tree species abundance across 101 subtropical forest species in China. *Frontiers in Plant Science*, *12*, 541577. https://doi.org/10.3389/fpls.2021.541577.

Lyons, K. G., Brigham, C. A., Traut, B. H., & Schwartz, M. W. (2005). Rare species and ecosystem functioning. *Conservation Biology*, *19*(4), 1019–1024. https://doi.org/10.1111/j.1523-1739.2005.00106.x

Menges E. S. (2008). Restoration demography and genetics of plants: when is a translocation successful? *Australian Journal of Botany*, 56, 187–196.

Mi, X., Sun, Z., Song, Y., Liu, X., Yang, J., Wu, J., Ci, X., Li, J., Lin, L., Cao, M., & Ma, K. (2021). Rare tree species have narrow environmental but not functional niches. *Functional Ecology*, *35*(2), 511–520. https://doi.org/10.1111/1365-2435.13714.

Miedema Brown, L., & Anand, M. (2022). Plant functional traits as measures of ecosystem service provision. *Ecosphere*, *13*(2), e3930. https://doi.org/10.1002/ecs2.3930.

Murthy, H. N., Joseph, K. S., Paek, K. Y., & Park, S. Y. (2023). Bioreactor systems for micropropagation of plants: present scenario and future prospects. *Frontiers in Plant Science*, *14*, 1159588.

Murthy, H. N., Joseph, K. S., Paek, K. Y., & Park, S. Y. (2023). Bioreactor systems for micropropagation of plants: present scenario and future prospects. *Frontiers in Plant Science*, *14*, 1159588. https://doi.org/10.3389/fpls.2023.1159588

Nadarajan, J., & Pritchard, H. W. (2014). Biophysical characteristics of successful oilseed embryo cryoprotection and cryopreservation using vacuum infiltration vitrification: an innovation in plant cell preservation. *PLoS One*, *9*(5), e96169.

Pavlik, B. M. (1996). Defining and measuring success. In: Falk, D. A., Millar, C. I., Olwell, M. (eds.) *Restoring Diversity. Strategies for Reintroduction of Endangered Plants*. Island Press, Washington, DC, pp. 127–155.

Popova, E., Kulichenko. I., & Kim, H.-H. 2023. Critical role of regrowth conditions in post-cryopreservation of in vitro plant germplasm. *Biology*, *12*(4), 542. https://doi.org/10.3390/biology12040542

Streczynski, R., Clark, H., Whelehan, L., Ang, S.-T., Hardstaff, L., Funnekotter, B., Bunn, E., Offord, C. A., Sommerville, K., & Mancera, R. 2019. Current issues in plant cryopreservation and importance for ex situ conservation of threatened Australian native species. *Australian Journal of Botany*, 67, 1–15

Thakur, J., Abbas, N. S., Bhardwaj, S., & Kaula, B. C. (2021). Biotechnological tools in the propagation and conservation of threatened species: an overview. *Indian Journal of Biotechnology*, 20(3).

Thakur, J., Abbas, N. S., Bhardwaj, S., & Kaula, B. C. (2022). Biotechnological tools in the propagation and conservation of threatened species: an overview.

Velmurugan, A. V., Jaisankar, I., & Singh, A. (2018). *Hotspots—An Introduction and Role in Conservation* (pp. 1–21).

Volis, S. (2023). Living collections of threatened plants in botanic gardens: when is ex situ cultivation less appropriate than quasi in situ cultivation? *Journal of Zoological and Botanical Gardens*, *4*(2), 462–475. https://doi.org/10.3390/jzbg4020034

Walters, C., & Pence, V. C. (2021). The unique role of seed banking and cryobiotechnologies in plant conservation. *Plants, People, Planet*, *3*(1), 83–91.

White, P. S., & Walker, J. L. (1997). Approximating nature's variation: selecting and using reference information in restoration ecology. *Restoration Ecology*, 5, 338–349.

Zhang, S., Zang, R., & Sheil, D. (2022). Rare and common species contribute disproportionately to the functional variation within tropical forests. *Journal of Environmental Management*, *304*, 114332. https://doi.org/10.1016/j.jenvman.2021.114332

14 Enhancing Resilience
Harnessing Traits for Coping with Disturbances

Naresh Singh, R. S. Bali, Pankaj Kumar, and Priyanka Thakur

14.1 INTRODUCTION

Resilience within forest ecosystems is characterized as the inherent ability of these systems to endure disturbances and recuperate from them while preserving their fundamental functions and structures. This idea emphasizes how important biodiversity is on many levels, from genetic diversity to landscape heterogeneity, as a key factor that makes things more resilient. Forests exhibiting resilience are capable of withstanding environmental stressors, including climate change, wildfires, and pest infestations, without experiencing considerable degradation. These ecosystems demonstrate a capacity for adaptability, enabling them to reorganize and maintain their ecological functions despite external challenges. The implementation of effective forest management strategies that aim to enhance biodiversity and reduce competition among tree species is crucial for cultivating resilience, thereby guaranteeing that forests persist in providing essential ecosystem services and sustaining their health over extended periods (Kumar et al., 2021d; Messier et al., 2019; Forzieri et al., 2022; Yan et al., 2011).

14.1.1 Importance of Coping with Disturbances (Natural and Anthropogenic)

Addressing both natural and anthropogenic perturbations is essential for preserving ecological integrity and resilience within ecosystems. Such perturbations, whether instigated by natural phenomena like wildfires and storms or by anthropogenic actions including urbanization and environmental contamination, profoundly influence biodiversity, ecosystem functions, and the overall health of the environment.

14.1.2 Natural Disturbances

Natural disturbances are essential for fostering biodiversity and ecological processes. They create variability in habitats, which is crucial for species coexistence. For instance:

DOI: 10.1201/9781003545231-14

Biodiversity Enhancement: Wildfires, floods, and storms can enhance the variety of life forms by forming fresh environments and living spaces. These natural occurrences can revitalize ecosystems by removing established plants, thereby making room for fresh development and a wide range of species to flourish (Kumar et al., 2023).

Ecosystem Resilience: In natural settings, natural disturbances more often than not give rise to a resilient structure that improves recovery from disturbances with better efficiency. In this regard, resilience is very critical for the sustenance of functions and services that humans use (Meyer et al., 2021).

14.1.3 Anthropogenic Disturbances

Human-induced disturbances often disrupt natural processes and can lead to significant ecological consequences:

Biodiversity Loss: Anthropogenic activities significantly impact habitats, reducing their diversity and potentially leading to a decline in species populations. As a result, this homogenizes habitats and decreases available niches for various organisms and thereby threatens biodiversity (Prakash and Verma, 2022).

Altered Disturbance Regimes: The human-related alterations of natural disturbance regimes make ecosystems vulnerable to invasive species and other stressors. For example, fire suppression may result in larger fuel loads, thereby making fires more severe when they occur (Thom and Seidl, 2016).

14.1.4 Coping Mechanisms

To effectively cope with these disturbances, both ecosystems and human communities must adopt adaptive management strategies:

Natural Disturbance-Based Management: This practice applies the role of natural disturbances in forest management. The focus has been on biodiversity and sustainable resource use. This therefore implies that such management practice requires ecological health with economic values (Kuuluvainen et al., 2021).

Restorative Environments: Studies have shown that exposure to natural environments reduces stress in human populations, highlighting the psychological value of healthy ecosystems. These ecosystems offer significant restorative effects that improve well-being and cognition (Berto, 2014).

14.2 UNDERSTANDING DISTURBANCES

14.2.1 Types of Disturbances

Natural Disturbances: Natural disturbances in forests play a crucial role in shaping ecosystems, influencing biodiversity, and driving ecological processes. The primary types of natural disturbances include

Fire: Wildfires are common and can vary in intensity, affecting vast areas of forest. They help maintain ecosystem health by clearing underbrush, promoting new growth, and facilitating nutrient cycling.

Insect Outbreaks: Insects such as bark beetles can devastate tree populations, leading to significant changes in forest structure and composition. These outbreaks often occur in response to environmental stressors, including drought and climate change.

Wind Disturbances: Strong winds can uproot trees or break branches, creating gaps in the canopy that allow sunlight to reach the forest floor, fostering new plant growth and enhancing biodiversity.

Drought: Extended periods of low precipitation can weaken trees, making them more susceptible to pests and diseases. Drought conditions can also alter species composition, as some trees may not survive.

Flooding: Heavy rains or snowmelt can lead to flooding, which may drown trees but also enrich the soil with nutrients as water recedes, supporting the growth of flood-adapted species.

Disease: Tree diseases caused by fungi, bacteria, or viruses can reduce tree populations and alter forest dynamics. Other disturbances, like drought or insect infestations, may exacerbate some diseases.

14.2.2 Anthropogenic Disturbances

Anthropogenic disturbances, including logging, land conversion, and climate change, significantly impact forest ecosystems globally. These activities lead to deforestation, biodiversity loss, and alterations in carbon dynamics.

14.2.2.1 Logging and Deforestation

Logging is the most significant cause of forest degradation and deforestation. It usually involves the harvesting of some very valuable timber, which makes the health of the forests deteriorate gradually until they are completely converted to other uses. The World Wildlife Fund believes that logging contributes to climate change by producing emissions of carbon from soil disturbances and the decay of biomass in places that are cleared.

14.2.2.2 Land Conversion

Land conversion is the change of a natural forest to agricultural use, or to urban use, etc. The demand for agricultural commodities such as palm oil, soy, beef, and timber primarily drives this change. According to ClientEarth, forest conversion is currently the primary cause of global deforestation, with the majority of this activity occurring in the tropics due to extensive agricultural expansion.

14.2.2.3 Climate Change Effects

Climatic change exacerbates anthropogenic disturbances in forests. Increases in temperature and changes in precipitation patterns influence the health and resilience of forests. Deforestation is considered a significant contributor on climate change, accounting for approximately 20% of global greenhouse gas emissions. The

Intergovernmental Panel on Climate Change comments on the amount of carbon input to the atmosphere because of land-use changes that approximate 1.6 ± 0.8 gigatons per year.

14.2.2.4 Regulatory Challenges

Overlapping laws and governance issues complicate the regulation of logging and land conversion. In many countries, weak legal frameworks allow illegal logging and conversion practices to flourish. Illegal activities accounted for nearly half of tropical deforestation from 2000 to 2012. Effective governance is crucial for balancing economic needs with environmental protection.

14.2.3 Impacts of Disturbances on Forest Health and Biodiversity

Through logging, land conversion, and climate change, humans have fundamentally altered the health and biodiversity of forests. In most cases, they alter the structure and composition of forest ecosystems, leading to a decline in ecosystem services such as carbon storage, water regulation, and habitat provision. For instance, increased logging and land conversion cause habitat fragmentation that threatens the survival and biodiversity of species. Studies show that although disturbances at times increase species richness by producing heterogeneous habitats, they generally reduce aggregate ecosystem carbon storage by around 38.5% on average in events of stand-replacement disturbance (Thom and Seidl, 2016). These interactions are made even more complicated by climate change, which makes disturbances in forests like wildfires and insect outbreaks worse. These can be very bad for the health of the forest in a chain reaction (Kumar et al., 2020a,b; Sanchez et al., 2021). These interact in such a manner that there is a need for the effective management of forests that can support ecological integrity and meet human needs.

14.2.4 The Role of Disturbances in Shaping Forest Dynamics

Natural disturbances are crucial drivers of forest dynamics, influencing biodiversity, ecosystem structure, and recovery processes. Various types of disturbances, including fire, windthrow, and insect outbreaks, play significant roles in shaping forest ecosystems across different biomes.

Fire: Fire is a dominant disturbance in many forest ecosystems, particularly in boreal regions. It varies in severity and frequency, leading to diverse ecological outcomes. According to Shorohova et al. (2023), crown fires can kill a lot of trees, while mixed-severity fire regimes may lead to complicated recovery patterns that depend on the climate and the plants that are there.

Wind Throw: Wind disturbances can uproot trees or break their trunks, significantly altering forest structure. The return intervals for windthrow events can vary widely, impacting the regeneration dynamics and species composition within affected areas (Viljur et al., 2022).

Insect Outbreaks: Insect infestations can lead to widespread tree mortality and changes in species composition. These disturbances often interact with other environmental factors, such as climate change, which may exacerbate their effects (Viljur et al., 2022).

14.2.5 Ecological Impacts of Disturbances

Biodiversity: Disturbances can enhance biodiversity by creating heterogeneous habitats and promoting the growth of disturbance-adapted species. The intermediate disturbance hypothesis suggests that moderate levels of disturbance can maximize species richness by preventing dominant species from monopolizing resources (Viljur et al., 2022).

Forest Recovery: Pre-disturbance forest composition and post-disturbance environmental conditions influence the recovery trajectory following disturbances. Factors such as local topography and community functional composition play critical roles in determining recovery rates (Viljur et al., 2022).

Nutrient Cycling: Disturbances often lead to increased nutrient availability through the decomposition of dead organic matter. This process can enhance soil fertility and influence subsequent plant growth and carbon dynamics within the ecosystem (Hérault and Piponiot, 2018).

14.3 PHENOTYPIC PLASTICITY AND ADAPTABILITY

Phenotypic plasticity is an important biological concept describing the organism's ability to change its phenotype in conformity with environmental changes. This flexibility can be expressed in various ways, such as alterations in the morphology, physiology, and also the behavior, which aids in the survival and proper growth of an organism under changing conditions. Phenotypic plasticity can be described as the ability of a single genotype to give rise to different phenotypes upon exposure to different environmental pressures (Fusco and Minelli, 2010; Kumar et al., 2021a,b). It comprises two basic aspects:

Developmental Plasticity: Changes that occur during an organism's development due to environmental cues. For instance, some insects may develop different wing lengths based on temperature variations.

Phenotypic Flexibility: Short-term adjustments that occur in response to immediate environmental changes, such as behavioral modifications or metabolic adjustments (Forsman, 2015).

These mechanisms are not mutually exclusive; rather, they often interact. For example, genetic changes can lead to similar phenotypes as those produced by plastic responses, emphasizing the interconnectedness of plasticity and adaptation (Sabino-Pinto et al., 2019).

14.3.1 Types of Phenotypic Plasticity

1. **Active Plasticity**: This refers to anticipatory changes that frequently integrate and respond to environmental signals. An example is the behavioral adaptations of prey in response to predator presence.
2. **Passive Plasticity**: When direct environmental factors, such as resource constraints on growth, cause unanticipated changes, passive plasticity occurs (Forsman, 2015).

14.3.2 Role in Adaptation

Phenotypic plasticity plays a significant role in how species adapt to new or changing environments. It allows organisms to maintain their genetic integrity while producing alternative phenotypes suited for different conditions. This flexibility can be crucial for survival during rapid environmental changes, such as climate shifts (Gibert et al., 2019; Ghalambor et al., 2007).

14.3.3 Adaptive vs. Nonadaptive Plasticity

Adaptive Plasticity: Enhances survival and reproductive success under specific conditions. For instance, certain fish species can alter their body shape based on water flow to optimize swimming efficiency.

Nonadaptive Plasticity: There is a type of plasticity called nonadaptive plasticity that might not help an organism survive if its environment changes too quickly for it to adapt (Ghalambor et al., 2007).

14.3.4 Evolutionary Implications

The relationship between phenotypic plasticity and evolution is complex. While traditionally viewed as a stabilizing force against speciation, recent studies suggest that plasticity can facilitate evolutionary divergence by exposing traits to natural selection (Sabino-Pinto et al., 2019; Gibert et al., 2019). This process, known as plastic acclimatization, allows for the gradual evolution of traits that were initially plastic but become fixed through genetic adaptation over time.

14.3.5 Significance of Phenotypic Plasticity

Phenotypic plasticity is defined as the ability of the same genotype to produce a range of phenotypes according to the fluctuating environmental conditions. This phenomenon is important for organisms, as it enables them to adapt to changing and different environments, thus improving their chances of survival and reproduction. The importance of phenotypic plasticity lies in its role within ecologic and evolutionary processes; it permits species to cope with environmental stressors, such as climate change or habitat alteration, via immediate adaptations. For instance, plant leaves modify their size and shape according to the availability of light while animals may be forced to change behavior or physiology in response to differences in their habitats intended to maximize resource use. Additionally, phenotypic plasticity can result in genetic change because it serves as a buffer against environmental variation, meaning that the propensity of populations to be persistent can result in the emergence of new traits through time because of natural selection (Kumar et al., 2021c; Sommer, 2020; Fusco and Minelli, 2010; Miras, 2024) (Table 14.1).

14.3.6 Drought Tolerance Strategies

Drought tolerance strategies in trees encompass a range of physiological and morphological adaptations that enable them to cope with water scarcity. These strategies can be broadly categorized into drought avoidance and drought tolerance mechanisms.

14.3.6.1 Drought Avoidance Strategies

Stomatal Regulation: Trees can control water loss by regulating stomatal closure. Early closure (isohydry) minimizes water loss but can limit photosynthesis, while late closure (anisohydry) allows for continued photosynthesis at the risk of embolism in the xylem (Markesteijn, 2010).

Leaf Shedding: Many species, especially in tropical dry forests, shed leaves during prolonged droughts to reduce transpiration and conserve water (Polle et al., 2019).

Root Growth: Increased root depth and biomass allocation allow trees to access deeper soil moisture, enhancing their ability to survive during dry periods (Ravn et al., 2022).

14.3.6.2 Drought Tolerance Strategies

Osmotic Adjustment: Trees accumulate solutes such as sugars and amino acids to maintain cell turgor pressure during drought stress, thus enabling continued cellular function (Polle et al., 2019).

Hydraulic Safety Margin: Species with a high hydraulic safety margin can handle lower water potentials without getting xylem embolism, which lets them keep moving water even when there is a drought (Markesteijn, 2010).

Water-Use Efficiency (WUE): Some trees enhance their WUE by optimizing the ratio of photosynthesis to transpiration, which helps mitigate the effects of carbon starvation during drought (Ravn et al., 2022).

14.3.7 Responses to Nutrient Availability

Some tree responses to nutrient levels are so significant for influencing the growth, biomass allocation, and health of trees. When nutrient availability is high, above-ground growth is augmented with increased leaf area and photosynthetic capacity. High leaf mass induces a higher number of light capture; it enhances carbon assimilation and will, therefore, promote more rapid growth. In nutrient-poor environments, trees can favor root growth to reach deeper layers of soil for water and nutrients, thus increasing the root–shoot ratio. Thereby, these trees survive unfavorable environments by maximizing resource acquisition from soil. The trees modify their physiological processes, including nitrogen uptake efficiency and leaf nutrient composition, to enhance growth in nutrient variability. For example, it has been proven that the nitrogen content in leaves increases in fertile soils and directly correlates with the increased growth rate and biomass accumulation observed (Kumar et al., 2019, 2022). Such adaptation strategies emphasize the role of nutrient availability in tree physiology and community dynamics within forest ecosystems (Singh et al., 2023).

14.3.8 Role of Genetic Diversity in Enhancing Plasticity

It is thought that genetic variation helps improve the phenotypic plasticity of tree populations, which gives them the ability to adapt to changing environmental conditions. Most of the time, having more genetic diversity means that you are more likely to have beneficial alleles that make you more flexible in environments that change

TABLE 14.1
Examples of Plastic Responses to Environmental Changes

Environmental Change	Tree Species	Plastic Response
Drought conditions	*Balsam Fir*	Increased root:shoot ratio to access deeper moisture and enhanced water-use efficiency
Temperature fluctuations	*Quercu srobur* (Oak)	Altered leaf morphology and photosynthetic rates to optimize energy capture under varying temperatures
Nutrient availability	*Populus deltoides* (Cottonwood)	Changes in biomass allocation, favoring root growth when nutrients are limited
Light availability	*Betula pendula* (Birch)	Leaf thickness and size adjustments based on light exposure, with thicker leaves in high light
Soil pH variations	*Pinus sylvestris* (Pine)	Altered root structure and nutrient uptake efficiency in response to soil acidity changes
Wind exposure	*Salix alba* (Willow)	Development of thicker stems and flexible branches to withstand strong winds

in temperature, moisture, or nutrient availability. Because of this genetic diversity, the trees show a range of phenotypes in response to stressors, which improves their chances of survival and reproduction. Trees with more genetic diversity are better able to adapt to changes in drought and nutrient levels, for example, by changing their growth patterns, root architecture, or physiological processes. Since one genotype corresponds to any one specific combination of two alleles and because there might exist in a population of many individuals multiple genotypes, the presence of several genotypes within a population may ensure overall fitness by ensuring that at least some individuals are especially well-suited to particular environmental conditions. In this way, genetic diversity not only protects tree populations from short-term bad changes in their environment, but it also makes them more resilient and helps them evolve in the long term (de la Mata et al., 2022; Forsman, 2015).

14.4 HARNESSING TRAITS FOR FOREST MANAGEMENT

14.4.1 Integrating Trait-Based Approaches into Forest Management Practices

Using the useful features of different tree species to make ecosystems more resilient and able to adapt to changes in the environment is what trait-based approaches are all about. The approach mainly focuses on understanding how specific traits, such as drought tolerance, growth rates, and nutrient acquisition, affect species performance and interactions within forest ecosystems. When forest managers look at these traits, they can make the right choice of species for reforestation and afforestation projects. This way, the chosen species will be able to thrive in the current environment and in the climates of the future.

14.4.2 Key Components of Trait-Based Approaches

Functional Trait Assessment: Identifying and measuring traits that influence tree growth, survival, and ecosystem functions is critical. For example, traits related to WUE and root morphology can inform management strategies in water-limited environments.

Species Selection: By understanding the functional traits of various species, managers can select trees that are more resilient to specific stresses, such as drought or nutrient-poor soils. This selection process can enhance forest productivity and biodiversity.

Ecosystem Dynamics: Trait-based approaches allow for a better understanding of how tree species interact with other organisms in the ecosystem, including competition for resources and responses to herbivory. This understanding can help predict how changes in species composition may affect overall forest health.

Adaptation to Climate Change: As climate change continues to impact forest ecosystems, integrating trait-based strategies can facilitate the selection of species that exhibit high phenotypic plasticity. This plasticity enables trees to adapt to varying conditions, such as increased temperatures or altered precipitation patterns.

Long-Term Monitoring: Implementing trait-based management requires ongoing monitoring of tree performance and ecosystem responses over time. This data can inform adaptive management practices, allowing for adjustments based on observed outcomes.

14.4.3 Benefits of Integrating Trait-Based Approaches

Enhanced Resilience: By focusing on functional traits, forest management can promote resilience against climate variability and other stressors.

Improved Biodiversity: Selecting a diverse range of species based on their traits can enhance ecosystem stability and function.

Sustainable Management Practices: Trait-based approaches support sustainable forestry by promoting practices that align with ecological principles rather than solely economic goals.

14.4.4 Selection and Breeding for Resilient Traits

The selection and breeding for resilient traits in trees is a critical aspect of forest management, particularly in the context of climate change and environmental stressors. Two primary methodologies, MAS and traditional breeding methods, can effectively implement this process.

14.4.4.1 Marker-Assisted Selection (MAS)

MAS is a modern breeding technique that utilizes molecular markers to identify and select for desirable traits in trees. This approach involves the following key steps:

Identification of Genetic Markers: Techniques like quantitative trait locus mapping identify specific DNA sequences associated with traits of interest, such as drought tolerance or disease resistance. These markers serve as indicators for the presence of these traits in breeding populations.

Genotyping: Trees are genotyped to determine the presence of these markers, allowing breeders to select individuals that carry favorable alleles without needing to wait for the traits to manifest phenotypically.

Accelerated Breeding Cycles: By using MAS, breeders can significantly speed up the breeding process. This is because they can select desired traits at the seedling stage rather than waiting for mature trees to exhibit them, thus increasing efficiency and reducing costs associated with traditional breeding methods.

Enhanced Precision: MAS provides a more precise approach to breeding compared to traditional methods, as it allows for the selection of specific genetic combinations that enhance resilience traits while minimizing undesirable characteristics.

14.4.4.2 Traditional Breeding Methods

Traditional breeding methods encompass a range of techniques that have been used for centuries to improve tree species. These methods include:

Cross-Pollination and Hybridization: Breeders cross different tree varieties to combine desirable traits. To produce hybrids that may exhibit improved resilience traits, breeders can deliberately transfer pollen from one tree to another through controlled pollination techniques.

Selection: Traditional methods often rely on phenotypic selection, where individuals with desirable characteristics are chosen for further breeding. Mass selection is a method that selects the best-performing individuals based on their observable traits.

Backcrossing: In this method, a hybrid tree is crossed back with one of its parent species to reinforce specific desirable traits while retaining some genetic material from both parents. This technique helps in stabilizing traits like pest resistance or drought tolerance over successive generations.

Tissue Culture and Clonal Propagation: By propagating genetically identical plants from a single parent tree, these techniques effectively preserve and propagate desirable traits.

Although traditional breeding methods can be time-consuming and may require multiple generations to achieve desired outcomes, they remain essential due to their cost-effectiveness and ability to utilize existing genetic diversity within populations. They also play a crucial role in integrating local adaptations into breeding programs, ensuring that selected trees are well-suited to their specific environments.

14.4.5 Restoration Strategies that Prioritize Resilient Species and Traits

Restoration strategies that focus on resilient species and traits improve the stability and functionality of an ecosystem when considering environmental changes. Mainly, these strategies are species selection with traits attributed to resilience, such as drought, pest resistance, and plastic phenotypes. For example, the incorporation of functional traits into restoration practice allows managers to select species that can easily adapt to changing conditions, thereby improving biodiversity and ecosystem services. Also, studies have shown that reintroducing species based on their ecological role and how they interact with other organisms in the ecosystem greatly improves the chances of recovery. Restoration approaches focusing on species network-based restoration emphasize critical species whose maintenance would guarantee mutualistic relationships, subsequently supporting recovery in overall biodiversity. Using genetic diversity at the population level can also help in adapting to environmental stressors. This makes sure that restored systems are not only diverse but also strong enough to handle future problems. These strategies can provide stronger ecosystems at the point of restoration that can endure the stresses associated with climate and habitat degradation.

14.5 CONCLUSION

In conclusion, harnessing specific traits in order to strengthen resilience is the key to success in an increasingly complex world. We looked at how resilience may be developed via intentional activity and appear as an active rather than passive response to adversity. The three major key traits include adaptability, optimism, and emotional regulation as the fundamental pillars of strength for both persons and communities in order to handle their challenges with better confidence and creativity. By cultivating such, one can not only survive the disturbances but emerge much stronger and more resourceful from them. Therefore, we cannot undervalue the importance of social support and community engagement. Resilience grows in a setting where people feel mutual and can share experiences and solve problems together; this method of gathering smoothes out individual means of coping as well as develops the social structure. Second, with a growth mindset in which failure may help one learn and grow, setbacks are positive learning milestones in turning reality into the philosophy that resilience is a path rather than a destination. As we move into a bright future full of opportunities, it is important to remember that resilience is a way of adapting to new situations and is closely linked to our ability to live with the challenges that come with social, economic, and environmental problems. By focusing on cultivating resilient properties at both the individual and the community level, we build for a future that will be more sustainable and equitable. And finally, by boosting resilience, we enable ourselves with the means to confront disturbances with courage and creativity, and hence, with this hope, we construct a context that nudges us toward innovation and collaboration in the face of adversity. In this way, we are getting ourselves ready to face all the eventualities that lie ahead and are adding toward a more resilient world for the many decades to come.

REFERENCES

Berto, R. (2014). The role of nature in coping with psycho-physiological stress: a literature review on restorativeness. *Behavioral Sciences*, *4*(4), 394–409.

de la Mata, R., Zas, R., Bustingorri, G., Sampedro, L., Rust, M., Hernandez-Serrano, A., & Sala, A. (2022). Drivers of population differentiation in phenotypic plasticity in a temperate conifer: a 27-year study. *Evolutionary Applications*, *15*(11), 1945–1962.

Forsman, A. (2015). Rethinking phenotypic plasticity and its consequences for individuals, populations and species. *Heredity*, *115*(4), 276–284.

Forzieri, G., Dakos, V., McDowell, N. G., Ramdane, A., & Cescatti, A. (2022). Emerging signals of declining forest resilience under climate change. *Nature*, *608*(7923), 534–539.

Fusco, G., & Minelli, A. (2010). Phenotypic plasticity in development and evolution: facts and concepts. *Philosophical Transactions of the Royal Society B: Biological Sciences*, *365*(1540), 547–556.

Ghalambor, C. K., McKay, J. K., Carroll, S. P., & Reznick, D. N. (2007). Adaptive versus non-adaptive phenotypic plasticity and the potential for contemporary adaptation in new environments. *Functional Ecology*, *21*(3), 394–407.

Gibert, P., Debat, V., & Ghalambor, C. K. (2019). Phenotypic plasticity, global change, and the speed of adaptive evolution. *Current Opinion in Insect Science*, *35*, 34–40.

Hérault, B., & Piponiot, C. (2018). Key drivers of ecosystem recovery after disturbance in a neotropical forest. *Forest Ecosystems*, *5*(1), 1–15.

Kumar, A., Dwivedi, G. K., Tewari, S., Paul, J., Anand, R., Kumar, N., Kumar, P., Singh, H., & Kaushal, R. (2020a). Carbon mineralization and inorganic nitrogen pools under *Terminalia chebula* Retz.-Based agroforestry system in Himalayan foothills, India. *Forest Science*, *66*(5), 634–643.

Kumar, A., Dwivedi, G. K., Tewari, S., Paul, J., Sah, V. K., Singh, H., Kumar, P., Kumar, N., & Kaushal, R. (2020b). Soil organic carbon pools under *Terminalia chebula* Retz. based agroforestry system in Himalayan foothills, India. *Current Science*, *118*(7), 1098–1103.

Kumar, A., Kumar, P., Singh, H., Bisht, S., & Kumar, N. (2021a). Relationship of physiological plant functional traits with soil carbon stock in temperate forest of Garhwal Himalaya. *Current Science*, *120*(8), 1368–1373

Kumar, A., Kumar, P., Singh, H., & Kumar, N. (2021b). Modulation of plant functional traits under essential plant nutrients during seasonal regime in natural forests of Garhwal Himalayas. *Plant and Soil*, *465*, 197–212.

Kumar, A., Kumar, P., Singh, H., & Kumar, N. (2021c). Impact of plant functional traits on near saturated hydraulic conductivity of soil under different forests of Kempty watershed in Garhwal Himalaya. *Indian Journal of Soil Conservation*, *49*(1), 38–44.

Kumar, A., Singh, H., Kumari, G., Bisht, S., Malik, A., Kumar, N., Singh, M., Raturi, A., Barthwal, S., Thakur, A., & Kaushal, R. (2022). Adaptive resilience of roadside trees to vehicular emissions via leaf enzymatic, physiological, and anatomical trait modulations. *Environmental Pollution*, *313*, 120191.

Kumar, N., Jeena, N., Kumar, A., Khairakpam, R., & Singh, H. (2021d). Comparative response of rice cultivars to elevated air temperature in Bhabar region of Indian Himalaya: status on yield attributes. *Heliyon*, *7*, e07474.

Kumar, N., Jeena, N., & Singh, H. (2019). Elevated temperature modulates rice pollen structure: a study from foothill Himalayan agro-ecosystem in India. *3 Biotech*, *9*, 175.

Kumar, R., Bhardwaj, A. K., & Chandra, K. K. (2023). Levels of natural and anthropogenic disturbances and assessment of their impact on plant community functional diversity. *Forestist*, *73*(1), 35–42.

Kuuluvainen, T., Angelstam, P., Frelich, L., Jõgiste, K., Koivula, M., Kubota, Y., & Macdonald, E. (2021). Natural disturbance-based forest management: moving beyond retention and continuous-cover forestry. *Frontiers in Forests and Global Change*, *4*, 629020.

Markesteijn, L. (2010). *Drought Tolerance of Tropical Tree Species: Functional Traits, Trade-Offs and Species Distribution*. Wageningen University and Research, Netherlands: Wageningen.

Messier, C., Bauhus, J., Doyon, F., Maure, F., Sousa-Silva, R., Nolet, P., & Puettmann, K. (2019). The functional complex network approach to foster forest resilience to global changes. *Forest Ecosystems*, *6*(1), 1–16.

Meyer, S. E., Callaham, M. A., Stewart, J. E., & Warren, S. D. (2021). Invasive species response to natural and anthropogenic disturbance. In *Invasive Species in Forests and Rangelands of the United States: A Comprehensive Science Synthesis for the United States Forest Sector* (pp. 85–110), Cham: Springer International Publishing.

Miras, K. (2024). Exploring the costs of phenotypic plasticity for evolvable digital organisms. *Scientific Reports*, *14*(1), 108.

Polle, A., Chen, S. L., Eckert, C., & Harfouche, A. (2019). Engineering drought resistance in forest trees. *Frontiers in Plant Science*, *9*, 1875.

Prakash, S., & Verma, A. K. (2022). Anthropogenic activities and biodiversity threats. *International Journal of Biological Innovations, IJBI*, *4*(1), 94–103.

Ravn, J., D'Orangeville, L., Lavigne, M. B., & Taylor, A. R. (2022). Phenotypic plasticity enables considerable acclimation to heat and drought in a cold-adapted boreal forest tree species. *Frontiers in Forests and Global Change*, *5*, 1075787.

Sabino-Pinto, J., Goedbloed, D. J., Sanchez, E., Czypionka, T., Nolte, A. W., &Steinfartz, S. (2019). The role of plasticity and adaptation in the incipient speciation of a fire salamander population. *Genes*, *10*(11), 875.

Sanchez, J. J., Marcos-Martinez, R., Srivastava, L., & Soonsawad, N. (2021). Valuing the impacts of forest disturbances on ecosystem services: an examination of recreation and climate regulation services in US national forests. *Trees for People 5*, 100123.

Shorohova, E., Aakala, T., Gauthier, S., Kneeshaw, D., Koivula, M., Ruel, J. C., & Ulanova, N. (2023). Natural disturbances from the perspective of forest ecosystem-based management. In Miguel Montoro Girona, Hubert Morin, Sylvie Gauthier, Yves Bergeron (Eds), *Boreal Forests in the Face of Climate Change: Sustainable Management* (pp. 89–121). Cham: Springer International Publishing.

Singh Ramesh, A., Cheesman, A. W., Flores-Moreno, H., Preece, N. D., Crayn, D. M., & Cernusak, L. A. (2023). Temperature, nutrient availability, and species traits interact to shape elevation responses of Australian tropical trees. *Frontiers in Forests and Global Change*, *6*, 1089167.

Sommer, R. J. (2020). Phenotypic plasticity: from theory and genetics to current and future challenges. *Genetics*, *215*(1), 1–13.

Thom, D., & Seidl, R. (2016). Natural disturbance impacts on ecosystem services and biodiversity in temperate and boreal forests. *Biological Reviews*, *91*(3), 760–781.

Viljur, M. L., Abella, S. R., Adámek, M., Alencar, J. B. R., Barber, N. A., Beudert, B., & Thorn, S. (2022). The effect of natural disturbances on forest biodiversity: an ecological synthesis. *Biological Reviews*, *97*(5), 1930–1947.

Yan, H., Zhan, J., & Zhang, T. (2011). Resilience of forest ecosystems and its influencing factors. *Procedia Environmental Sciences*, *10*, 2201–2206.

15 Technological Innovations in Functional Trait Research

Anand Kumar, Laxmidas Verma, and Anamika Yadav

15.1 INTRODUCTION

Forests play a vital role in supplying timber for human use, regulating essential ecosystem services such as carbon sequestration and soil erosion control, providing medicinal resources and recreational opportunities, and contributing to the rehabilitation of degraded lands. The association between plant traits and environmental effects is very necessary to detect the actual effect of traits for characters, especially for biotic and abiotic characters. Environmental changes such as changing climate in atmosphere, land use, and biotic exchange are triggering unprecedented ecosystem change. Because of their ability to fix atmospheric CO_2 and convert it to organic carbon, forests have been identified as a key mitigating factor for climate change. Restoring the same land is time consuming and many techniques are needed to restore the same. According to Mashizi and Sharafatmandrad (2023), functional trait diversity enables plants to enhance functional efficiency while minimizing building costs. Understanding the plant functional traits (PFTs) determines the performance of plants for the distribution along environments. Knowledge of plants' functional traits helps plants potential with stand to environment change. Global warming affects plants and forces species to track their climatic niche and shift distribution of plants upward and downward (Maharjan et al., 2021).

The prediction of plant species' responses to environmental change has never been more vital. Elevation-dependent warming happens when temperatures rise at a faster pace at higher elevations. In tropical montane landscapes, elevation dependent on warming can force species to shift their distribution ranges rapidly upward (Liu et al., 2018). Understanding the mechanism of species dispersal when elevation gradients rise is more important for predicting species responses to global warming. We can select the suitable genotype species for future planting activities. Díaz et al. (2004) studied and analyzed the variation in 12 traits among four floras in three continents. According to his research, species with a conservative strategy have small, thick, and tough leaves with slow turnover, while species with an acquisitive strategy have large, thin, and soft leaves with rapid turnover. One more study by Diaz et al. (2016) analyzed the variation in six major traits critical to growth, survival, and reproduction of vascular plant species across the globe. They revealed two major

DOI: 10.1201/9781003545231-15

axes of evolutionary specialization, namely, the plant size spectrum (that runs from short species with small diasporas to tall species with large diaspores) and the leaf economics spectrum (that runs from species with cheaply constructed acquisitive leaves to species with conservative leaves). These two axes together accounted for 74% of the trait variation across the globe. Despite increasing knowledge on how these plant strategies and their underlying functional trait's structure plant communities along environmental gradients, studies testing these concepts in species-rich tropical montane landscapes are limited (Díaz et al., 2016).

Many ecologists have used to detect the performance of functional traits among species to assess the extent to which these differences have adaptive value, thus resulting in differences in relative abundances of co-existing species. For example, studies in a water- and nitrogen-limited grassland in Greece found that low specific leaf area (SLA), leaf water content, mineral content, photosynthetic carbon assimilation, and high water-use efficiency led to higher species abundances. In US grasslands, there were positive connections between species abundance, biomass, and nitrogen concentration. In an eastern Chinese forest, it was discovered that species abundance and wood density were positively connected, but that species abundance and leaf nitrogen content were negatively connected.

While many studies have attempted to use functional traits to predict species' relative abundances in plant communities, empirical findings often show weak correlations between individual traits and species abundance. Kumar et al. (2021b) argued that traits affect species performance based on both abiotic factors (such as climate and resource availability) and biotic factors (such as pathogens and competition). For instance, traits related to pathogen resistance can lead to varying levels of demographic success depending on pathogen prevalence. Similarly, the relationship between drought tolerance traits and species abundance is expected to vary across climate regions with different levels of aridity or water deficit. Moreover, many studies have focused on a limited set of easily measured traits that may not accurately reflect the abundance of coexisting plant species. In this study, we aimed to measure the nature and strength of the association between functional characteristics and species abundance in a mature subtropical evergreen forest. We measured physiological traits, including photosynthetic carbon assimilation capacity, stem hydraulic conductivity, and drought tolerance, all of which directly impact species performance and fitness.

Functional traits, which are usually assessed at the species level, are crucial for comprehending plant function. The selection of species for trait measurement is determined by the research objectives, and species may be chosen based on phylogenetic groups or other relevant evolutionary criteria. On the other hand, the environment shapes vegetation, and vegetation characteristics, in turn, influence the local flows of matter and energy. Thus, local abundance becomes an important factor for selecting vegetation for study, especially in forests or places with woody plants, where the most numerous species are prioritized for undisturbed collection. Lavorel also makes the distinction between "variable traits" and "stable traits." Variable traits can differ within a species in response to the same environmental conditions and are assessed across multiple sites within a habitat range. Examples include vegetative and reproductive plant height, nutrient concentration in leaves, and branching patterns.

Stable traits, on the other hand, remain consistent within a representative population and can include leaf and stem dry matter content or leaf toughness. The distinction is significant because species exhibit varying degrees of stability or variability in their properties across environmental gradients. Therefore, it's necessary to test traits before assuming their stability for a given species, as noted by Freschet et al. (2021). This emphasizes the complexity of plant trait variability and the need for careful selection and testing in ecological studies. In this chapter, we define functional traits as morphological, physiological, or phenological features that can be measured at the individual plant level, directly influencing growth, biochemical changes, and overall fitness. The growing need to understand and predict these traits has revitalized a long-standing tradition of studying plant characteristics, shedding light on how plants respond to environmental changes.

15.2 PLANT FUNCTIONAL TRAIT MEASUREMENT

Kumar et al. (2022) selected 10 functional traits to reflect significant axes of plant functional diversity, with an emphasis on the trade-offs between fast-growing and slow-growing species. For each species, they sampled three to five individuals whose diameter at breast height (DBH) was similar to the mean DBH of that species in the plot. Data for 48 of the 101 species had been previously reported (Bigler, 2016). They assessed the carbon, phosphorus, and nitrogen economy of leaves by measuring SLA, leaf phosphorus content per unit mass, and leaf nitrogen content per unit mass. These traits are part of the leaf economics spectrum, which describes how plants balance growth and resource use (Liu et al., 2010). For leaf area (LA) measurements, 20 fully expanded, sun-exposed leaves were collected from the canopy of three to five individuals per species. The LA was measured using a LA meter (Li et al., 2021). These measurements provide critical insights into species' resource allocation strategies and overall growth patterns.

They assessed light capture strategies by measuring the maximum CO_2 assimilation rate per unit mass and stomatal conductance per unit mass. Additionally, maximum CO_2 assimilation rate per unit area and stomatal conductance per unit area were measured using a Li-6400 portable photosynthesis system on sunny days. The photosynthetic photon flux density was adjusted based on preliminary trials to ensure that leaves of all species were light-saturated. For each species, five to ten fully expanded, sun-exposed leaves were measured to ensure accurate representation of light capture and gas exchange (Chytyk, 2010).

The study evaluated sapwood density and hydraulic conductivity to assess biomass allocation and hydraulic capacity. Healthy, leaf-bearing branches (6–8 mm in diameter and 40–60 cm in length) were collected from three to five individuals per species in the early morning. Each branch was placed in black plastic bags with moist towels and transported to the laboratory without delay. Prior to initiating measurements, branch samples were cut underwater, and their ends were carefully retrimmed with a razor blade to ensure a clean cut. To quantify leaf drought tolerance, they measured the leaf turgor loss point through a series of pressure–volume relationship assessments. Early in the morning, when leaf water potentials were at their peak, we harvested leaf-bearing branches from three to five individuals per species.

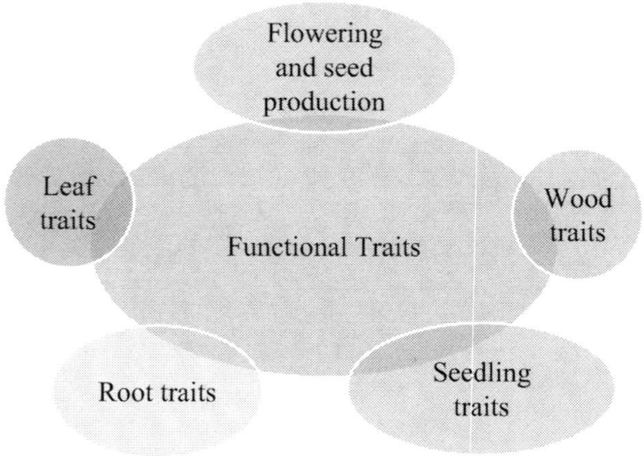

FIGURE 15.1 Schematic representation of functional plant traits.

The branches were then transferred to the laboratory, where we recut the basal ends and immersed them in distilled water to promote rehydration, while covering the leafy shoots with plastic to maintain humidity (Zhu et al., 2019) (Figure 15.1).

15.2.1 Structural Traits

Using a measuring tape, the two structural traits that characterize plant size canopy diameter and height were measured in the field. Plant height was recorded using a vertical measurement of the tree from the soil to the top of the canopy. Canopy diameter was established by measuring two perpendicular canopy widths directed north and west, respectively. Both canopy measurements were subsequently averaged to calculate the canopy diameter (Duan et al., 2016).

15.2.2 Leaf Traits

To assess LA and SLA, three fully developed, sun-exposed leaves were collected from each tree and stored in plastic bags to preserve moisture prior to scanning. The fresh LAs were measured using a HP Officejet 4500 desktop scanner, ensuring that the petiole was excluded from measurements to avoid influencing the leaf weight and misrepresenting the photosynthetic area (Khan et al., 2020). The leaves were then dried for 48 hours at 60°C and subsequently weighed. LA was calculated using ImageJ software, while SLA was determined by dividing the fresh LA by the corresponding dry weight (Khan et al., 2020).

For the carbon-to-nitrogen (C:N) ratio and leaf isotope analysis, three leaves per individual were collected. These leaves were dried for 48 hours at 60°C, then macerated and encapsulated before being sent to UC Davis Laboratory for chemical and isotopic analysis. Data on leaf phenological patterns, specifically the duration of dormancy in which tree species were leaf-free, were gathered from the literature (Klingler et al., 2020). In cases where a species was reported in multiple studies, an average value was calculated for that species.

15.2.3 Wood Traits

To minimize damage to the plants, they selected the largest branch above the main stem, averaging 50 mm in diameter, for sampling. Upon removal of the entire branch, we prepared two subsamples. The first subsample was placed in a plastic bag to preserve humidity for subsequent measurements of bark density, wood density, bark and wood water storage capacity, and bark thickness (Huynh et al., 2021).

To obtain the necessary measurements, they detached the bark from the wood. The fresh volumes of both bark and wood were measured using the water displacement method. Each sample was then submerged for 48 hours, after which we re-weighed them to determine their saturated weights. Following this, the samples were dried at 70°C for one week and weighed again to acquire their dry weights. Bark and wood densities were calculated by dividing the dry weight by their respective fresh volumes. Water storage capacities for both figures were determined by subtracting the dry weight from the saturated weight and dividing this difference by the dry weight. Bark thickness was assessed by measuring two perpendicular lengths from the vascular cambium to the outer surface of the fresh bark and calculating the average of these measurements (Kupfernagel, 2024).

15.2.4 Root Traits

To investigate root traits, we excavated a cylindrical area around the stem of each tree, measuring 40 cm in radius and 20 cm in depth, to expose the roots. The specific root area, defined as the fresh root surface area per unit of root dry mass (cm^2/g), was calculated by sampling roots with diameters smaller than 1 mm, following the methods outlined by Bordron et al. (2019). Root samples were preserved in ziplock bags inside a cooler bag to prevent moisture loss. The fresh root surface area was scanned using an HP Office jet 4500 desktop scanner and analyzed with ImageJ software. The roots were then dried at 70°C for one week and weighed to determine their dry mass. For roots larger than 1 mm, we measured the largest diameter (D) and length (L) to calculate their volume (V) using the cylindrical volume formula ($V = \pi r^2 L$), as described by Fagundes (2020). The total root volume for each individual tree was computed by summing the volumes of all measured roots, with the average root volume for each species calculated from all individual measurements. Root density was determined by dividing the root dry mass by the fresh root volume ($g\ cm^{-3}$). To obtain the fresh root volume for this calculation, one root with a diameter larger than 10 mm was sampled, and its volume was measured using the water displacement method. After drying the samples at 70°C for one week, their dry mass was recorded. To understand the physiological significance of these traits, we referred to the classifications by Cornelissen et al. (2003), outlining the trait names, units, and their respective functions.

15.2.5 Flowering and Seed Production

The reproductive processes of tree species are significantly influenced by cyclical patterns of flowering and seed production, commonly referred to as masting. There are several ways to define masting, but in general, it refers to a population's synchronous production of seeds at irregular intervals, which is frequently characterized by the episodic development of significant seed crops (Burley, 2004). At the population level, masting occurs when individuals synchronize their seed production. This phenomenon exhibits variability across years and tends to synchronize among different populations (Lamontagne and Boutin, 2007). For wind-pollinated species, such as those found in boreal environments, masting's principal evolutionary benefit is increased pollination efficiency (Pesendorfer et al., 2021). Environmental factors play a crucial role in driving both pollen and seed production. Research by Schimleck et al. (2019) examined the pollen production of several boreal species, including *Picea*, *Larix*, *Abies*, *Betula*, *Populus*, and *Alnus*, categorizing them into masting and non-masting pollen producers (Klisz et al., 2023). Masting species tend to produce a significant amount of pollen primarily before a masting event to coincide with favorable weather conditions, whereas non-masting species release pollen annually. Notably, some boreal tree species in North America and Europe exhibit intraspecific spatiotemporal synchrony in masting over large geographical areas (Vacchiano et al., 2021).

15.2.6 Wood Properties

The variability in wood properties arises from a complex interplay of genetic factors, stand-level conditions, and short-term weather events. Key wood properties, including tracheid or fiber length, strength, stiffness, and density, are critical not only for the vitality of the tree but also for the performance of wood as a material (Schimleck et al., 2019). Tracheid length varies between different trees and species, as well as within a single tree ring and between juvenile and mature wood. While it is primarily under genetic control, silvicultural practices can also exert influence. Notably, tracheid length tends to increase from the pith outward to the bark (Dahlen et al., 2021). Additionally, variations in growth rates, alongside the available time for growth, lead to intra-annual differences, resulting in longer tracheid lengths in latewood. This increase in tracheid length significantly affects various hydraulic parameters of living trees (Bryukhanova and Fonti, 2013).

Adequate mechanical properties of wood are essential for both supporting tree architecture and ensuring the functionality of wood products. Various characteristics, including hardness, tensile strength, bending strength, compression strength, and impact resistance, are all closely related to wood density. Among these, wood density and ring width are the most extensively studied parameters. Wood density represents the total amount of cell wall material per unit volume (Ramage et al., 2017). Ring width plays a significant role in determining the percentage of latewood, with smaller tree rings typically exhibiting a higher proportion of latewood and consequently greater wood density (Pritzkow et al., 2014). However, it's important to note that there are exceptions to this general trend. The impact of climate on earlywood density is particularly significant due to its close relationship with plant hydraulics. Low earlywood density can lead to negative water pressure that exceeds the wood's fracture limits, potentially resulting in radial or stem cracks. However, data on earlywood density obtained through X-ray densitometry is scarce in dendrochronological studies. In high-altitude or high-latitude sites, maximum density during the growing season is primarily influenced by summer temperatures, which coincide with the period of wood formation (Klisz et al., 2023).

In addition to environmental factors, wood formation and quality are strongly governed by genetic factors. Studies on various *Abies* species and provenances, particularly *Abies alba*, in eastern Austria have shown that species and provenance have a considerable impact on wood-quality characteristics (Segura et al., 2016). Variability in wood traits and their responses to drought indicated that the genus *Abies* accounts for approximately 10%–20% of the variance in wood density, while it contributes only minimally to the variance in ring width. Conversely, the provenance of *A. alba* influences about 10%–15% of the variation in ring width, but has a negligible effect on ring density (Klisz et al., 2023).

15.2.7 Seedling Functional Traits

The study conducted by Metz et al. (2023) focused on assessing various functional traits of young seedlings, aiming to better understand the ecological dynamics within the Yasuní Forest Dynamics Plot (FDP). A total of 747 seedlings belonging to 138 species were collected, highlighting the site's rich biodiversity. However, the research noted the limited overlap in species across different sites, with only a few species appearing at multiple locations.

To ensure standardization in the developmental stage of the seedlings, visuals of the first fully expanded photosynthetic organ were prioritized. The research took a detailed approach by collecting seedlings over several years, accounting for variations in germination times and growth rates influenced by environmental factors (Connell and Green, 2000). The cotyledon types of each species were categorized following Caplin's (2019) framework, distinguishing between photosynthetic and non-photosynthetic storage cotyledons. The latter encompassed various forms of cotyledons, from those storing energy in fleshy structures to those using specialized haustorial mechanisms.

In addition to analyzing cotyledon morphology, a series of quantitative measurements were recorded for each seedling. These included variables such as cotyledon and leaf count, plant height, root length, and fresh mass across various plant sections, as well as stem dimensions and volumetric parameters. A comprehensive analysis of the plants' dry mass was also performed, providing insights into their physiological characteristics. The methodology adopted, including scanning techniques for cotyledons and leaves, suggests a thorough approach aimed at capturing the complexity of seedling traits in the diverse setting of the Yasuní FDP (Caplin, 2019).

The findings from this study contribute to the understanding of seedling dynamics in tropical forests, offering insights into how functional traits may influence species survival and recruitment in varying ecological contexts. SLA is a key metric reflecting the dry matter content of leaves and is often associated with species growth rates, leaf longevity, and the investment in secondary metabolites for herbivore defense. Similarly, stem-specific density influences growth and survival, as variations in stem vessel width can impact hydraulic efficiency and mechanical resilience against herbivores and pathogens (Hidalgo-Triana et al., 2023).

In addition, specific root length (SRL), akin to SLA in leaves, indicates how resources are allocated within the plant, playing a significant role in nutrient uptake, relative growth rates, and root lifespan. However, SRL can exhibit different responses to growth conditions compared to leaf traits (Fort et al., 2013). The root-to-shoot biomass ratio serves as an index of a plant's allocation of dry mass between belowground and aboveground tissues. This ratio is believed to correlate with demographic performance, as it can vary among species and along environmental gradients. Factors influencing this variation may include species shade tolerance and the trade-offs associated with investing in mechanical tissue strength, expanding photosynthetic areas, carbon storage, and chemical defenses (Metz et al., 2023).

15.3 CONCLUSION

The recognition of microclimate's role in forest ecosystems has advanced significantly, providing a robust framework for investigating its impact on forest functioning. Understanding how microclimates influence biodiversity is crucial for effective forest management, sustainable use, and ensuring ecosystem services in a warming climate. Consequently, microclimates should be regarded as a vital ecosystem service. PFTs are essential for understanding forest ecology and ecosystem functioning. They help elucidate how forest plants adapt to various environments and respond to climate change. By optimizing resource efficiency and enhancing the resilience of plants to biotic stresses, particularly pest and disease resistance, these traits play a crucial role in developing climate-smart forestry practices.

REFERENCES

Bigler, C., 2016. Trade-offs between growth rate, tree size and lifespan of mountain pine (*Pinus montana*) in the Swiss National Park. *PloS one*, *11*(3), p. e0150402.

Bordron, B., Robin, A., Oliveira, I.R., Guillemot, J., Laclau, J.P., Jourdan, C., Nouvellon, Y., Abreu-Junior, C.H., Trivelin, P.C.O., Gonçalves, J.L.M. and Plassard, C., 2019. Fertilization increases the functional specialization of fine roots in deep soil layers for young *Eucalyptus grandis* trees. *Forest Ecology and Management*, *431*, pp. 6–16.

Bryukhanova, M. and Fonti, P., 2013. Xylem plasticity allows rapid hydraulic adjustment to annual climatic variability. *Trees*, *27*, pp. 485–496.

Burley, J., 2004. *Encyclopedia of Forest Sciences*. Netherlands: Elsevier.

Caplin, N.M., 2019. Developmental, morphological and physiological effects of chronic low doses of ionising radiation on plants on earth and in space (*Doctoral dissertation*, University of the West of England).

Chytyk, C.J., 2010. Leaf photosynthesis in wheat (*Triticum* spp.) under conditions of low temperature and CO_2 enrichment (*Doctoral dissertation*, University of Saskatchewan).

Connell, J.H. and Green, P.T., 2000. Seedling dynamics over thirty-two years in a tropical rain forest tree. *Ecology*, *81*(2), pp. 568–584.

Cornelissen, J.H., Lavorel, S., Garnier, E., Díaz, S., Buchmann, N., Gurvich, D.E., Reich, P.B., Ter Steege, H., Morgan, H.D., Van Der Heijden, M.G.A. and Pausas, J.G., 2003. A handbook of protocols for standardised and easy measurement of plant functional traits worldwide. *Australian Journal of Botany*, *51*(4), pp. 335–380.

Dahlen, J., Nabavi, M., Auty, D., Schimleck, L. and Eberhardt, T.L., 2021. Models for predicting the within-tree and regional variation of tracheid length and width for plantation loblolly pine. *Forestry: An International Journal of Forest Research*, *94*(1), pp. 127–140.

Díaz, S., Hodgson, J.G., Thompson, K., Cabido, M., Cornelissen, J.H., Jalili, A., Montserrat-Martí, G., Grime, J.P., Zarrinkamar, F., Asri, Y. and Band, S.R., 2004. The plant traits that drive ecosystems: evidence from three continents. *Journal of Vegetation Science*, *15*(3), pp. 295–304.

Díaz, S., Kattge, J., Cornelissen, J.H., Wright, I.J., Lavorel, S., Dray, S., Reu, B., Kleyer, M., Wirth, C., Colin Prentice, I. and Garnier, E., 2016. The global spectrum of plant form and function. *Nature*, *529*(7585), pp. 167–171.

Duan, T., Chapman, S.C., Holland, E., Rebetzke, G.J., Guo, Y. and Zheng, B., 2016. Dynamic quantification of canopy structure to characterize early plant vigour in wheat genotypes. *Journal of Experimental Botany*, *67*(15), pp.4523–4534.

Fagundes, M.V., 2020. Atributosfuncionais de caatinga, umaflorestasazonalmenteseca: estratégias, distribuiçãoespacial, interações entre plantas e restauração de comunidades.

Fagundes, M.V., Souza, A.F., Oliveira, R.S. and Ganade, G., 2022. Functional traits above and below ground allow species with distinct ecological strategies to coexist in the largest seasonally dry tropical forest in the Americas. *Frontiers in Forests and Global Change*, *5*, p. 930099.

Fort, F., Jouany, C. and Cruz, P., 2013. Root and leaf functional trait relations in *Poaceae* species: implications of differing resource-acquisition strategies. *Journal of Plant Ecology*, *6*(3), pp. 211–219.

Freschet, G.T., Pagès, L., Iversen, C.M., Comas, L.H., Rewald, B., Roumet, C., Klimešová, J., Zadworny, M., Poorter, H., Postma, J.A. and Adams, T.S., 2021. A starting guide to root ecology: strengthening ecological concepts and standardising root classification, sampling, processing and trait measurements. *New Phytologist*, *232*(3), pp. 973–1122.

Hidalgo-Triana, N., Pérez-Latorre, A.V., Adomou, A.C., Rudner, M. and Thorne, J.H., 2023. Adaptations to the stressful combination of serpentine soils and Mediterranean climate drive plant functional groups and trait richness. *Frontiers in Plant Science*, *14*, p. 1040839.

Huynh, T., Lee, D.J., Applegate, G. and Lewis, T., 2021. Field methods for above and belowground biomass estimation in plantation forests. *MethodsX*, *8*, p. 101192.

Khan, A., Sun, J., Zarif, N., Khan, K., Jamil, M.A., Yang, L., Clothier, B. and Rewald, B., 2020. Effects of increased N deposition on leaf functional traits of four contrasting tree species in northeast China. *Plants*, *9*(9), p. 1231.

Klingler, A., Schaumberger, A., Vuolo, F., Kalmár, L.B. and Pötsch, E.M., 2020. Comparison of direct and indirect determination of leaf area index in permanent grassland. *PFG–Journal of Photogrammetry, Remote Sensing and Geoinformation Science*, *88*(5), pp. 369–378.

Klisz, M., Chakraborty, D., Cvjetković, B., Grabner, M., Lintunen, A., Mayer, K., George, J.P. and Rossi, S., 2023. Functional traits of boreal species and adaptation to local conditions. In Miguel Montoro Girona, Hubert Morin, Sylvie Gauthier, Yves Bergeron (Eds), *Boreal Forests in the Face of Climate Change: Sustainable Management* (pp. 323–355). Cham: Springer International Publishing.

Kupfernagel, C., 2024. Understanding the influence of wood species on resin modification and property enhancement in resin modified wood.

Lamontagne, J.M. and Boutin, S., 2007. Local-scale synchrony and variability in mast seed production patterns of *Picea glauca*. *Journal of Ecology*, *95*(5), pp. 991–1000.

Li, R., Zhu, S., Lian, J., Zhang, H., Liu, H., Ye, W. and Ye, Q., 2021. Functional traits are good predictors of tree species abundance across 101 subtropical forest species in China. *Frontiers in Plant Science*, *12*, p. 541577.

Liu, B., Liang, E., Liu, K. and Camarero, J.J., 2018. Species-and elevation-dependent growth responses to climate warming of mountain forests in the Qinling Mountains, central China. *Forests*, *9*(5), p. 248.

Liu, F., Yang, W., Wang, Z., Xu, Z., Liu, H., Zhang, M., Liu, Y., An, S. and Sun, S., 2010. Plant size effects on the relationships among specific leaf area, leaf nutrient content, and photosynthetic capacity in tropical woody species. *Acta Oecologica*, *36*(2), pp. 149–159.

Maharjan, S.K., Sterck, F.J., Dhakal, B.P., Makri, M. and Poorter, L., 2021. Functional traits shape tree species distribution in the Himalayas. *Journal of Ecology*, *109*(11), pp. 3818–3834.

Mashizi, A.K. and Sharafatmandrad, M., 2023. Dry forests conservation: a comprehensive approach linking ecosystem services to ecological drivers and sustainable management. *Global Ecology and Conservation*, *47*, p. e02652.

Metz, M.R., Wright, S.J., Zimmerman, J.K., Hernandéz, A., Smith, S.M., Swenson, N.G., Umaña, M.N., Valencia, L.R., Waring-Enriquez, I., Wordell, M. and Zambrano, M., 2023. Functional traits of young seedlings predict trade-offs in seedling performance in three neotropical forests. *Journal of Ecology*, *111*(12), pp. 2568–2582.

Pesendorfer, M.B., Ascoli, D., Bogdziewicz, M., Hacket-Pain, A., Pearse, I.S. and Vacchiano, G., 2021. The ecology and evolution of synchronized reproduction in long-lived plants. *Philosophical Transactions of the Royal Society B*, 376(1839), p. 20200369.

Pritzkow, C., Heinrich, I., Grudd, H. and Helle, G., 2014. Relationship between wood anatomy, tree-ring widths and wood density of *Pinus sylvestris* L. and climate at high latitudes in northern Sweden. *Dendrochronologia*, *32*(4), pp. 295–302.

Ramage, M.H., Burridge, H., Busse-Wicher, M., Fereday, G., Reynolds, T., Shah, D.U., Wu, G., Yu, L., Fleming, P., Densley-Tingley, D. and Allwood, J., 2017. The wood from the trees: the use of timber in construction. *Renewable and Sustainable Energy Reviews*, *68*, pp. 333–359.

Schimleck, L., Dahlen, J., Apiolaza, L.A., Downes, G., Emms, G., Evans, R., Moore, J., Pâques, L., Van den Bulcke, J. and Wang, X., 2019. Non-destructive evaluation techniques and what they tell us about wood property variation. *Forests*, *10*(9), p. 728.

Segura, V., Descauses, M.C.L., Charpentier, J.P., Kinkel, K., Mandin, C., Ader, K., Belmokhtar, N., Boizot, N., Buret, C., Dejardin, A. and Guérin, V., 2016, May. A systems biology approach for identifying candidate genes involved in the natural variability of biomass yield and chemical properties in black poplar. In *IUFRO Genomics and Forest Tree Genetics* (p. np), HAL Open Science.

Vacchiano, G., Pesendorfer, M.B., Conedera, M., Gratzer, G., Rossi, L. and Ascoli, D., 2021. Natural disturbances and masting: from mechanisms to fitness consequences. *Philosophical transactions of the Royal Society B*, 376(1839), p. 20200384.

Zhu, S.D., Li, R.H., He, P.C., Siddiq, Z., Cao, K.F. and Ye, Q., 2019. Large branch and leaf hydraulic safety margins in subtropical evergreen broadleaved forest. *Tree Physiology*, *39*(8), pp. 1405–1415.

16 Future Avenues
Advancing Research and Implementation of Functional Traits

Laxmidas Verma, Anand Kumar, and Anamika Yadav

16.1 INTRODUCTION

Plant functional traits refer to the inherent characteristics or attributes of plants that exert influence on their ecological roles and interactions within the environment (Violle et al., 2007). These traits play a pivotal role in comprehending the adaptive mechanisms of plants to diverse ecological niches, their competitive dynamics with other species, and their responses to environmental fluctuations. Researchers use plant functional traits to categorize and analyze plant species, evaluate ecosystem functions, and predict ecosystem reactions to impacts or climatic shifts. In the realm of crop enhancement, these traits prove highly advantageous (Kumar et al., 2021d, 2022; Malik et al., 2024). Moreover, plant functional traits are increasingly utilized to create mechanistic models that can predict how ecological communities respond to abiotic and biotic perturbations, as well as how species influence ecosystem functions and services in our rapidly changing world. However, significant challenges persist in this area of research (Kumar et al., 2021a; Funk et al., 2017).

Forest microclimates differ significantly from the surrounding environment, influencing forest biodiversity and functions. To enhance our understanding and predictions regarding how forests respond to climate change, it is essential to integrate microclimates into ecological research. Despite their potential influence, our knowledge of how microclimates beneath tree canopies affect species, community dynamics, and ecosystem processes remains inadequate. De et al. emphasize the necessity of considering microclimates to comprehend biodiversity patterns and ecosystem functions across diverse forest landscapes. Additionally, we discuss the bidirectional relationship between macroclimate warming and microclimates, highlighting how changes in the broader climate can influence microclimates and how land-use alterations may further modify these interactions across different biomes (Kumar et al., 2020a; De Frenne et al., 2021).

DOI: 10.1201/9781003545231-16

16.2 HYBRIDIZATION IN TREE SPECIES

The phenomenon of hybridization among tree species, as highlighted in the aforementioned studies, underscores the complex interplay between genetic variation and environmental adaptation. Bawa and Holliday (2017) emphasize that hybridization can facilitate local adaptation in transitional environments, supporting the notion that gene flow among closely related species can be beneficial in the face of changing conditions (Bawa & Holliday, 2017). The insights from the *Populus trichocarpa* × *P. balsamifera* hybrid zone illustrate how one species can gain adaptive traits from another that thrive in different environmental conditions. Such directional introgression enables species like *P. trichocarpa* to expand their habitats into colder areas, demonstrating fluidity in species ranges driven by hybridization (Kumar et al., 2020b). Similarly, Suarez-Gonzalez et al. (2018) investigate the hybridization between *Pinus strobiformis* and *Pinus flexilis*, revealing how this genetic exchange may facilitate the northward expansion of *P. strobiformis* in response to climate changes. Their extensive research, which includes morphological assessments and genetic studies, demonstrates that while morphological traits may suggest one hybrid zone location, genomic data can disclose a different story, revealing continuing shifts in hybrid zones.

Importantly, this study points to the potential of natural hybridization to contribute positively to biodiversity, contrary to views that regard it as detrimental. Introgression can lead to the emergence of intermediate phenotypes and facilitate adaptive variation that can be crucial for species resilience in an era of climate change. Menon et al.'s findings are especially noteworthy because they provide a genomic perspective on these processes, emphasizing the importance of hybridization in evolutionary dynamics and the need to incorporate these mechanisms into predictive models for species' future distributions and adaptations. This research may guide conservation strategies, ensuring that biodiversity is understood as a dynamic and adaptive trait in response to shifting environmental conditions (Kumar et al., 2015, 2016; Menon et al., 2020).

16.3 GENOMIC TOOLS

Genomics alone cannot fully address the complexities of how populations adapt to their environments. To better understand adaptation to both biotic and abiotic stresses throughout the lifespan of tree populations, it is essential to assess phenotypes across large cohorts. The phenomics era, defined by technologies such as unmanned aerial vehicles equipped with hyperspectral sensors, has transformed our ability to monitor phenology and responses to environmental stresses such as drought and pest invasions by generating leaf spectral indices (e.g., Calderon et al., 2015). Precision phenotyping not only enhances our understanding of tree responses but also opens new avenues for researching specific issues like resistance to various stresses.

Despite the emergence of low-cost, high-throughput technologies capable of producing vast genomic and phenotypic data, forest biologists studying long-lived tree populations encounter significant challenges. As noted by Wegrzyn et al. (2020), the current state of data integration for non-model species lags far behind that for model species.

To avoid redundancy and inefficiencies in data storage and retrieval while keeping pace with technological advancements, the development of robust cyberinfrastructures is imperative. These infrastructures should support all facets of the data life cycle—from acquisition and storage to integration and visualization (Kumar et al., 2017, 2019; Wegrzyn et al., 2020).

To effectively implement and sustain such cyberinfrastructures, substantial improvements must be made in four critical areas: data standards, ontologies, analytic workflows, and integrated databases. Wegrzyn et al. (2020) provide insights into these areas, highlighting best practices currently employed by the genomics community. Among these, establishing data standards is paramount. Initiatives such as FAIR (findability, accessibility, interoperability, and reusability) should be emulated and endorsed by academic journals and funding agencies alike. Adopting these best practices in cyberinfrastructures will ultimately benefit all research communities studying non-model species, as well as advance our understanding of ecological and evolutionary dynamics (Wegrzyn et al., 2020; Isabel et al., 2020).

16.4 GENOMIC EDITING

Genetic engineering (GE) has indeed revolutionized agricultural practices by introducing desired traits into crops, thereby improving their resilience and productivity. As outlined by Fernandez-Cornejo et al. (2014), the introduction of herbicide tolerance allows farmers to effectively manage weeds without damaging crops, providing a more efficient and sustainable approach to weed control. Similarly, engineering crops for resistance to pests and diseases reduces the need for chemical pesticides, benefiting both the environment and the health of consumers (Fernandez-Cornejo et al., 2014). The development of abiotic stress tolerances, such as drought and salinity, is especially important in the context of climate change, as extreme weather patterns are increasingly threatening crop viability. Improved quality attributes, including nutritional enhancements and better flavor profiles, also cater to changing consumer preferences, thereby creating additional market opportunities (Chan et al., 2020).

However, as observed by Chan et al. (2020), the complexities of genetic characteristics, particularly those related to yield, provide significant challenges. Yield is governed by multiple genes and environmental factors, making it difficult to achieve consistent improvements through GE alone. The interaction of different traits and environmental factors might result in unanticipated consequences, complicating the breeding and commercialization processes. These challenges underline the need for continued research and development in GE, as well as an integrated approach that combines traditional breeding practices with modern biotechnological methods to achieve the desired agricultural outcomes effectively (Chan et al., 2020).

Overall, while GE has revealed considerable potential for crop improvement, continued efforts are required to overcome the inherent complexity and ensure that these technologies can be successfully implemented on a commercial scale. The application of GE in forestry reflects a growing recognition of its potential to address critical challenges faced in forest management and conservation. Among the primary objectives of GE in this context are species rescue and restoration, which are increasingly vital due to habitat loss, climate change, and disease pressures that threaten many tree species. By employing genetic tools, breeders can develop trees that are better suited to survive in changing environments, resist pests and diseases, and recover lost populations (Harfouche et al., 2011).

Additionally, enhancing the performance of forest plantations is another key goal of GE. Breeders strive to enhance growth rates, product yield, and overall quality of wood and non-wood forest products, which is critical for addressing rising global resource demand while optimizing land-use efficiency. Achieving these objectives often involves incorporating desirable traits from wild relatives or other species into commercial lines, ensuring that the resulting varieties are not only productive but also resilient to environmental stresses (Chang et al., 2018). An in-depth knowledge of the current gene pools and breeding lines is necessary for the strategic incorporation of GE into conventional breeding programs. It requires careful planning to ensure that new traits are effectively incorporated without disrupting the genetic diversity that is crucial for the long-term health and adaptability of forest ecosystems.

Furthermore, while GE offers substantial advantages, it also comes with ethical considerations and regulatory challenges that must be navigated. Stakeholder engagement and transparent communication regarding the benefits and risks associated with genetically engineered trees are essential for gaining public trust and acceptance (Kumar et al., 2021b; Harfouche et al., 2011; Chang et al., 2018).

The application of GE in forestry indeed mirrors agricultural practices in various ways, especially with the focus on trait enhancement and performance improvement of specific tree species in managed environments. Through the use of GE, researchers are working to address a number of forestry-related issues, including growth rates, disease resistance, and environmental adaptation. As noted, during the GE development phase, the elimination of natural regeneration opportunities is critical. This is achieved through methods such as removing reproductive structures or employing genetic modifications that lead to sterility strategies designed to minimize unintended crossbreeding with wild populations. This is particularly important for maintaining the integrity of altered features in the target species while limiting the escape of changed genes into wild relatives, which could disturb ecosystems or have unforeseen consequences. The use of sterile or single-genotype planting ensures tight control over the genetic characteristics of the cultivated trees. Research, including that by Brunner et al. (2007), Zhang et al. (2012), has emphasized these methods as effective ways to safeguard native biodiversity while leveraging the advantages of GE.

However, this approach also raises important ecological and ethical considerations. The long-term impacts of genetically engineered trees on natural forest ecosystems, potential loss of genetic diversity, and the need for strict regulation and monitoring are aspects that must be thoroughly examined. Balancing the benefits of GE, such as increased productivity and resilience to environmental stressors, against ecological risks is crucial for sustainable forestry practices. As research advances, ongoing discussions among scientists, policymakers, and the public will be essential to guide the responsible deployment of GE in forestry.

16.5 EPIGENETIC

Epigenetics enhances our understanding of gene influence on traits, a complex issue for scientists. It goes beyond traditional genetics by exploring how genes and their products interact to shape plant characteristics (Waddington, 2012). Contemporary molecular biology focuses on DNA-related processes that dictate phenotype, extending beyond traditional DNA sequences. A key area of advancement is the application of epigenetics in plant breeding, which allows for the tracing of heritable epigenetic modifications that enhance crop traits (Slotkin, 2016). Modifications such as DNA methylation, histone alterations, chromatin remodeling, and small RNA activity influence phenotypic variation and crop fitness (Varotto et al., 2020). DNA methylation plays a crucial role in essential traits like seed dormancy, flowering time, and yield, revealing its significance in plant evolution and domestication (Song et al., 2017). Understanding DNA methylation enables breeders to promote beneficial traits while minimizing adverse epigenetic changes resulting from breeding strategies (Lin et al., 2019).

Epigenetic mechanisms in eukaryotes are primarily supported by chromatin structure, which consists of DNA wrapped around nucleosomes (histone octamers). Chromatin exists in two states: euchromatin, characterized by a decondensed state that is gene-rich and transcriptionally active, and heterochromatin, which is condensed, enriched in repetitive sequences, and transcriptionally silent. These chromatin states are regulated by epigenetic marks that define various chromatin domains, which influence gene expression and cellular functions (Roudier et al., 2011; Bannister and Kouzarides, 2011; Amaral et al., 2020).

Research on forest trees is poised for significant advancement due to improved genomic resources and decreasing costs. The sequencing of genomes like *Populus trichocarpa* (Tuskan et al., 2006), *Picea abies* (Nystedt et al., 2013), and *Eucalyptus grandis* (Myburg et al., 2014) has likely driven a rise in forest epigenetic studies, which depend on high-quality genome assemblies. Furthermore, the term "epigenetic memory" describes the long-term impacts of maternal environmental factors on seed development and offspring performance in forest tree species, especially Picea abies. Johnsen et al. (2005) identified that temperature and photoperiod during seed production interact to create this memory, influencing phenology, frost hardiness, and bud burst timing through changes in gene expression (Johnsen et al., 2005; Kvaalen and Johnsen, 2008; Kumar et al., 2021c). Supportingly, Besnard et al. (2008) found that these maternal conditions primarily affect phenotypic traits rather than driving significant genotypic selection (Besnard et al., 2008). This method explains why Central European *P. abies* can quickly adapt in Norway, exhibiting bud set characteristics comparable to local seedlings rather than their own origin (Skroppa et al., 2010).

16.6 CONCLUSION

Long-term forest health is crucial for global ecosystem functions and is vital in addressing climate change. Healthy forests regulate climate, support biodiversity, and provide essential resources, making their maintenance essential for sustainability and ecological balance. Forest ecosystem architects are exposed to various abiotic and biotic stresses throughout their lifespan. Their poor adaptability to climate change may make them more vulnerable to both native and foreign pests, as well as other natural disturbances. Genomics is enhancing our understanding and tools for forest management and conservation, enabling quicker adaptations in breeding programs to address climate change, even for non-model species. In addition, the integration of physiology and gene editing is revolutionizing plant trait discovery, offering significant advancements for agriculture and environmental sustainability. This synergy helps in the identification of intricate connections between genetic elements, physiological systems, and environmental effects that promote development, stress resilience, and productivity. However, this progress raises important policy concerns balancing open access for scientific advancement with the protection of intellectual property is crucial. Future research should prioritize the collection of functional trait data to better understand species dynamics, functional types, and their relationships during land-use transitions in tropical ecosystems.

REFERENCES

Amaral, J., Ribeyre, Z., Vigneaud, J., Sow, M.D., Fichot, R., Messier, C., Pinto, G., Nolet, P. and Maury, S., 2020. Advances and promises of epigenetics for forest trees. *Forests*, *11*(9), p. 976.

Bannister, A.J. and Kouzarides, T., 2011. Regulation of chromatin by histone modifications. *Cell Research*, *21*(3), pp. 381–395.

Bawa, R. and Holliday, J.A., 2017. Genomics of speciation in temperate and boreal angiosperm trees. In Bawa, R., Holliday, J.A. (Eds), *Comparative and Evolutionary Genomics of Angiosperm Trees* (pp. 159–177).

Besnard, G., Acheré, V., Jeandroz, S., Johnsen, Ø., Rampant, P.F., Baumann, R., Müller-Starck, G., Skrøppa, T. and Favre, J.M., 2008. Does maternal environmental condition during reproductive development induce genotypic selection in *Picea abies*? *Annals of Forest Science*, *65*(1), pp. 1–6.

Brunner, A.M., Li, J., DiFazio, S.P., Shevchenko, O., Montgomery, B.E., Mohamed, R., Wei, H., Ma, C., Elias, A.A., VanWormer, K. and Strauss, S.H., 2007. Genetic containment of forest plantations. *Tree Genetics & Genomes*, *3*, pp. 75–100.

Chan, R.L., Trucco, F. and Otegui, M.E., 2020. Why are second-generation transgenic crops not yet available in the market? *Journal of Experimental Botany*, *71*(22), pp. 6876–6880.

Chang, S., Mahon, E.L., MacKay, H.A., Rottmann, W.H., Strauss, S.H., Pijut, P.M., Powell, W.A., Coffey, V., Lu, H., Mansfield, S.D. and Jones, T.J., 2018. Genetic engineering of trees: progress and new horizons. *In Vitro Cellular & Developmental Biology-Plant*, *54*, pp. 341–376.

De Frenne, P., Lenoir, J., Luoto, M., Scheffers, B.R., Zellweger, F., Aalto, J., Ashcroft, M.B., Christiansen, D.M., Decocq, G., De Pauw, K. and Govaert, S., 2021. Forest microclimates and climate change: importance, drivers and future research agenda. *Global Change Biology*, *27*(11), pp. 2279–2297.

Fernandez-Cornejo, J., Wechsler, S., Livingston, M. and Mitchell, L., 2014. Genetically engineered crops in the United States. *USDA-ERS Economic Research Report*, (162).

Funk, J.L., Larson, J.E., Ames, G.M., Butterfield, B.J., Cavender-Bares, J., Firn, J., Laughlin, D.C., Sutton-Grier, A.E., Williams, L. and Wright, J., 2017. Revisiting the Holy Grail: using plant functional traits to understand ecological processes. *Biological Reviews*, *92*(2), pp. 1156–1173.

Harfouche, A., Meilan, R. and Altman, A., 2011. Tree genetic engineering and applications to sustainable forestry and biomass production. *Trends in Biotechnology*, *29*(1), pp. 9–17.

Isabel, N., Holliday, J.A. and Aitken, S.N., 2020. Forest genomics: advancing climate adaptation, forest health, productivity, and conservation. *Evolutionary Applications*, *13*(1), pp. 3–10.

Johnsen, O, Dæhlen, O.G., Østreng, G. and Skrøppa, T., 2005. Daylength and temperature during seed production interactively affect adaptive performance of Picea abies progenies. *New Phytologist*, *168*(3), pp. 589–596.

Kumar, A., Dwivedi, G. K., Tewari, S., Paul, J., Anand, R., Kumar, N., Kumar, P., Singh, H. and Kaushal, R., 2020a. Carbon mineralization and inorganic nitrogen pools under *Terminalia chebula* Retz.-Based agroforestry system in Himalayan foothills, India. *Forest Science*, *66*(5), pp. 634–643.

Kumar, A., Dwivedi, G. K., Tewari, S., Paul, J., Sah, V. K., Singh, H., Kumar, P., Kumar, N. and Kaushal, R., 2020b. Soil organic carbon pools under *Terminalia chebula* Retz. based agroforestry system in Himalayan foothills, India. *Current Science*, *118*(7), pp. 1098–1103.

Kumar, A., Kumar, P., Singh, H., Bisht, S. and Kumar, N., 2021a. Relationship of physiological plant functional traits with soil carbon stock in temperate forest of Garhwal Himalaya. *Current Science*, *120*(8), pp. 1368–1373.

Kumar, A., Kumar, P., Singh, H. and Kumar, N., 2021b. Modulation of plant functional traits under essential plant nutrients during seasonal regime in natural forests of Garhwal Himalayas. *Plant and Soil*, *465*, pp. 197–212.

Kumar, A., Kumar, P., Singh, H. and Kumar, N., 2021c. Impact of plant functional traits on near saturated hydraulic conductivity of soil under different forests of Kempty watershed in Garhwal Himalaya. *Indian Journal of Soil Conservation*, *49*(1), pp. 38–44.

Kumar, A., Singh, H., Kumari, G., Bisht, S., Malik, A., Kumar, N., Singh, M., Raturi, A., Barthwal, S., Thakur, A. and Kaushal, R., 2022. Adaptive resilience of roadside trees to vehicular emissions via leaf enzymatic, physiological, and anatomical trait modulations. *Environmental Pollution*, *313*, pp. 120191.

Kumar, N., Jeena, N. and Singh, H., 2019. Elevated temperature modulates rice pollen structure: a study from foothill Himalayan agro-ecosystem in India. *3 Biotech*, *9*, pp. 175.

Kumar, N., Jeena, N., Kumar, A., Khairakpam, R. and Singh, H., 2021d. Comparative response of rice cultivars to elevated air temperature in Bhabar region of Indian Himalaya: status on yield attributes. *Heliyon*, *7*, pp. e07474.

Kumar, N., Kumar, N., Shukla, A., Shankhdhar, S.C. and Shankhdhar, D., 2015. Impact of terminal heat stress on pollen viability and yield attributes of rice (*Oryza sativa* L.). *Cereal Research Communications*, *43*(4), pp. 616–626.

Kumar, N., Shankhdhar, S.C. and Shankhdhar, D. 2016. Impact of elevated temperature on antioxidant activity and membrane stability in different genotypes of rice (*Oryza sativa* L.). *Indian Journal of Plant Physiology (Now Plant Physiology Report)*, *21*(1), pp. 37–43.

Kumar, N., Suyal, D.C., Sharma, I.P., Verma, A. and Singh, H., 2017. Elucidating stress proteins in rice (*Oryza sativa* L.) genotype under elevated temperature: a proteomic approach to understand heat stress response. *3 Biotech*, *7*, p. 205.

Kvaalen, H. and Johnsen, Ø., 2008. Timing of bud set in Picea abies is regulated by a memory of temperature during zygotic and somatic embryogenesis. *New Phytologist*, *177*(1), pp. 49–59.

Lin, W., Xiao, X., Zhang, H., Li, Y., Liu, S., Sun, W., Zhang, X. and Wu, Q., 2019. Whole-genome bisulfite sequencing reveals a role for DNA methylation in variants from callus culture of pineapple (*Ananas comosus* L.). *Genes*, *10*, pp. 877.

Malik, A., Kumar, K., Kumari, G., Kumar, N. and Singh, H., 2024. Application of functional traits in modelling productivity and resilience under climate change. In Kumar. N. and Singh, H. (Eds), *Plant Functional Traits for Improving Productivity* (pp. 77–96). Singapore: Springer Nature Singapore.

Menon, M., Landguth, E., Leal-Saenz, A., Bagley, J.C., Schoettle, A.W., Wehenkel, C., Flores-Renteria, L., Cushman, S.A., Waring, K.M. and Eckert, A.J., 2020. Tracing the footprints of a moving hybrid zone under a demographic history of speciation with gene flow. *Evolutionary Applications*, *13*(1), pp. 195–209.

Myburg, A.A., Grattapaglia, D., Tuskan, G.A., Hellsten, U., Hayes, R.D., Grimwood, J., Jenkins, J., Lindquist, E., Tice, H., Bauer, D. and Goodstein, D.M., 2014. The genome of *Eucalyptus grandis*. *Nature*, *510*(7505), pp. 356–362.

Nystedt, B., Street, N.R., Wetterbom, A., Zuccolo, A., Lin, Y.C., Scofield, D.G., Vezzi, F., Delhomme, N., Giacomello, S., Alexeyenko, A. and Vicedomini, R., 2013. The Norway spruce genome sequence and conifer genome evolution. *Nature*, *497*(7451), pp. 579–584.

Roudier, F., Ahmed, I., Bérard, C., Sarazin, A., Mary-Huard, T., Cortijo, S., Bouyer, D., Caillieux, E., Duvernois-Berthet, E., Al-Shikhley, L. and Giraut, L., 2011. Integrative epigenomic mapping defines four main chromatin states in Arabidopsis. *The EMBO Journal*, *30*(10), pp. 1928–1938.

Skroppa, T., Tollefsrud, M.M., Sperisen, C. and Johnsen, O., 2010. Rapid change in adaptive performance from one generation to the next in *Picea abies*—Central European trees in a Nordic environment. *Tree Genetics & Genomes*, *6*, pp. 93–99.

Slotkin, R.K., 2016. Plant epigenetics: from genotype to phenotype and back again. *Genome Biology*, *17*, p. 57.

Song, Q., Zhang, T., Stelly, D.M. and Chen, Z.J., 2017. Epigenomic and functional analyses reveal roles of epialleles in the loss of photoperiod sensitivity during domestication of allotetraploid cottons. *Genome Biol*ogy, *18*, p. 99.

Suarez-Gonzalez, A., Hefer, C.A., Lexer, C., Douglas, C.J. and Cronk, Q.C., 2018. Introgression from Populus balsamifera underlies adaptively significant variation and range boundaries in *P. trichocarpa*. *New Phytologist*, *217*(1), pp. 416–427.

Tuskan, G.A., Difazio, S., Jansson, S., Bohlmann, J., Grigoriev, I., Hellsten, U., Putnam, N., Ralph, S., Rombauts, S., Salamov, A. and Schein, J., 2006. The genome of black cottonwood, *Populus trichocarpa* (Torr. & Gray). *Science*, *313*(5793), pp. 1596–1604.

Varotto, S., Tani, E., Abraham, E., Krugman, T., Kapazoglou, A., Melzer, R., Radanović, A. and Miladinović, D., 2020. Epigenetics: possible applications in climate-smart crop breeding. *Journal of Experimental Botany*, *71*, pp. 5223–5236.

Violle, C., Navas, M.L., Vile, D., Kazakou, E., Fortunel, C., Hummel, I. and Garnier, E., 2007. Let the concept of trait be functional! *Oikos*, *116*(5), pp. 882–892.

Waddington, C.H., 2012. The epigenotype. *International Journal of Epidemiology*, *41*, pp. 10–13.

Wegrzyn, J.L., Falk, T., Grau, E., Buehler, S., Ramnath, R. and Herndon, N., 2020. Cyberinfrastructure and resources to enable an integrative approach to studying forest trees. *Evolutionary Applications*, *13*(1), pp. 228–241.

Zhang, C., Norris-Caneda, K.H., Rottmann, W.H., Gulledge, J.E., Chang, S., Kwan, B.Y.H., Thomas, A.M., Mandel, L.C., Kothera, R.T., Victor, A.D. and Pearson, L., 2012. Control of pollen-mediated gene flow in transgenic trees. *Plant Physiology*, *159*(4), pp. 1319–1334.

17 Climate-Resilient Approaches for Sustainable Water Management Through Forest Resilience and Plant Functional Traits

Amit Bijlwan, Ankita Jha, Shivani Kothiyal, Roomesh Kumar Jena, and Rajeev Ranjan

17.1 INTRODUCTION

Greenhouse gas emissions are currently at unprecedented levels. Even if nations adhere to their existing commitments outlined in the nationally determined contributions under the Paris Agreement, scientists are virtually certain that the global average temperature will exceed pre-industrial levels by at least 1.5°C sometimes after 2030 (IPCC (Intergovernmental Panel on Climate Change), 2018a). Currently, approximately half of the world's population, around 8 billion people, faces severe water scarcity for at least part of the year. This crisis stems from a combination of climatic factors, such as prolonged droughts and intensified hydrological cycles, and non-climatic factors, including rapid population growth, unsustainable water use, and inadequate infrastructure (WMO (World Meteorological Organization), 2019). Since the 1970s, nearly 44% of all documented natural disasters have been related to floods, underscoring the vulnerability of communities to water-related hazards. It is estimated that about 60% of global adaptation measures are designed specifically to address water-related challenges, highlighting the critical role water plays in resilience strategies. Human-induced climate change has accelerated the intensification of the global hydrological cycle, disrupting the physical availability and distribution of water resources. This has compounded pre-existing water-related vulnerabilities that are often tied to socioeconomic inequalities. Over the past two decades, glaciers worldwide have experienced a dramatic loss of mass, with rates exceeding 0.5 m of water equivalent per year. This significant decline has far-reaching consequences for ecosystems and human societies, particularly for high mountain and polar communities that depend on glacial meltwater for their cultural practices, agriculture, and

DOI: 10.1201/9781003545231-17

daily water needs (Caretta et al., 2023; Majumder, 2015; United Nations Environment Programme, 2022).

Climate change has also led to observable shifts in global water systems, and streamflow has increased in many northern high-latitude regions, primarily driven by climatic changes rather than direct human intervention. Meanwhile, tropical aquifers have shown episodic recharge during intense precipitation and flooding events, a phenomenon that brings both opportunities and challenges for groundwater-dependent sectors (de Graaf et al., 2019; Ndehedehe et al., 2023a). Water is increasingly recognized as a critical "climate connector," linking environmental, social, and economic dimensions of resilience. The integration of water resource management and climate adaptation is essential to mitigating the impacts of climate change on the water cycle. Failure to act could lead to severe consequences, including threats to human health, economic instability, and potential conflicts over shared water resources (Kumar et al., 2021d; Ndehedehe et al., 2023b; Nimma et al., 2025).

Developing countries face unique and pressing challenges in water management, compounded by unpredictable rainfall patterns, population growth, and rising sea levels that result in saltwater intrusion into freshwater systems. Irregular and extreme weather events such as unseasonal rainfall, droughts, and flood-degrade soil disrupt agricultural production and threaten food security, leaving millions of farmers vulnerable to income losses (Brotherton & Joyce, 2015; Kumar et al., 2015; Joyce et al., 2016). Recent record-breaking floods and droughts serve as stark reminders of the cascading impacts of climate change on societies worldwide.

Water originating from forest ecosystems is often of exceptional quality, particularly in undisturbed areas. However, various forestry-related activities, such as timber harvesting, conservation management, and the construction of forest roads, can significantly impact water production, quality, and availability. These impacts arise due to increased sedimentation, nutrient loading, and disruptions in natural hydrological processes. Research indicates that while forestry interventions aimed at boosting water production may alter ecosystem evaporation rates and temporarily increase water flow, they often result in elevated levels of sediment and nutrients in nearby water bodies. In recent years, the expansion of agriculture and urbanization, coupled with the growing demand for timber and raw wood products, has led to a shift toward fast-growing exotic plantations over natural forest ecosystems. This conversion offers several economic and ecological benefits, such as higher timber yields and carbon sequestration. However, it also introduces challenges for water resources. Fast-growing tree species typically have higher water demands, which can reduce surface water flow and deplete underground water reserves. These effects are particularly pronounced in regions where natural forests are replaced with monoculture plantations, altering the balance of local water cycles (Abildtrup & Strange, 2000; Keleş, 2019; Núñez et al., 2006).

Furthermore, plantation forestry often relies on intensive silvicultural practices, including the use of fertilizers, herbicides, and pesticides, which can negatively impact nearby water resources. The leaching of chemicals into the soil and water systems poses risks to aquatic ecosystems and drinking water quality. Additionally, interventions like clear-cutting, frequently employed in plantation management, can exacerbate soil erosion, increase runoff, and disrupt groundwater recharge, further compromising the hydrological stability of forested landscapes. While afforestation and reforestation efforts can play an essential role in mitigating climate change and restoring degraded lands, careful consideration must be given to their design and implementation. Selecting tree species that align with local water availability, minimizing chemical inputs, and adopting sustainable forest management practices can help balance the economic benefits of plantations with the need to preserve water resources and ecosystem health (Futter et al., 2016; Sun et al., 2017; Trisurat et al., 2016). By integrating these approaches, it is possible to maximize the positive impacts of forestry while mitigating its potential downsides for water security.

Plant functional traits are the physio-morphological and phenological characteristics that indirectly influence a plant's fitness and performance by shaping its life history strategies (Kumar et al., 2016; Kumar et al., 2017; Tyagi & Kumar, 2024). These traits play a crucial role in determining how plants interact with their environment and contribute to ecosystem functioning. However, the increasing stressors brought about by climate change, such as rising temperatures, altered precipitation patterns, and more frequent extreme weather events, are reducing the resilience of plants (Lavorel & Garnier, 2002). This not only affects the structural composition of forests but also alters the ecosystem services they provide, driving shifts in functional traits over time. Functional traits are a cornerstone of ecological research, particularly in two critical areas. First, they are used to understand how plant communities respond to environmental changes, such as droughts, floods, and nutrient limitations. Second, they are instrumental in assessing the impacts of changes in community composition on broader ecosystem processes, including carbon sequestration, water cycling, and nutrient dynamics (Bonan et al., 2002; Cleland et al., 2007; Kumar et al., 2019; Kumar et al., 2020a; Kumar et al., 2020b).

In semi-arid regions, where water scarcity is a significant concern, the efficient use of water has become a pressing issue due to the expansion of irrigated agricultural areas and the increasing water demands of crops (Gupta et al., 2024). Plants in these regions have evolved a suite of adaptive traits to cope with drought stress, which often necessitate highly conservative survival and recovery strategies. Among these traits, deep root systems stand out as a critical adaptation. These systems enable plants to access water stored in deeper soil layers, thereby improving their water-use efficiency (WUE). Stomatal regulation is another essential trait for drought resistance and enhanced WUE. By controlling the opening and closing of stomata, plants can optimize water loss through transpiration while maintaining sufficient carbon dioxide uptake for photosynthesis. This fine balance is vital for survival in water-limited environments (Ilyas et al., 2021; Kumar et al., 2021a; Twalla et al., 2022).

Additionally, the trait of succulence, which is associated with water storage and utilization, is particularly relevant in arid ecosystems. Succulent plants, such as those belonging to the Cactaceae and Crassulaceae families, have specialized tissues that allow them to store significant amounts of water. This adaptation enables them to thrive in environments with prolonged dry periods. Beyond its ecological significance, the study of succulence offers valuable insights into strategies for enhancing WUE in crops and other plant species, especially as agricultural systems face increasing water scarcity (Sharma et al., 2023; Tardieu, 2022). By better understanding these functional traits and their responses to environmental stressors, researchers can develop strategies to improve plant resilience and water management in both natural and managed ecosystems. This knowledge is particularly critical in the context of climate change, as it can inform efforts to sustain ecosystem services, conserve biodiversity, and support agricultural productivity in water-limited regions.

To address these challenges, an integrated approach to water management, known as Integrated Water Resources Management, combined with comprehensive climate adaptation planning, is critical. This dual strategy aims to optimize the equitable distribution of water resources, ensure sustainable use, and build resilience against the growing disruptions caused by climate change. By implementing suitable measures, nations can work toward maintaining stable and prosperous societies in the face of intensifying water-related challenges.

17.2 CLIMATE CHANGE IMPACTS ON WATER RESOURCES

Climate change exerts a profound impact on global water resources, manifesting in intricate spatiotemporal patterns and interactions between physical processes and human activities. These effects compound existing challenges in the sustainable management of water resources, which are already under immense stress in many parts of the world. The consequences of climate change extend to the availability, quality, and quantity of water. Hydrological changes driven by a warming climate pose significant risks to society. These risks arise not only from direct disruptions to hydrometeorological processes, such as changes in precipitation patterns, increased frequency of extreme weather events, and shifts in evaporation and runoff rates, but also from indirect effects. For example, altered water cycles can jeopardize energy production by reducing the reliability of hydropower, exacerbate food insecurity by disrupting irrigation and agricultural water supply, and hinder economic development by increasing the costs of water management and infrastructure (Schewe et al., 2014; Wang et al., 2011; Kumar et al., 2021c, 2022). Additionally, these impacts can deepen social inequalities, disproportionately affecting vulnerable populations that are least equipped to adapt to water-related challenges.

Moreover, the interconnected nature of water systems amplifies these risks. Changes in one aspect of the water cycle can cascade through ecosystems and human systems, creating feedback loops that further strain water resources. For instance, reduced water availability can compromise agricultural yields, leading to economic losses and increased food prices, which in turn exacerbate poverty and inequality. Similarly, declining water quality can intensify health risks, increase treatment costs, and limit water usability for industrial and domestic purposes. Addressing these multifaceted challenges requires integrated water management strategies that consider the interplay of physical and human dimensions (Milly et al., 2008; UN-Water, 2013; Van Loon et al., 2016). Climate adaptation measures, such as enhancing water storage infrastructure, improving WUE, restoring ecosystems, and developing early warning systems for extreme events, are critical for building resilience. Additionally, policies must prioritize equitable access to water resources, ensuring that the needs of marginalized communities are met.

The impacts of climate change on the cryosphere are extensive, with a significant global decline in snow and ice cover. Snowfields, glaciers, and permafrost are projected to diminish in almost all regions throughout the 21st century, a trend supported by high confidence in scientific projections. Accelerated glacier melting is expected to negatively affect water resources, particularly in mountain regions and their surrounding lowlands, with tropical mountain areas being among the most vulnerable. While the rapid melting of glaciers may temporarily increase local streamflow, this phenomenon leads to long-term consequences such as reduced glacier cover, more variable river flows, decreased baseflows, and altered seasonal timing of peak streamflow (Hock et al., 2022; Meredith et al., 2022).

Over the last century, water use has grown at more than twice the rate of population increase. Combined with a more erratic and unpredictable water supply driven by climate change, this trend is expected to exacerbate water scarcity in already stressed regions and create new challenges in areas that currently have abundant water resources. Approximately four billion people experience severe physical water scarcity for at least one month annually. By 2050, an estimated 685 million people in over 570 cities could face at least a 10% decline in freshwater availability due to climate change, with some cities, such as Amman, Cape Town, and Melbourne, potentially seeing reductions of 30%–49% (C40 Cities, 2018). The societal and economic impacts of water scarcity are likely to be profound, with some regions projected to lose up to 6% of their gross domestic product, while migration and conflict risks rise as water shortages intensify (FAO/World Bank Group, 2018).

In addition to scarcity, freshwater resources worldwide are increasingly polluted by organic waste, pathogens, agricultural fertilizers, pesticides, heavy metals, and emerging contaminants. The discharge of untreated municipal and industrial wastewater, coupled with intensified agricultural practices, including livestock farming, is contributing to higher levels of organic pollution. Decreased runoff and extensive water extractions further reduce the natural dilution capacity of rivers, worsening water quality (Baraer et al., 2015; Huss et al., 2017; IPCC, 2019; Mekonnen & Hoekstra, 2012). The warming of water bodies due to climate change has also intensified the prevalence of harmful algal blooms. Many lakes and estuaries, essential sources of drinking water and ecosystem services, are now experiencing toxic cyanobacterial blooms that disrupt food webs, generate hypoxia, and threaten both human and environmental health. Freshwater and coastal wetlands are similarly under pressure from human activities, including altered water flow regimes and declining water quality. Climate change is expected to further strain these vital ecosystems, with adverse implications for fisheries and aquaculture, which depend on healthy wetland environments for sustainability (Buytaert et al., 2014; WHO, 2017; Zandaryaa & Mateo-Sagasta, 2018). The combined effects of climate change, pollution, and human interference underscore the urgent need for integrated water management strategies to safeguard the planet's freshwater and aquatic ecosystems for future generations (Kumar et al., 2021b).

Global warming is expected to intensify the hydrological cycle, leading to an increase in the intensity of extreme precipitation events and a heightened risk of flooding (Allen & Ingram, 2002). However, the changes observed often deviate from the theoretical expectation based on the increased water-holding capacity of a warmer atmosphere, particularly in regions where water availability is limited. Extreme precipitation events are projected to become more intense across many regions as atmospheric water vapor concentrations rise, in line with the thermodynamic Clausius–Clapeyron relationship, which predicts an increase of approximately 6%–7% in saturation vapor concentrations per degree Celsius of temperature rise (Ingram, 2016). Nonetheless, atmospheric dynamics, such as the poleward expansion of the tropical Hadley circulation, can either amplify or mitigate these thermodynamic effects on a regional scale. These dynamic changes can alter the patterns and intensity of extreme precipitation, highlighting the complex interplay between temperature, atmospheric circulation, and moisture transport. Water availability also significantly influences the relationship between temperature and moisture, with notable variability observed across different climatic zones (Norris et al., 2019).

Recent studies have explored how daily extreme precipitation changes correlate with water availability. Findings suggest that 30-year averaged annual precipitation maxima, when aggregated over both wet and dry regions, are likely to increase as the climate warms. Notably, rarer and more extreme precipitation events are expected to be particularly sensitive to climate change. These events are influenced more by vertical moisture transport processes than by horizontal moisture advection, further emphasizing the importance of regional and local factors in shaping precipitation responses to global warming (Scheff & Frierson, 2012; Zhang et al., 2017).

17.3 FOREST RESILIENCE: A CLIMATE-ADAPTATION STRATEGY

Forests span approximately 41 million square kilometers, covering about 30% of the Earth's land surface. They play a vital role in the global carbon cycle by absorbing roughly 33% of human-caused carbon emissions, making them a cornerstone in strategies to mitigate future climate change (McDowell et al., 2020). Beyond their role in carbon sequestration, forests provide essential ecosystem services, including regulating water flows, protecting soils, and conserving biodiversity, all of which are crucial for societal well-being. The resilience of forest ecosystems and their ability to withstand and recover from both natural and human-induced disturbances is critical to maintaining these benefits. However, tropical, arid, and temperate forests are experiencing significant declines in resilience, likely driven by increasing water scarcity and climate variability. Resilience enables vegetation communities to adapt and recover after disturbances, but its decline signals vulnerability to environmental changes. In contrast, boreal forests exhibit varying local trends, with many showing increasing resilience due to the potential benefits of warming temperatures and CO_2 fertilization, which may offset some adverse climate impacts (Dakos et al., 2015; Scheffer et al., 2015). These resilience trends are evident across both managed and untouched forests, underscoring the influence of large-scale climate drivers.

Diminished forest resilience has been statistically linked to abrupt reductions in primary productivity, which occurs as ecosystems approach critical resilience thresholds. Alarmingly, about 23% of undisturbed forests representing 3.32 petagrams of carbon in gross primary productivity have already reached these critical thresholds, leading to further degradation of their resilience (Friedlingstein et al., 2022). Forests with low resilience are more susceptible to external disturbances and face an increased risk of abrupt and potentially irreversible changes. As global efforts to combat climate change intensify, forest-based mitigation strategies that maintain and enhance carbon sinks and stocks are becoming increasingly critical. However, human activities such as forest management practices and land-use changes can profoundly affect the capacity of forests to recover from disturbances. These activities influence key ecological characteristics, including tree species composition, age distribution, canopy density, rooting depth, and primary productivity (Canadell et al., 2007). Maintaining and enhancing forest resilience is essential for sustaining their role in mitigating climate change and supporting biodiversity and ecosystem services. Addressing the challenges posed by declining resilience requires an integrated approach to forest conservation, management, and restoration that considers both global climate drivers and local ecological dynamics.

Trees function as natural pumps within ecosystems, drawing water from the soil and releasing it into the atmosphere through transpiration. A single mature tree can transpire hundreds of liters of water daily, significantly influencing local humidity levels. Additionally, tree canopies intercept rainfall, slowing its descent and facilitating gradual absorption into the soil, while their extensive root systems absorb groundwater and stabilize soil structure. Forests are not merely passive participants in the water cycle; they actively drive it, playing a critical role in regulating climate and sustaining the hydrological cycle through biophysical processes. These processes are intricately connected to land surface properties, including albedo, surface roughness, and canopy conductance. These factors influence the exchange of energy and water between the land and atmosphere. Research has highlighted the direct biophysical effects of forests on surface temperature, showing a transition from cooling effects in tropical regions to warming effects in boreal zones (Hohenegger & Stevens, 2018; Perugini et al., 2017; Taylor et al., 2012).

Global climate models (GCMs) predict that large-scale deforestation will reduce precipitation and cloud cover, particularly in tropical areas. However, high-resolution regional climate models and empirical analyses based on satellite imagery indicate a more nuanced picture. For instance, small-scale deforestation in the Amazon has been linked to increased cloud formation and precipitation, driven by surface heterogeneity. In contrast, certain regions, such as West Africa, exhibit inhibited cloud formation over forested areas, which appears to challenge the predictions of GCMs. Conversely, other regions like Western Europe and Central America have shown enhanced cloud formation over forests (EcoMatcher, 2024; Schwärzel et al., 2020). These disparities between modeling studies and observational data underscore the complexities of forest–cloud interactions. The mechanisms driving these interactions differ across spatial scales and regions, reflecting the challenges in accurately representing clouds and convection processes in climate models. This highlights the need for further research to unravel the intricate dynamics between forests, clouds, and the broader climate system, which are critical for understanding and managing the global impacts of deforestation and forest conservation.

Forest ecosystems exhibit a remarkable ability to adapt to changing climates, driven by their complexity and inherent mechanisms of resilience. Genetic diversity is a foundational element of this adaptability, as natural selection favors individuals with traits suited to evolving environmental conditions (Pakhira et al., 2024). Over time, these resilient genotypes become dominant, enhancing the forest's capacity to withstand climatic shifts. Additionally, trees demonstrate phenotypic plasticity, enabling them to modify their form, structure, and function in response to environmental signals such as temperature changes or water availability. This plasticity enhances their ability to cope with short-term stresses and contributes to long-term survival. Species migration is another critical component of forest adaptation, allowing ecosystems to relocate gradually to more favorable climates (de la Mata et al., 2022). However, natural migration often lags the rapid pace of climate change. Assisted migration, which involves the deliberate transplantation of tree species to regions with suitable climatic conditions, has emerged as a potential strategy to bridge this gap. Such proactive interventions could ensure the survival of species unable to migrate quickly enough to keep pace with climate change.

Trees also employ physiological mechanisms to cope with stressors like drought and heat. These include the production of protective enzymes, adjustments in photosynthetic activity, and metabolic shifts that help conserve energy and water during adverse conditions. These mechanisms are crucial for maintaining individual tree health and the broader forest ecosystem's stability (Seleiman et al., 2021). Adaptive forest management strategies play a pivotal role in enhancing forest resilience. These approaches include selective breeding programs to cultivate species with desirable traits, mimicking natural disturbances to maintain ecological balance, and promoting mixed-species forests to increase biodiversity and reduce vulnerability to pests and diseases. Integrating ecological interactions and functional traits into these management practices fosters a holistic approach, ensuring that forests can better adapt to climate uncertainties.

Habitat restoration is equally vital for fostering forest resilience. Restorative efforts aimed at recovering native species and ecosystems can help rebuild biodiversity and restore essential ecological functions. Soil and water conservation practices, such as contour planting, mulching, and reforestation, further enhance forest health and resilience, ensuring that these ecosystems can thrive under changing climatic conditions (Gao et al., 2023; Liu et al., 2024b). Climate-smart forestry has emerged as a forward-looking approach to bolster forest ecosystems' robustness. By integrating sustainable practices with climate adaptation strategies, climate-smart forestry not only enhances resilience but also provides valuable ecosystem services, including carbon sequestration, water regulation, and economic benefits for local communities. The combined efforts of adaptive management, habitat restoration, and climate-smart forestry are essential for navigating the uncertainties of a changing climate. These strategies empower forests to thrive in the face of adversity while continuing to deliver vital ecosystem services.

This chapter aims to contribute valuable insights into the scientific community and policymakers, equipping them with the knowledge needed to understand forest ecosystem resilience and develop effective strategies to mitigate climate change impacts on these critical ecosystems. Through a deeper understanding and implementation of these practices, we can ensure that forests remain vital and resilient components of our planet's ecological balance.

17.4 PLANT FUNCTIONAL TRAITS AND THEIR ROLE IN WATER MANAGEMENT

In water-limited ecosystems, plants adapt to severe water scarcity by modulating internal physiological processes, characterized by reduced photosynthetic activity and transpiration rates, coupled with enhanced WUE. Episodic rainfall events induce changes in water availability, which stimulate plant recruitment, promote substrate synthesis, and facilitate biomass accumulation, enabling the completion of life cycles. These adaptive responses are governed by intricate physiological mechanisms. For instance, fluctuations in soil moisture near root systems initiate the movement of water through vascular tissues, following a hydraulic gradient from roots to leaves. This process is accompanied by stomatal opening, enabling transpiration and the uptake of CO_2 for carbon assimilation via photosynthesis. The synthesized photosynthates are subsequently translocated to various plant tissues, including roots, while sustained water uptake supports cellular respiration, leading to an increase in soil respiration activity belowground. Beyond water-use strategies, the physiological responses of plants to soil moisture variations are influenced by functional traits shaped by long-term environmental adaptation. Functional traits, such as specific leaf area and relative water content, serve as indicators of plant physiological status, while morphological traits, including leaf shape and the distribution, length, and surface area of fine roots, reflect the integrated role of these processes in adaptation. For instance, leaf anatomical and physiological characteristics regulate the balance between water conservation and carbon dioxide assimilation by modulating stomatal conductance. Additionally, photosynthetic efficiency is determined by the interplay of various foliar attributes, such as leaf nitrogen content, thickness, and associated rates of photosynthesis and transpiration. These traits collectively mediate plant responses to environmental water dynamics and contribute to their overall adaptive capacity in arid and semi-arid ecosystems (Bussotti et al., 2015; Feng et al., 2022; Liu et al., 2024a; Zirbel et al., 2017).

Tropical rainforests, known for their structural and biological diversity, host various plant functional groups exhibiting unique adaptations for managing wet leaf conditions. These adaptations include trichome coverage, surface roughness, and water-repellent features such as cuticular waxes. Such features facilitate efficient water shedding, minimizing water-induced stomatal interference and ensuring photosynthetic activity during and after rain, fog, or dew events. In addition to leaf surface wettability, canopy structure and leaf arrangements play critical roles in determining the duration of leaf wetness and its effect on photosynthetic efficiency. While tropical forests experience frequent wet leaf conditions due to high rainfall, plants in semiarid regions encounter sporadic rain interspersed with extended dry periods, requiring different strategies. In these environments, dew formation on plant surfaces often serves as an alternative water source through foliar uptake, improving both leaf and soil moisture content and subsequently enhancing photosynthetic conditions (Aparecido et al., 2017; Zhang et al., 2019).

For example, certain species within the *Quercus* genus, common to semiarid regions, exhibit specific adaptations. These include trichomes on the abaxial surface to shield against intense radiation, regulate leaf temperature, and prevent dehydration. Additionally, these plants often have highly water-repellent abaxial surfaces and, in some species, the ability to absorb water directly through the adaxial surface.

17.4.1 Morphological Traits of Selected Plant Species

A detailed examination of the leaf or leaflet traits of 10 selected plant species reveals diverse adaptations to their respective microclimates:

- *Stachytarpheta jamaicensis*: Found in partially shaded environments, this plant grows to a height of 1–2 m with leaf sizes averaging 40.6 cm^2. The leaves have rough textures with multiple veins, sparse trichomes on both sides, acute drip tips, and thin, serrated margins.
- *Tibouchina heteromalla*: Growing in full sun to ~3 m, this species has large leaves (248.2 cm^2) with five midribs and multiple veins. Dense trichomes cover both surfaces, especially the abaxial side, with acute drip tips and intermediate thickness.
- *Zamia skinneri*: A shade-adapted species (1–1.5 m tall), it features 20.8 leaflets with an average size of 127.8 cm^2. Leaves have parallel, dentated veins, are thick and rigid, and possess distinctive drip tips.
- *Calathea crotalifera*: In partially shaded environments, this species reaches 1–2 m with smooth, waxy leaves (238.0 cm^2). It lacks trichomes, has small drip tips, and exhibits thin and flexible leaf structures.
- *Costus laevis*: A shade-tolerant plant growing to 2–3 m, it has thick, firm, waxy leaves (149.3 cm^2) with short trichomes on the abaxial surface, drip tips, and robust morphology.
- *Carapa guianensis*: A late-successional species growing to 15–35 m, it has smooth leaves averaging 79.6 cm^2 with undulated margins. Trichomes are absent, and the leaf tips are rounded with intermediate thickness.
- *Senna alata*: Thriving in full sun at 5–10 m, this species has small, smooth, water-repellent leaves (21.2 cm^2). It features numerous short trichomes on the abaxial side, rounded drip tips, and thin texture.
- *Quercus stellata*: Found in full sun (~6 m tall), this species has moderately coarse leaves (72.6 cm^2) with intermediate density trichomes and thin, lobed margins.
- *Quercus muehlenbergii*: Like PO, this species grows in full sun (~6 m), with smooth leaves (77.4 cm^2), dense abaxial trichomes, and thin, sinuated margins.
- *Quercus macrocarpa*: Also adapted to full sun (~4 m), it features coarse leaves (73.3 cm^2) with intermediate density trichomes on the abaxial side, absent drip tips, and thin, lobed margins.

Species in tropical and temperate regions of India exhibit distinct morphological adaptations to thrive in their respective climates. Tropical species, which experience hot and humid conditions, often have large leaves, smooth textures, and sparse trichomes. For instance, the Banyan tree (*Ficus benghalensis*) has broad, glossy leaves and aerial roots that help it absorb humidity and provide structural support. Similarly, the Neem tree (*Azadirachta indica*) features pinnate leaves with minimal trichomes and a deep root system, enabling it to access groundwater during dry spells. The Sal tree (*Shorea robusta*) has leathery, fire-resistant leaves and a thick trunk, while the Teak tree (*Tectona grandis*) displays rough-textured, dense trichome leaves for efficient water transport in its deciduous lifecycle. In contrast, temperate species in cooler climates adapt with features such as needle-like or thick, leathery leaves, dense trichomes, and waxy coatings to retain water and resist cold. The Deodar cedar (*Cedrus deodara*) exemplifies this with its smooth, waxy needles and conical shape that sheds snow effectively. The Oak (*Quercus leucotrichophora*) has oblong leaves with coarse textures and dense trichomes on the abaxial surface, while the Rhododendron (*Rhododendron arboreum*) showcases thick, leathery leaves with trichomes on young shoots to minimize water loss. Maple trees (*Acer oblongum*) feature broad, lobed leaves and are deciduous, shedding foliage to conserve energy during cold seasons. These adaptations highlight the ecological specialization of species in India, reflecting their resilience to the environmental challenges of their habitats.

These observations underline the intricate interplay between plant morphology and environmental conditions, showcasing how diverse leaf traits contribute to water management, resilience, and survival across varying ecosystems.

17.5 STRATEGIES FOR CLIMATE-RESILIENT WATER MANAGEMENT

17.5.1 Nature-Based Solutions: Reforestation, Afforestation, and Agroforestry

Trees are significant water consumers, and large-scale afforestation efforts in arid and semi-arid regions can adversely impact hydrological systems. These effects include substantial declines in groundwater levels, depletion of soil moisture, and reductions in annual runoff and water yields. Among various land-use changes, the conversion of grasslands and shrublands into forests is particularly associated with decreased annual runoff, especially when tree species with high water consumption rates are introduced. To evaluate the hydrological consequences of land-use and land-cover changes, along with the effects of climate change, various hydrological models have been utilized. Models such as Hydrologiska ByrånsVattenbalansavdelning, MIKE-Système HydrologiqueEuropéen, Water Flow and Balance Simulation Model, and the Soil and Water Assessment Tool (SWAT) are widely recognized in this domain (Khadka et al., 2023). Among these, SWAT stands out due to its capacity for multi-objective analysis, including the evaluation of forest cover dynamics on water balance components. It integrates multiple environmental processes and supports watershed management practices, providing a robust foundation for informed policymaking.

Forests and trees play a critical role in modulating global, regional, and local hydrological cycles. However, large-scale afforestation in drylands frequently leads to negative hydrological impacts, such as soil moisture depletion, reduced runoff, and decreased water yields. These effects necessitate the adoption of water-saving strategies, particularly in irrigation-dependent regions like downstream areas of Central Asia. Here, water consumption rates are exceptionally high, while irrigation WUE and overall water productivity remain notably low, underscoring the need for sustainable water management interventions.

Nature-based solutions (NBS) are instrumental in sustaining ecological systems by mitigating risk factors or enhancing the resilience of systems to external disturbances. In arid regions affected by desertification, NBS provide a transformative approach by converting deserts into grasslands, shrublands, and forests through the establishment of vegetation. This transformation reduces ecological risks and promotes ecosystem stability. Enhanced biodiversity increases food web complexity, stabilizing various ecosystem functions and processes such as nutrient recycling, energy flux regulation, and biomass production. Reforestation plays a critical role in improving soil water-retention capacity and stability, which are essential for maintaining forest ecosystem integrity. The water retention capacity of reforested areas can be quantified by measuring water yield, the volume of water supplied by specific land cover types while soil stability can be assessed by calculating soil erosion mass, which reflects soil loss under different land covers. Reforestation also serves as a powerful climate change mitigation strategy by sequestering atmospheric carbon dioxide into aboveground biomass, root systems, and soil organic matter (Kim et al., 2021; Zuazo & Pleguezuelo, 2009).

Additionally, forest ecosystems contribute significantly to rural livelihoods and national economies by generating income and employment opportunities through the sale of forest products. Agroforestry, which integrates agricultural and forestry practices, has been identified as a viable alternative income source for livestock farmers. By combining trees and farmlands, agroforestry supports sustainable land use and enhances ecosystem services, offering a holistic solution to environmental and socioeconomic challenges.

17.5.2 Enhancing Functional Trait Diversity in Managed Forests

Plant functional traits serve as key indicators of individual species' adaptive capacities and simultaneously reflect the functional attributes of the broader ecosystem. Climate change exerts profound impacts on ecosystems by disrupting their stability and altering biodiversity (Cui et al., 2020). Functional traits in plants demonstrate a capacity to respond and adapt to environmental changes, with their variations providing insight into the adaptive strategies employed by species confronting climate change challenges.

As environmental or altitudinal gradients shift from humid to arid conditions, notable changes occur in plant functional traits: leaf area tends to decrease, nitrogen and phosphorus concentrations increase, and the carbon-to-nitrogen ratio declines. Plants exhibit adaptive responses to environmental changes by modifying both their external morphological features and internal physiological mechanisms, thereby influencing ecosystem processes. Alterations in traits such as plant architecture, stomatal dynamics, photosynthetic pathways, photosynthesis and respiration rates, water-use physiology, and phenological timing are critical indicators of regional climate shifts. These morphological and physiological adjustments underscore the intricate relationship between plant functional traits and their adaptive responses to climatic variability (Andrew et al., 2022). The study highlights the varying vulnerability of *Rhododendron arboreum* across temperate forests in the western Himalayas, with maximum susceptibility in ban oak forests and lower vulnerability in moist temperate deciduous forests due to adaptive traits like leaf stoichiometry modulation. Key sensitivity traits such as light harvesting, nutrient flus, and gas flux indicate responses to stress, while adaptive traits like resource efficiency, architecture and defense, resource limitation, and facilitation enhance resilience. Effective conservation strategies should integrate these traits to guide management practices like selective felling, afforestation, and ecological monitoring, ensuring sustainable outcomes and ecosystem resilience (Pandey et al., 2024).

17.6 CONCLUSION

The findings presented in this chapter emphasize the critical role of plant functional traits in driving ecosystem resilience and adaptation to climate change. Traits such as light harvesting, nutrient and gas fluxes, resource efficiency, and defense mechanisms serve as pivotal indicators of plant responses to environmental stressors, guiding targeted conservation strategies. Effective management practices, including selective afforestation, agroforestry, and ecological monitoring, can mitigate hydrological and ecological challenges while enhancing biodiversity and soil–water conservation. The vulnerability assessment of *Rhododendron arboreum* across forest types in the western Himalayas highlights the necessity of trait-based approaches to conservation. Integrating hydrological models and NBS provides actionable insights into sustaining ecosystem services and achieving climate resilience. By fostering a nuanced understanding of trait co-variation dynamics and ecosystem functionality, this chapter advocates for adaptive management frameworks to safeguard biodiversity and ensure sustainable development in the face of evolving environmental challenges.

REFERENCES

Abildtrup, J., & Strange, N. (2000). The option value of non-contaminated forest watersheds. *Forest Policy and Economics*, 1(2), 115–125. https://doi.org/10.1016/S1389-9341(99)00005-2

Allen, M. R., & Ingram, W. J. (2002). Constraints on future changes in climate and the hydrologic cycle. *Nature*, 419(6903), 224–232. https://doi.org/10.1038/nature01092

Andrew, S. C., Gallagher, R. V., Wright, I. J., & Mokany, K. (2022). Assessing the vulnerability of plant functional trait strategies to climate change. *Global Ecology and Biogeography*, 31(6), 1194–1206. https://doi.org/10.1111/geb.13501

Aparecido, L. M. T., Miller, G. R., Cahill, A. T., & Moore, G. W. (2017). Leaf surface traits and water storage retention affect photosynthetic responses to leaf surface wetness among wet tropical forest and semiarid savanna plants. *Tree Physiology*, 37(10), 1285–1300. https://doi.org/10.1093/treephys/tpx092

Baraer, M., McKenzie, J., Mark, B. G., Gordon, R., Bury, J., Condom, T., Gomez, J., Knox, S., & Fortner, S. K. (2015). Contribution of groundwater to the outflow from ungauged glacierized catchments: A multi-site study in the tropical Cordillera Blanca, Peru. *Hydrological Processes*, 29(11), 2561–2581. https://doi.org/10.1002/hyp.10386

Bonan, G. B., Levis, S., Kergoat, L., & Oleson, K. W. (2002). Landscapes as patches of plant functional types: An integrating concept for climate and ecosystem models. *Global Biogeochemical Cycles*, 16(2). https://doi.org/10.1029/2000GB001360

Brotherton, S. J., & Joyce, C. B. (2015). Extreme climate events and wet grasslands: Plant traits for ecological resilience. *Hydrobiologia*, 750(1), 229–243. https://doi.org/10.1007/s10750-014-2129-5

Bussotti, F., Pollastrini, M., Holland, V., & Brüggemann, W. (2015). Functional traits and adaptive capacity of European forests to climate change. *Environmental and Experimental Botany*, 111, 91–113. https://doi.org/10.1016/j.envexpbot.2014.11.006

Buytaert, W., Zulkafli, Z., Grainger, S., Acosta, L., Alemie, T. C., Bastiaensen, J., De BiÃ¨vre, B., Bhusal, J., Clark, J., Dewulf, A., Foggin, M., Hannah, D. M., Hergarten, C., Isaeva, A., Karpouzoglou, T., Pandeya, B., Paudel, D., Sharma, K., Steenhuis, T., … Zhumanova, M. (2014). Citizen science in hydrology and water resources: Opportunities for knowledge generation, ecosystem service management, and sustainable development. *Frontiers in Earth Science*, 2. https://doi.org/10.3389/feart.2014.00026

C40 Cities. (2018). Restoring the flow. C40 Cities website. www.c40.org/other/the-future-we-don-t-want-restoring-the-flow.

Canadell, J. G., Le Quéré, C., Raupach, M. R., Field, C. B., Buitenhuis, E. T., Ciais, P., Conway, T. J., Gillett, N. P., Houghton, R. A., & Marland, G. (2007). Contributions to accelerating atmospheric CO_2 growth from economic activity, carbon intensity, and efficiency of natural sinks. *Proceedings of the National Academy of Sciences*, 104(47), 18866–18870. https://doi.org/10.1073/pnas.0702737104

Caretta, M. A., Mukherji, A., Arfanuzzaman, M., Betts, R.A., Gelfan, A., Hirabayashi, Y., Lissner, T. K., Liu, J., Lopez Gunn, E., Morgan, R., Mwanga, S., & Supratid, S. (2023). Water. In *Climate Change 2022 – Impacts, Adaptation and Vulnerability* (pp. 551–712). Cambridge University Press. https://doi.org/10.1017/9781009325844.006

Cleland, E., Chuine, I., Menzel, A., Mooney, H., & Schwartz, M. (2007). Shifting plant phenology in response to global change. *Trends in Ecology & Evolution*, 22(7), 357–365. https://doi.org/10.1016/j.tree.2007.04.003

Cui, E., Weng, E., Yan, E., & Xia, J. (2020). Robust leaf trait relationships across species under global environmental changes. *Nature Communications*, 11(1), 2999. https://doi.org/10.1038/s41467-020-16839-9

Dakos, V., Carpenter, S. R., van Nes, E. H., & Scheffer, M. (2015). Resilience indicators: Prospects and limitations for early warnings of regime shifts. *Philosophical Transactions of the Royal Society B: Biological Sciences*, 370(1659), 20130263. https://doi.org/10.1098/rstb.2013.0263

de Graaf, I. E. M., Gleeson, T., (Rens) van Beek, L. P. H., Sutanudjaja, E. H., & Bierkens, M. F. P. (2019). Environmental flow limits to global groundwater pumping. *Nature*, 574(7776), 90–94. https://doi.org/10.1038/s41586-019-1594-4

de la Mata, R., Zas, R., Bustingorri, G., Sampedro, L., Rust, M., Hernandez-Serrano, A., & Sala, A. (2022). Drivers of population differentiation in phenotypic plasticity in a temperate conifer: A 27-year study. *Evolutionary Applications*, 15(11), 1945–1962. https://doi.org/10.1111/eva.13492

EcoMatcher, B. F. C. (2024, November 15). The importance of forests in hydrological balance - Bas Fransen, CEO EcoMatcher - Medium. https://ecomatcher.medium.com/the-importance-of-forests-in-hydrological-balance-d2798e9a520f.

FAO/World Bank Group. (2018). Water management in fragile systems: Building resilience to shocks and protracted crises in the Middle East and North Africa. Cairo. Rome/Washington, DC, FAO/World Bank Group. openknowledge.worldbank.org/handle/10986/30307.

Feng, X., Liu, R., Li, C., Li, M., Wang, Y., & Li, Y. (2022). Multi-level physiological and morphological adjustment of Haloxylonammodendron related to groundwater drawdown in a desert ecosystem. *Agricultural and Forest Meteorology*, 324, 109096. https://doi.org/10.1016/j.agrformet.2022.109096

Friedlingstein, P., Jones, M. W., O'Sullivan, M., Andrew, R. M., Bakker, D. C. E., Hauck, J., Le Quéré, C., Peters, G. P., Peters, W., Pongratz, J., Sitch, S., Canadell, J. G., Ciais, P., Jackson, R. B., Alin, S. R., Anthoni, P., Bates, N. R., Becker, M., Bellouin, N., … Zeng, J. (2022). Global carbon budget 2021. *Earth System Science Data*, 14(4), 1917–2005. https://doi.org/10.5194/essd-14-1917-2022

Futter, M. N., Högbom, L., Valinia, S., Sponseller, R. A., & Laudon, H. (2016). Conceptualizing and communicating management effects on forest water quality. *Ambio*, 45(S2), 188–202. https://doi.org/10.1007/s13280-015-0753-6

Gao, L., Kantar, M. B., Moxley, D., Ortiz-Barrientos, D., & Rieseberg, L. H. (2023). Crop adaptation to climate change: An evolutionary perspective. *Molecular Plant*, 16(10), 1518–1546. https://doi.org/10.1016/j.molp.2023.07.011

Gupta, S., Ranjan, R., Bijlwan, A., Jha, A., & Jena, R. K. (2024). Enhancing water use efficiency through trait-based approaches. In *Plant Functional Traits for Improving Productivity* (pp. 133–147). Springer Nature Singapore. https://doi.org/10.1007/978-981-97-1510-7_8

Hock, R., Rasul, G., Adler, C., Cáceres, B., Gruber, S., Hirabayashi, Y., Jackson, M., Kääb, A., Kang, S., Kutuzov, S., Al. Milner, Molau, U., Morin, S., Orlove, B., & Steltzer, H. (2022). High mountain areas. In *The Ocean and Cryosphere in a Changing Climate* (pp. 131–202). Cambridge University Press. https://doi.org/10.1017/9781009157964.004

Hohenegger, C., & Stevens, B. (2018). The role of the permanent wilting point in controlling the spatial distribution of precipitation. *Proceedings of the National Academy of Sciences*, 115(22), 5692–5697. https://doi.org/10.1073/pnas.1718842115

Huss, M., Bookhagen, B., Huggel, C., Jacobsen, D., Bradley, R. S., Clague, J. J., Vuille, M., Buytaert, W., Cayan, D. R., Greenwood, G., Mark, B. G., Milner, A. M., Weingartner, R., & Winder, M. (2017). Toward mountains without permanent snow and ice. *Earth's Future*, 5(5), 418–435. https://doi.org/10.1002/2016EF000514

Ilyas, M., Nisar, M., Khan, N., Hazrat, A., Khan, A. H., Hayat, K., Fahad, S., Khan, A., & Ullah, A. (2021). Drought tolerance strategies in plants: A mechanistic approach. *Journal of Plant Growth Regulation*, 40(3), 926–944. https://doi.org/10.1007/s00344-020-10174-5

Ingram, W. (2016). Increases all round. *Nature Climate Change*, 6(5), 443–444. https://doi.org/10.1038/nclimate2966

IPCC. (2019). IPCC special report on the ocean and cryosphere in a changing climate.

IPCC (Intergovernmental Panel on Climate Change). (2018). Summary for Policymakers. Global Warming of 1.5°C. An IPCC Special Report on the impacts of global warming of 1.5°C above pre-industrial levels and related global greenhouse gas emission pathways, in the context of strengthening the global response to the threat of climate change, sustainable development, and efforts to eradicate poverty. Geneva, IPCC.

Joyce, C. B., Simpson, M., & Casanova, M. (2016). Future wet grasslands: Ecological implications of climate change. *Ecosystem Health and Sustainability*, 2(9). https://doi.org/10.1002/ehs2.1240

Keleş, S. (2019). An assessment of hydrological functions of forest ecosystems to support sustainable forest management. *Journal of Sustainable Forestry*, 38(4), 305–326. https://doi.org/10.1080/10549811.2018.1547879

Khadka, D., Babel, M. S., & Kamalamma, A. G. (2023). Assessing the impact of climate and land-use changes on the hydrologic cycle using the SWAT model in the Mun River Basin in Northeast Thailand. *Water*, 15(20), 3672. https://doi.org/10.3390/w15203672

Kim, G., Kim, J., Ko, Y., Eyman, O. T. G., Chowdhury, S., Adiwal, J., Lee, W., & Son, Y. (2021). How do nature-based solutions improve environmental and socio-economic resilience to achieve the sustainable development goals? Reforestation and afforestation cases from the Republic of Korea. *Sustainability*, 13(21), 12171. https://doi.org/10.3390/su132112171

Kumar, A., Dwivedi, G. K., Tewari, S., Paul, J., Anand, R., Kumar, N., Kumar, P., Singh, H., & Kaushal, R. (2020a). Carbon mineralization and inorganic nitrogen pools under *Terminalia chebula* Retz.-Based agroforestry system in Himalayan foothills, India. *Forest Science*, 66(5), 634–643.

Kumar, A., Dwivedi, G. K., Tewari, S., Paul, J., Sah, V. K., Singh, H., Kumar, P., Kumar, N., & Kaushal, R. (2020b) Soil organic carbon pools under *Terminalia chebula* Retz. based agroforestry system in Himalayan foothills, India. *Current Science*, 118(7), 1098–1103.

Kumar, A., Kumar, P., Singh, H., Bisht, S., & Kumar, N. (2021a). Relationship of physiological plant functional traits with soil carbon stock in temperate forest of Garhwal Himalaya. *Current Science*, 120(8), 1368–1373.

Kumar, A., Kumar, P., Singh, H., & Kumar, N. (2021b). Modulation of plant functional traits under essential plant nutrients during seasonal regime in natural forests of Garhwal Himalayas. *Plant and Soil*, 465, 197–212.

Kumar, A., Kumar, P., Singh, H., & Kumar, N. (2021c). Impact of plant functional traits on near saturated hydraulic conductivity of soil under different forests of Kempty watershed in Garhwal Himalaya. *Indian Journal of Soil Conservation*, 49(1), 38–44.

Kumar, A., Singh, H., Kumari, G., Bisht, S., Malik, A., Kumar, N., Singh, M., Raturi, A., Barthwal, S., Thakur, A., & Kaushal, R. (2022). Adaptive resilience of roadside trees to vehicular emissions via leaf enzymatic, physiological, and anatomical trait modulations. *Environmental Pollution*, 313, 120191.

Kumar, N., Suyal, D. C., Sharma, I.P., Verma, A., & Singh, H. (2017). Elucidating stress proteins in rice (*Oryza sativa* L.) genotype under elevated temperature: A proteomic approach to understand heat stress response. *3 Biotech*, 7, 205.

Kumar, N., Jeena, N., & Singh, H. (2019). Elevated temperature modulates rice pollen structure: A study from foothill Himalayan agro-ecosystem in India. *3 Biotech*, 9, 175.

Kumar, N., Jeena, N., Kumar, A., Khairakpam, R., & Singh, H. (2021d). Comparative response of rice cultivars to elevated air temperature in Bhabar region of Indian Himalaya: Status on yield attributes. *Heliyon*, 7, e07474.

Kumar, N., Kumar, N., Shukla, A., Shankhdhar, S. C., & Shankhdhar, D. (2015). Impact of terminal heat stress on pollen viability and yield attributes of rice (*Oryza sativa* L.). *Cereal Research Communications*, 43(4), 616–626.

Kumar, N., Shankhdhar, S. C., & Shankhdhar, D. (2016). Impact of elevated temperature on antioxidant activity and membrane stability in different genotypes of rice (*Oryza sativa* L.). *Indian Journal of Plant Physiology (Now Plant Physiology Report)*, 21(1), 37–43.

Lavorel, S., & Garnier, E. (2002). Predicting changes in community composition and ecosystem functioning from plant traits: Revisiting the Holy Grail. *Functional Ecology*, 16(5), 545–556. https://doi.org/10.1046/j.1365-2435.2002.00664.x

Liu, H., Yin, D., He, P., Cadotte, M. W., & Ye, Q. (2024a). Linking plant functional traits to biodiversity under environmental change. *Biological Diversity*, 1(1), 22–28. https://doi.org/10.1002/bod2.12004

Liu, J., Dou, Y., & Chen, H. (2024b). Stepwise ecological restoration: A framework for improving restoration outcomes. *Geography and Sustainability*, 5(2), 160–166. https://doi.org/10.1016/j.geosus.2024.02.003

Majumder, M. (2015). *Impact of Urbanization on Water Shortage in Face of Climatic Aberrations*. Springer.

McDowell, N. G., Allen, C. D., Anderson-Teixeira, K., Aukema, B. H., Bond-Lamberty, B., Chini, L., Clark, J. S., Dietze, M., Grossiord, C., Hanbury-Brown, A., Hurtt, G. C., Jackson, R. B., Johnson, D. J., Kueppers, L., Lichstein, J. W., Ogle, K., Poulter, B., Pugh, T. A. M., Seidl, R., … Xu, C. (2020). Pervasive shifts in forest dynamics in a changing world. *Science*, 368(6494). https://doi.org/10.1126/science.aaz9463

Mekonnen, M. M., & Hoekstra, A. Y. (2012). A global assessment of the water footprint of farm animal products. *Ecosystems*, 15(3), 401–415. https://doi.org/10.1007/s10021-011-9517-8

Meredith, M., Sommerkorn, M., Cassotta, S., Derksen, C., Ekaykin, A., Hollowed, A., Kofinas, G., Mackintosh, A., Melbourne-Thomas, J., Muelbert, M. M. C., Ottersen, G., Pritchard, H., & Schuur, E. A. G. (2022). Polar Regions. In *The Ocean and Cryosphere in a Changing Climate* (pp. 203–320). Cambridge University Press. https://doi.org/10.1017/9781009157964.005

Milly, P. C. D., Betancourt, J., Falkenmark, M., Hirsch, R. M., Kundzewicz, Z. W., Lettenmaier, D. P., & Stouffer, R. J. (2008). Stationarity is dead: Whither water management? *Science*, 319(5863), 573–574. https://doi.org/10.1126/science.1151915

Ndehedehe, C. E., Adeyeri, O. E., Onojeghuo, A. O., Ferreira, V. G., Kalu, I., &Okwuashi, O. (2023a). Understanding global groundwater-climate interactions. *Science of The Total Environment*, 904, 166571. https://doi.org/10.1016/j.scitotenv.2023.166571

Ndehedehe, C. E., Ferreira, V. G., Adeyeri, O. E., Correa, F. M., Usman, M., Oussou, F. E., Kalu, I., Okwuashi, O., Onojeghuo, A. O., Getirana, A., & Dewan, A. (2023b). Global assessment of drought characteristics in the Anthropocene. *Resources, Environment and Sustainability*, 12, 100105. https://doi.org/10.1016/j.resenv.2022.100105

Nimma, D., Devi, O. R., Laishram, B., Ramesh, J. V. N., Boddupalli, S., Ayyasamy, R., Tirth, V., &Arabil, A. (2025). Implications of climate change on freshwater ecosystems and their biodiversity. *Desalination and Water Treatment*, 321, 100889. https://doi.org/10.1016/j.dwt.2024.100889

Norris, J., Chen, G., & Neelin, J. D. (2019). Thermodynamic versus dynamic controls on extreme precipitation in a warming climate from the community earth system model large ensemble. *Journal of Climate*, 32(4), 1025–1045. https://doi.org/10.1175/JCLI-D-18-0302.1

Núñez, D., Nahuelhual, L., & Oyarzún, C. (2006). Forests and water: The value of native temperate forests in supplying water for human consumption. *Ecological Economics*, 58(3), 606–616. https://doi.org/10.1016/j.ecolecon.2005.08.010

Pakhira, R., Biswas, S., Manoj, C., & Singh, H. (2024). Biological insights into forest adaptation to climate change: Ecosystem resilience and adaptation. In *Urban Forests, Climate Change and Environmental Pollution* (pp. 307–329). Springer Nature Switzerland. https://doi.org/10.1007/978-3-031-67837-0_15

Pandey, R., Rawat, M., Singh, R., Kumar, M., Chaturvedi, R., Cui, L., & Li, T. (2024). Plant trait approach to assess the vulnerability of *Rhododendron arboreum* in western Himalayas. *Environmental and Sustainability Indicators*, 23, 100415. https://doi.org/10.1016/j.indic.2024.100415

Perugini, L., Caporaso, L., Marconi, S., Cescatti, A., Quesada, B., de Noblet-Ducoudré, N., House, J. I., & Arneth, A. (2017). Biophysical effects on temperature and precipitation due to land cover change. *Environmental Research Letters*, 12(5), 053002. https://doi.org/10.1088/1748-9326/aa6b3f

Scheffer, M., Carpenter, S. R., Dakos, V., & van Nes, E. H. (2015). Generic indicators of ecological resilience: Inferring the chance of a critical transition. *Annual Review of Ecology, Evolution, and Systematics*, 46(1), 145–167. https://doi.org/10.1146/annurev-ecolsys-112414-054242

Scheff, J., & Frierson, D. M. W. (2012). Robust future precipitation declines in CMIP5 largely reflect the poleward expansion of model subtropical dry zones. *Geophysical Research Letters*, 39(18). https://doi.org/10.1029/2012GL052910

Schewe, J., Heinke, J., Gerten, D., Haddeland, I., Arnell, N. W., Clark, D. B., Dankers, R., Eisner, S., Fekete, B. M., Colón-González, F. J., Gosling, S. N., Kim, H., Liu, X., Masaki, Y., Portmann, F. T., Satoh, Y., Stacke, T., Tang, Q., Wada, Y., … Kabat, P. (2014). Multimodel assessment of water scarcity under climate change. *Proceedings of the National Academy of Sciences*, 111(9), 3245–3250. https://doi.org/10.1073/pnas.1222460110

Schwärzel, K., Zhang, L., Montanarella, L., Wang, Y., & Sun, G. (2020). How afforestation affects the water cycle in drylands: A process-based comparative analysis. *Global Change Biology*, 26(2), 944–959. https://doi.org/10.1111/gcb.14875

Seleiman, M. F., Al-Suhaibani, N., Ali, N., Akmal, M., Alotaibi, M., Refay, Y., Dindaroglu, T., Abdul-Wajid, H. H., & Battaglia, M. L. (2021). Drought stress impacts on plants and different approaches to alleviate its adverse effects. *Plants*, 10(2), 259. https://doi.org/10.3390/plants10020259

Sharma, N., Raman, H., Wheeler, D., Kalenahalli, Y., & Sharma, R. (2023). Data-driven approaches to improve water-use efficiency and drought resistance in crop plants. *Plant Science*, 336, 111852. https://doi.org/10.1016/j.plantsci.2023.111852

Sun, X., Onda, Y., Kato, H., Gomi, T., & Liu, X. (2017). Estimation of throughfall with changing stand structures for Japanese cypress and cedar plantations. *Forest Ecology and Management*, 402, 145–156. https://doi.org/10.1016/j.foreco.2017.07.036

Tardieu, F. (2022). Different avenues for progress apply to drought tolerance, water use efficiency and yield in dry areas. *Current Opinion in Biotechnology*, 73, 128–134. https://doi.org/10.1016/j.copbio.2021.07.019

Taylor, C. M., de Jeu, R. A. M., Guichard, F., Harris, P. P., & Dorigo, W. A. (2012). Afternoon rain more likely over drier soils. *Nature*, 489(7416), 423–426. https://doi.org/10.1038/nature11377

Trisurat, Y., Eawpanich, P., &Kalliola, R. (2016). Integrating land use and climate change scenarios and models into assessment of forested watershed services in Southern Thailand. *Environmental Research*, 147, 611–620. https://doi.org/10.1016/j.envres.2016.02.019

Twalla, J. T., Ding, B., Cao, G., Bao, S., Li, M., Chen, X., Xie, X., & Wang, J. (2022). Roles of stomata in gramineous crops growth and biomass production. *Cereal Research Communications*, 50(4), 603–616. https://doi.org/10.1007/s42976-021-00216-3

Tyagi, J., & Kumar, M. (2024). Plant functional trait: Concept and significance. In *Plant Functional Traits for Improving Productivity* (pp. 1–22). Springer Nature Singapore. https://doi.org/10.1007/978-981-97-1510-7_1

United Nations Environment Programme. (2022). Where water management meets climate change adaptation to boost resilience. UNEP - UN Environment Programme. https://www.unep.org/technical-highlight/where-water-management-meets-climate-change-adaptation-boost-resilience

UN-Water. (2013). What is water security? Infographic. UN-Water. https://www.unwater.org/publications/what-water-security-infographic

Van Loon, A. F., Gleeson, T., Clark, J., Van Dijk, A. I. J. M., Stahl, K., Hannaford, J., Di Baldassarre, G., Teuling, A. J., Tallaksen, L. M., Uijlenhoet, R., Hannah, D. M., Sheffield, J., Svoboda, M., Verbeiren, B., Wagener, T., Rangecroft, S., Wanders, N., & Van Lanen, H. A. J. (2016). Drought in the Anthropocene. *Nature Geoscience*, 9(2), 89–91. https://doi.org/10.1038/ngeo2646

Wang, A., Lettenmaier, D. P., & Sheffield, J. (2011). Soil moisture drought in China, 1950–2006. *Journal of Climate*, 24(13), 3257–3271. https://doi.org/10.1175/2011JCLI3733.1

WHO. (2017). Climate-resilient water safety plans: Managing health risk associated with climate variability and change. WHO. www.who.int/globalchange/publications/climate-resilient-water-safety-plans/en/

WMO (World Meteorological Organization). (2019). WMO statement on the state of the global climate in 2018. Geneva, WMO. library.wmo.int/doc_num.php?explnum_id=5789

Zandaryaa, S., & Mateo-Sagasta, J. (2018). Organic matter, pathogens and emerging pollutants. In J. Mateo-Sagasta, S. Marjani Zadeh and H. Turral (eds.), *More People, More Food, Worse Water? A Global Review of Water Pollution from Agriculture*. Rome/Colombo, Food and Agriculture Organization of the United Nations/International Water Management Institute (FAO/IWMI), pp. 125–138. www.fao.org/3/ca0146en/CA0146EN.pdf

Zhang, X., Zwiers, F. W., Li, G., Wan, H., & Cannon, A. J. (2017). Complexity in estimating past and future extreme short-duration rainfall. *Nature Geoscience*, 10(4), 255–259. https://doi.org/10.1038/ngeo2911

Zhang, Z., Szota, C., Fletcher, T. D., Williams, N. S. G., & Farrell, C. (2019). Green roof storage capacity can be more important than evapotranspiration for retention performance. *Journal of Environmental Management*, 232, 404–412. https://doi.org/10.1016/j.jenvman.2018.11.070

Zirbel, C. R., Bassett, T., Grman, E., & Brudvig, L. A. (2017). Plant functional traits and environmental conditions shape community assembly and ecosystem functioning during restoration. *Journal of Applied Ecology*, 54(4), 1070–1079. https://doi.org/10.1111/1365-2664.12885

Zuazo, V. H. D., & Pleguezuelo, C. R. R. (2009). Soil-erosion and runoff prevention by plant covers: A review. In *Sustainable Agriculture* (pp. 785–811). Springer Netherlands. https://doi.org/10.1007/978-90-481-2666-8_48

18 Functional Traits for Predicting Climate-Driven Forest Changes

Kashish Walia, Yourmilla Kumari, Garima, and Aruna Mehta

18.1 INTRODUCTION

Climate is a crucial factor in shaping forest ecosystems and is deeply intertwined with forest health. Climate change impacts forest productivity both directly and indirectly, with significant effects on forests and the forest sector. These changes affect forests as both drivers and recipients of a changing climate, resulting in impacts such as prolonged growing seasons, shifts in insect species distributions, and increased forest degradation. Beyond these direct impacts, climate change can reshape the composition and dynamics of entire forest ecosystems. As a key determinant of forest tree distribution, climate drives shifts in tree species' ranges as a consequence of global warming (Kumar et al., 2019; Kumar et al., 2021d; Dyderski et al., 2018). To adapt, tree species migrate to coincide with favourable climatic conditions. Although trees cannot physically relocate like animals, forests migrate through the passive dispersal of seeds and the establishment of seedlings in suitable areas (Davis and Shaw, 2001). Evidence indicates that numerous tree species are already shifting toward higher latitudes and altitudes (Woodall et al., 2009). To manage these changes effectively, predictive models have become an essential tool by offering data-driven insights into forest growth, biodiversity conservation, wildfire risk, and climate change impacts. Predictive models help forest managers forecast forest growth, guiding decisions on sustainable timber harvesting, carbon sequestration, and simulating wildfire behaviour. These models also help protect biodiversity by predicting changes in habitats and supporting reforestation efforts. This makes forest management strategies more flexible and long-lasting (Peng, 2000).

A key part of understanding how forests respond to climate change lies in plant functional traits (FTs), heritable characteristics related to morphology, physiology, or phenology that influence species performance and environmental response (Kumar et al., 2016; Kumar et al., 2017; Garnier et al., 2016). The key uses of FTs include characterizing community responses to environmental changes, such as community assembly processes and quantifying how shifts in these communities affect ecosystem processes. The growth of trait-based ecology has shown that traits and environmental factors are consistently linked. This gives us a way to guess how biodiversity

 DOI: 10.1201/9781003545231-18

will react to changes in the climate. By combining insights from predictive models and trait-based ecology, forest management can better anticipate and respond to the ecological challenges posed by climate change.

18.1.1 Key Functional Traits Relevant to Climate Change

Global climate change presents substantial challenges to ecosystems, especially by impacting plant species and their FTs. These traits, including leaf morphology, photosynthetic efficiency, root depth, and reproductive strategies, play critical roles in how plants respond to changing conditions like temperature fluctuations, altered rainfall patterns, and increased atmospheric CO_2. Understanding these dynamics is crucial for sustainable development and achieving the Sustainable Development Goals. Changes in plant traits not only affect individual fitness but also influence ecosystem structure, community composition, and resilience (Violle et al., 2007; Kumar et al., 2020a; Kumar et al., 2020b). FTs are key indicators of how plants will adapt to climate change. To deal with more frequent droughts, traits that affect how efficiently plants use water (like stomatal conductance and leaf area) are very important. On the other hand, traits that are sensitive to temperature (like photosynthetic capacity and thermal tolerance) help scientists guess how plants will react to global warming. Research on these traits can guide forest management and conservation strategies, enhancing ecosystem resilience and adaptation to future climate scenarios (Lavorel & Garnier, 2002). Integrating observational and experimental approaches will further deepen our knowledge of plant responses to climate change.

18.1.2 Drought Tolerance-Associated Traits

Drought tolerance is important for plants in arid and semi-arid regions, where water is limited. Key traits include deep root systems for accessing water from deeper soil layers and thick cuticles to reduce water loss. Physiological adaptations like osmotic adjustment and increased water-use efficiency also help plants maintain function during drought (Chaves et al., 2003). Additionally, some species exhibit drought-deciduousness, shedding leaves to minimize water loss. These traits influence forest species composition, with drought-tolerant species more likely to dominate in water-stressed areas. Understanding these traits is vital for guiding species selection in reforestation efforts.

Milovancevic et al. (2024) describe how drought leads to tissue dehydration, inhibits cell division and growth, and induces oxidative stress by increasing reactive oxygen species. These effects slow important physiological processes such as photosynthesis, transpiration, and mineral uptake, reducing plant growth and yield (Anjum et al., 2017; Kumar et al., 2021a; Kumar et al., 2021b). In response, plants adopt resistance strategies such as drought escape (ephemeroid strategy), drought avoidance, and drought tolerance (Laxa et al., 2019).

18.1.3 Temperature Sensitivity-Associated Traits

Temperature sensitivity in plants is shaped by traits like leaf size and shape, which aid heat dissipation and heat-shock proteins that protect cells from high-temperature damage. Phenological traits, such as flowering and leaf-out timing, are also temperature-dependent, influencing growth and reproduction cycles (Korner, 2003).

In warmer climates, plants that can handle heat better—for example, those that can quickly release heat and keep photosynthesis going even when they're under a lot of stress—will do better. On the other hand, species that only survive in small thermal niches may go extinct, which could change the make-up of forests. Understanding these traits is crucial for predicting forest ecosystem responses to climate change.

18.1.4 Traits Influencing Carbon Sequestration and Storage

Forests are vital to the global carbon cycle, and traits influencing carbon sequestration and storage are crucial for climate change mitigation. Key traits include wood density, tree height, and growth rate, with high wood density species storing more carbon per unit volume (Chave et al., 2009; Kumar et al., 2021c). Additionally, traits affecting belowground carbon storage, such as root depth and turnover, are important. Deeper roots enhance long-term soil carbon storage, while high root turnover increases soil organic carbon content. Understanding these traits aids in selecting species and management practices to optimize carbon sequestration (Poorter et al., 2017).

18.1.5 Traits Affecting Species Distribution and Migration

Climate change is prompting shifts in species distributions as plants seek more suitable habitats. Key traits influencing these migrations include seed dispersal mechanisms, reproductive timing, and stress tolerance. Species with rapid dispersal and establishment traits are more likely to succeed in new environments, while those with limited dispersal may face local extinction (Thuiller et al., 2005). Traits related to biotic interactions, such as competitive ability and pest resistance, also affect migration. Understanding these traits is important for predicting climate change impacts on biodiversity and ecosystem functions (Parmesan and Yohe, 2003).

18.2 MECHANISMS OF TRAIT INFLUENCE ON FOREST DYNAMICS

The growth, recruitment, and death of trees, as shown by tree community biomass and density, shape forest dynamics. These processes are affected by biotic factors like species interactions and abiotic conditions like climate and soil properties (Poorter et al., 2017). Plant FTs, such as leaf area, wood density, and seed dispersal mechanisms, significantly affect species interactions and resource acquisition. Traits that enhance drought tolerance can determine species dominance in arid environments, influencing forest composition (Cornwell et al., 2008). Understanding these processes is vital for predicting forest responses to environmental changes and developing strategies for biodiversity conservation (Diaz et al., 2004).

18.3 TRAIT–CLIMATE INTERACTIONS: UNDERSTANDING THE LINKAGES

The interactions between FTs and climate are complex and interconnected, influencing plant performance and ecosystem processes. Temperature and soil moisture affect traits like drought tolerance, creating dynamic feedback loops between climate and plant responses. Trait–climate models create different scenarios to see how different combinations of traits and climate conditions affect the way a forest grows and changes over time. This helps managers find weak spots and improve the resilience of forests. Understanding these linkages is essential for forming strategies to mitigate climate change impacts on forest ecosystems and ensure their long-term sustainability (Suding et al., 2008).

18.4 IMPACT OF TRAITS ON FOREST STRUCTURE AND COMPOSITION

FTs significantly influence forest structure and composition, affecting species dominance, regeneration after disturbances, and evolutionary dynamics. Fast-growing, light-demanding species can quickly colonize open areas following disturbances, while shade-tolerant species establish in the understory (Pacala and Rees, 1998). Changing the distribution of traits can change the makeup of species and the way ecosystems work. For example, more species that store a lot of carbon could make carbon storage better, while more species that can survive in dry conditions could hurt biodiversity. Understanding these dynamics is vital for forest management to sustain ecosystem services and adapt to environmental changes (Wright et al., 2010; Kumar et al., 2022).

18.5 ROLE OF FUNCTIONAL TRAITS IN ECOSYSTEM PROCESSES

FTs are essential to ecosystem processes like nutrient cycling, energy flow, and carbon storage. Different factors, like the quality of the leaf litter, the structure of the roots, and the relationships between microbes, affect how nutrients move through food webs and how carbon is stored above and below ground (Chapin et al., 2011). For example, species with high-nitrogen leaves enhance nutrient cycling and soil fertility but may also increase greenhouse gas emissions. Additionally, root traits that promote deep soil penetration can enhance subsoil carbon storage, potentially offsetting aboveground carbon losses. Understanding these traits is crucial for forest management to sustain ecosystem services and adapt to environmental changes (Wardle et al., 2004).

18.6 TRAIT-BASED RESPONSES TO CLIMATE EXTREMES

Forests worldwide offer insights into how FTs influence responses to climate extremes. In Mediterranean forests, species with deep root systems and drought-tolerant leaves demonstrate greater resilience during prolonged dry periods (Klein et al., 2018). Conversely, tropical forests facing extreme heat waves show shifts in species composition, with temperature-sensitive species declining and heat-tolerant species becoming more dominant (Feeley et al., 2012). These examples underscore the significance of FTs in forest management and conservation planning. By prioritizing traits that enhance resilience to climate extremes, forest managers can support the long-term health and stability of ecosystems (Table 18.1).

TABLE 18.1
Trait-Based Responses to Climate Extremes

Climate Extreme	Functional Trait	Plant Response	Example Species	Ecosystem Impact
Drought	Deep root systems	Increased water uptake from deeper soil layers	*Quercus ilex* (Holm Oak)	Enhanced drought resilience; dominance in dry regions
	Leaf shedding (drought-deciduousness)	Reduced water loss during dry periods	*Prosopis glandulosa* (Mesquite)	Temporary reduction in photosynthesis; potential species shifts
Heatwaves	Heat-shock proteins	Protection of cellular structures from heat stress	*Arabidopsis thaliana*	Maintained cellular function; possible adaptation to heat
	Leaf size reduction	Reduced surface area for heat absorption	*Pinus pinea* (Stone Pine)	Altered canopy structure; reduced heat load
Frost	Frost tolerance (cell membrane stability)	Prevention of cell damage due to freezing temperatures	*Picea abies* (Norway Spruce)	Survival in cold environments; potential range expansion
	Delayed leaf-out	Avoidance of early frost damage	*Fagus sylvatica* (European Beech)	Synchronization with favourable growing conditions
Flooding	Aerenchyma formation in roots	Enhanced oxygen diffusion in waterlogged soils	*Taxodium distichum* (Bald Cypress)	Increased survival in flooded areas; wetland forest dominance
	Adventitious rooting	Development of roots above waterlogged soil	*Salix* spp. (Willows)	Stabilization of soil; recovery after flooding
Windstorms	Flexible stem structure	Reduction of wind damage by bending rather than breaking	*Acer rubrum* (Red Maple)	Lowered windthrow risk; potential changes in forest structure
	Shallow root systems	Increased vulnerability to windthrow due to reduced anchorage	*Picea sitchensis* (Sitka Spruce)	Increased gap formation; shifts in species composition

Extreme weather events like droughts and heat waves are killing more trees in forests across all biomes. Soil, topography, and disturbances from pests and wildfires are some of the things that are contributing to this. Human activities primarily drive current changes, despite historical data indicating previous drought-induced forest shifts. A global review revealed that shrubs replaced trees in 34% of cases and grasses in 10%, indicating potential forest collapse. Early warning systems utilizing long-term monitoring and remote sensing are vital for anticipating these shifts. Although global forest collapse is not widespread, ongoing events underscore future vulnerabilities (Lloret and Batllori, 2021).

18.7 PREDICTIVE MODELLING OF FOREST CHANGES USING FUNCTIONAL TRAITS

Predictive modelling of forest changes using FTs is an effective method for understanding how forests react to environmental changes like climate change, land-use alterations, and natural disturbances. FTs provide insights into organisms' growth, reproduction, and survival, which are crucial for ecosystem dynamics. Various modelling approaches, including statistical models (like regression and generalized additive models), analyse relationships between forest variables and environmental factors to forecast species distribution and forest health. Machine learning models, including random forests and neural networks, improve predictions by handling complex interactions and large datasets. Based on ecological processes, process-based models simulate how forests grow. Hybrid models, on the other hand, use both statistical and mechanistic methods to get more accurate results. Spatial models predict how geographic gradients influence forest patterns and time series models analyse changes over time. Collectively, these approaches create a comprehensive framework for managing forest ecosystems amid environmental changes.

18.8 INCORPORATING FUNCTIONAL TRAITS INTO CLIMATE MODELS

Integrating FTs into climate models greatly improves predictions of ecosystem responses to climate change, especially in forests. Traits like leaf size, wood density, and root depth reveal how species interact with their environment and affect ecosystem processes. In contrast to species-based approaches, which prioritize species identity, trait-based approaches focus on physiological, morphological, or life-history characteristics (Zakharova et al., 2019). These trait-based models provide insights into how species influence and are influenced by environmental changes. Integrating FTs into climate models can lead to more accurate forecasts of ecosystem dynamics under changing conditions.

18.8.1 Enhanced Predictive Accuracy

Incorporating FTs into climate models allows for more precise predictions of species responses to the changing climatic conditions. Traits that affect physiological performance, growth, and survival (e.g., drought tolerance and nutrient uptake efficiency) provide insights into how species will fare under different climate scenarios. For instance, models that include traits related to water-use efficiency can better predict which species are likely to survive prolonged droughts or increased temperatures.

18.8.2 Mechanistic Understanding

FTs offer a mechanistic foundation for comprehending ecosystem responses. By connecting traits to ecological processes like the cycling of carbon and nutrients, models can show how changes in the species composition caused by changes in the climate will impact the way ecosystems work (Loverel and Garnier, 2002). This approach helps predict changes in forest structure, productivity, and biodiversity with greater precision.

18.8.3 Improved Species Distribution Models

FTs can enhance species distribution models by adding another layer of information about species' ecological requirements and tolerances. When traits are added to these models, they can show how changes in climate will affect the distribution of species based on their abilities, not just whether they are present or not (Kearney and Porter, 2009a).

18.8.4 Adaptation and Resilience Insights

Including FTs in climate models can identify which species or communities are more likely to adapt to or resist climate change. FTs can be used to predict community responses and guide conservation strategies. Traits associated with resilience and adaptation—such as high seed dispersal rates or rapid growth under stress—can help in predicting how forests might shift or remain stable amidst climate changes (Lavorel and Garnier, 2002).

18.8.5 Management and Conservation Planning

By understanding how FTs affect species' responses to climate change, forest managers and conservationists can make better-informed decisions. For instance, they can focus on preserving or restoring species with traits that boost resilience to climate change or plan the introduction of species with beneficial traits that could thrive under future climate conditions (Penuelas and Boada, 2003).

18.8.6 Challenges and Considerations

The inclusion of FTs into climate models involves various challenges.

18.8.6.1 Trait Data Gaps

Incomplete trait data can limit model accuracy.

18.8.6.2 Trait Variability

Models must account for environmental variations in species traits.

18.8.6.3 Integration Challenges

Adding traits increases model complexity, which requires careful calibration.

18.9 TOOLS AND TECHNIQUES FOR TRAIT-BASED PREDICTIONS

Trait-based predictions use various tools and techniques to understand species responses to environmental changes.

18.9.1 Statistical Models

Methods like generalized linear models are used to link species traits with environmental factors, predicting how traits influence species distribution and ecosystem resilience. Breiman (2001) conducted a study on the impact of drought tolerance traits on species distribution and forest resilience in the context of climate change.

18.9.2 Machine Learning

Algorithms like support vector machines and neural networks help identify complex patterns between traits and environmental stressors like drought or temperature changes.

18.9.3 Functional Trait Databases

Large-scale databases like TRY and FLORES (a global database of plant traits) compile global plant traits (such as leaf area, wood density, and root depth), providing the necessary data for trait-based models and predictions.

18.9.4 Climate Models

Dynamic vegetation models (DVMs) and biogeochemical models integrate trait data with climate scenarios, allowing predictions of how species and ecosystems may shift under future climate conditions. DVMs may use traits like leaf area and growth rate to forecast changes in carbon storage capacity as species compositions shift due to climate impacts.

By adding trait data to these models, researchers can make better guesses about how forests will change over time and help management and conservation efforts work better.

18.10 MODEL APPLICATION IN DIFFERENT FOREST TYPES

Functioning trait-based models are useful for understanding how various FTs of plants (leaf area, seed size, wood density, etc.) influence ecosystem processes and dynamics in different forest types. Trait-based models provide valuable insights into forest ecosystem functioning, helping in predicting responses to environmental changes, management practices, and natural disturbances. By focusing on FTs, researchers can get a more profound understanding of the ecological roles of different species and the potential impacts of changing conditions on forest ecosystems (Table 18.2).

TABLE 18.2
Trait-Based Models for Monitoring Forest Health Across Different Forest Types

S. No.	Forest Types	Trait Used	Application	References
1.	Tropical rainforests	Specific leaf area (SLA), wood density, and maximum height	Predicting tree growth and carbon storage	Baraloto et al. (2010)
2.	Temperate forests	Leaf phenology (timing of leaf out and fall), cold tolerance, and drought tolerance	Understanding the response to climate change	Richardson et al. (2013)
3	Boreal forests	Bark thickness, resprouting ability, and seed dispersal mechanisms	Modelling fire resistance and recovery	Bond and Midgley (2001)
4	Mediterranean forests	Leaf size, stomatal conductance, and root depth	Water-use efficiency and drought resistance	Valladares et al. (2007)
5.	Mangrove forests	Root structure, salt tolerance, and growth form	Coastal protection and resilience	Krauss and Ball (2013)
6.	Urban forests	Leaf area index, air pollution tolerance, evapotranspiration rates	Mitigating urban heat islands and air pollution	Gill et al. (2007)
7.	Dry forests and savannas	Leaf toughness, seed size, and nitrogen fixation ability	Fire dynamics and nutrient cycling	Fernandez-Garcia et al. (2020)
8.	Alpine forests	Growth form (e.g., dwarfism), frost tolerance, phenological traits	Adaptation to high altitude and cold temperatures	Nagy and Grabherr (2009).
9.	Floodplain forests	Root architecture, anaerobic tolerance, nutrient uptake efficiency	Flood tolerance and nutrient cycling	Kozlowski (1997)

18.11 TRAIT-BASED INDICATORS FOR MONITORING FOREST HEALTH

Monitoring forest health is essential as climate change increases threats like droughts, pest infestations, and extreme weather events. Trait-based indicators provide a comprehensive approach to understand forest responses by focusing on FTs. It offers valuable insights into the functional roles of organisms within ecosystems, enabling informed decision-making for forest conservation and management. To make a monitoring plan work, combinations of traditional field methods with new technologies and ecological indicators are required. This will help forest ecosystems stay healthy and adaptable to climate change (Violle et al., 2007).

18.12 IDENTIFYING KEY INDICATORS OF CLIMATE-DRIVEN CHANGE

Essential indicators of forest health, including physiological, structural, ecosystem-level, and biodiversity metrics, are crucial for identifying stress and evaluating the effects of climate change.

18.12.1 Physiological Indicators

Things like tree growth rates, leaf water potential, chlorophyll content, and photosynthetic efficiency can show when plants are under stress because of things like drought, extreme temperatures, or not getting enough nutrients.

18.12.2 Structural Indicators

Things like canopy density, tree height, leaf area index, and biomass show how healthy and productive a forest is. They help us see how changes in climate affect the forest's structure and composition.

18.12.3 Ecosystem-Level Indicators

Rates of nutrient cycling, soil moisture content, and carbon sequestration are important functions that show how healthy and resilient a forest is.

18.12.4 Biodiversity Indicators

Species diversity, abundance, and distribution patterns are essential for ecosystem resilience. Monitoring shifts in biodiversity, including invasive species or keystone species decline, assesses overall forest health.

18.13 FUNCTIONAL TRAITS AS INDICATORS OF FOREST RESPONSE TO CLIMATE CHANGE

Identifying the key traits that respond to climate drivers is essential for developing effective monitoring systems.

18.13.1 Leaf Area and Specific Leaf Area (SLA)

Changes in the leaf area and SLA indicate resource availability and stress responses, such as drought tolerance. Higher SLA plants are more competitive in nutrient-rich, high-light environments but more susceptible to drought stress. Monitoring these traits offers insights into forest shifts under climate change (Wright et al., 2004).

18.13.2 Phenology

The timing of leaf-out, flowering, and senescence reveals climate-driven changes. Warming temperatures cause changes in plant phenology, such as earlier leaf-out in the spring and later senescence in the fall. These changes affect how plants interact with pollinators, herbivores, and decomposers, which in turn affects the way ecosystems work (Menzel et al., 2006).

18.13.3 Wood Density

Wood density affects a tree's growth rate, mechanical support, and resistance to stressors like drought. Trees with higher wood density grow more slowly but are more resistant to breakage and drought. Monitoring wood density changes can predict shifts in forest structure and composition due to climatic stress (Chave et al., 2009).

18.14 METHODS FOR MONITORING AND ASSESSING TRAIT VARIABILITY

Monitoring trait variability involves both in situ measurements and statistical modelling approaches. Here are some common methods:

18.14.1 Field Sampling

The traditional approach measures traits like leaf area, SLA, and wood density directly from plant samples in the field using standardized protocols. While providing detailed, site-specific data, these methods are labour-intensive and time-consuming, ensuring consistency and comparability across studies (Cornelissen et al., 2003).

18.14.1.1 Permanent Sample Plots (PSPs)

PSPs are set aside areas where tree growth, death, and regrowth can be measured over and over again. This gives long-term information about how forests change and how climate change affects them.

18.14.1.2 Tree Ring Analysis (Dendrochronology)

Looking at tree rings to figure out how trees grew and when they were droughty in the past. This helps us understand how changing climates affect forests over time.

18.14.1.3 Forest Health Monitoring Programs

Regular surveys of tree conditions, including foliage colour, canopy density, and pest infestations, provide data on stress factors affecting forest health.

18.14.2 Trait Databases and Meta-Analysis

Large-scale trait databases, such as TRY and FLORES, compile plant trait data from various studies worldwide, providing a valuable resource for understanding trait variability across species and regions. Researchers can use these databases to do meta-analyses, which help them find global patterns in the variation of traits and how they relate to environmental factors (Kattge et al., 2011). By comparing traits across different species and regions, scientists can better understand how climate change affects forests.

18.14.3 Functional Diversity Indices

Functional diversity indices assess the diversity, distribution, and abundance of traits within a community. Key metrics include functional richness (number of unique traits), functional evenness (distribution of traits), and functional divergence (differences from the community mean). These measurements help figure out how resilient and well an ecosystem is working. More functional diversity is usually a sign of more stability and resistance to changes (Pla et al., 2012).

18.15 USE OF REMOTE SENSING AND FIELD DATA IN MONITORING

Remote sensing is an effective tool for monitoring forest health across large spatial and temporal scales. It allows for the detection of changes in forest structure, composition, and function that may not be visible from the ground.

18.15.1 Spectral Imaging

Remote sensing via satellites and drones with spectral sensors measures vegetation indices like the Normalized Difference Vegetation Index. These indices offer insights into leaf area, chlorophyll content, and plant health; they detect stress responses like drought or nutrient deficiency before visible signs appear (Ollinger, 2011).

18.15.2 LiDAR (Light Detection and Ranging)

LiDAR technology uses laser pulses to generate high-resolution maps of forest structure, measuring canopy height, biomass, and leaf area index. This data is crucial for understanding the three-dimensional structure, health, and dynamics of forests (Lefsky et al., 2002).

18.15.3 Integration with Field Data

Combining remote sensing data with field-based measurements enhances the accuracy and reliability of forest health assessments. Field data provide ground-truthing for remote sensing models, ensuring that trait-based indicators reflect actual forest conditions (Asner et al., 2010).

18.16 PRACTICAL APPLICATIONS: EARLY WARNING SYSTEMS FOR FOREST DECLINE

Trait-based indicators can be used to develop early warning systems for forest decline, allowing for proactive management and conservation efforts:

18.16.1 Detecting Stress and Mortality

Monitoring key traits like leaf water content and chlorophyll fluorescence enables early detection of drought stress and nutrient deficiencies in forests. Early identification allows for timely interventions, reducing the risk of forest decline and supporting more effective management strategies to preserve forest health and resilience (Lausch et al., 2017).

18.16.2 Predicting Forest Resilience

Understanding trait variability helps in modelling forest responses to stressors like drought, fire, and pests. By identifying traits linked to resilience, researchers can predict which forests tend to withstand climate change and prioritize them for conservation (Morin et al., 2008).

18.16.3 Guiding Management Practices

Trait-based monitoring guide's reforestation and afforestation by identifying species and genotypes better suited to future climates, enhancing forest adaptability and maintaining ecosystem services amid climate change.

18.17 CHALLENGES IN USING FUNCTIONAL TRAITS FOR PREDICTION

Plant FTs are invaluable for understanding ecological interactions, yet the weak correlation between plant traits and environmental factors is contradictory. Although plant traits regulate interactions with the environment, they do not consistently predict plant distributions or performance in specific environments. Addressing this contradiction requires a more accurate examination of trait variation mechanisms across different scales, from within species to among communities. It is also important to combine what we know about physiological and evolutionary processes that lead to similar ecological outcomes through different traits and strategies. This also calls for reconsidering how FTs contribute to the overall plant phenotype, which may often function independently of environmental tolerances. In short, resolving this contradiction demands a more holistic view of how traits and environments interact across evolutionary and ecological contexts. The conceptual importance of FTs and their relationships with environmental factors such as climate and soil often explains less than 10%–20% of trait variation globally, raising questions about the interpretation of trait-environment dynamics (Violle et al., 2007). Thus, it highlights the need for a deeper understanding of trait variation across ecological scales.

18.18 IMPLICATIONS FOR FOREST CONSERVATION AND MANAGEMENT

Forest conservation and sustainable management are vital for mitigating climate change, preserving biodiversity and supporting ecosystem services like water regulation and soil protection. Forests serve as carbon sinks, help slow global warming, and sustain ecological resilience. Sustainable practices, including community-based approaches, protect indigenous rights and promote eco-friendly activities like agroforestry and ecotourism. International frameworks like the Paris Agreement and REDD+ enhance conservation efforts, ensuring that forests play an important role in climate mitigation and adaptation.

18.18.1 Adapting Forest Management Strategies to Climate Change

The adaptation of forests to climate change will impact global forestry practices, complicating sustainable management amid rising temperatures, changing precipitation, and extreme weather. Vulnerable ecosystems like boreal and tropical forests are vital carbon sinks but can emit greenhouse gases when degraded. Despite not knowing what the future climate effects will be, forest management and silviculture need to include climate adaptation measures to make ecosystems more resilient, make sure carbon is stored, and keep providing ecosystem services.

18.18.2 Policy Implications for Sustainable Forest Management

Sustainable forest management (SFM) gained global prominence after the 1992 Rio Earth Summit, shifting focus from timber production to broader ecosystem services like biodiversity conservation and climate regulation (Bulkeley and Newell, 2023). Legal frameworks, such as the Forest Conservation Act of 1980 and India's Biological Diversity Act of 2002, promote afforestation, ecosystem protection, and community engagement, emphasizing the importance of biodiversity for climate resilience (Thompson and Ian, 2009). Global forest policy trends now advocate for mixed-species plantations, responsible resource extraction, and inclusive decision-making (Sheppard et al., 2020). Integrating climate resilience into forest management is critical, as diverse ecosystems better withstand climate variability making effective community-driven conservation strategies essential for long-term sustainability (Berchoux et al., 2023).

18.18.3 Collaborative Approaches to Implementing Trait-Based Strategies for Forest Management and Conservation

Successfully implementing trait-based forest management requires synthesizing diverse data, identifying key FTs, and guiding species selection for restoration. Collaboration between researchers and practitioners is crucial to bridging knowledge gaps. A collaborative framework involves stakeholder engagement where research institutions, government agencies, local communities, and the private sector work together to customize management practices based on traits like growth rate, drought tolerance, fire resistance, and nutrient cycling. This approach enhances ecological outcomes while addressing socioeconomic goals. Additionally, integrating research and data is vital for studying FTs that influence ecosystem services like carbon storage and water regulation. Global initiatives like the Center for International Forestry Research (CIFOR) and the World Agroforestry Centre (ICRAF) support these efforts by focusing on tree species traits that provide both environmental and economic benefits.

18.19 EMERGING TRENDS IN TRAIT-BASED CLIMATE RESEARCH AND GENOMIC INTEGRATION

Over the past decade, trait-based approaches have revitalized community ecology by offering tools to monitor and predict biodiversity changes. Although these methods primarily concentrate on the direct effects of climate and land use changes, they overlook how biotic interactions, such as plant-animal relationships, affect species' climate responses. Though recent studies examine FTs in interactions like pollination and dispersal, there's still a need for frameworks addressing how species interactions shape adaptation to climate change (Schleuning et al., 2020). Recent advances in genomic resources and computational tools have significantly improved the study of forest ecosystem adaptation. Genomics is now integral to forest science, with next-generation sequencing enabling researchers to analyse genome-wide adaptation in key species. However, challenges remain due to polygenic traits and population dynamics, necessitating long-term field experiments. The integration of genomic data with phenotyping and computational methods offers promise for anticipating tree responses to environmental changes (Neophytou et al., 2022).

18.20 CONCLUSION

FTs are essential for understanding and predicting climate-driven forest changes. By examining traits related to drought tolerance, temperature sensitivity, carbon sequestration, and species distribution, researchers and forest managers can obtain valuable understanding of how forests might respond to shifting climatic conditions. These traits not only influence forest structure and composition but also affect ecosystem processes, making them crucial for forest dynamics. The incorporation of FTs into predictive models has enhanced our ability to forecast the impacts of climate change on forests, offering a more refined approach for conservation and management strategies. These models offer early warning systems for forest health, facilitating proactive management when paired with cutting-edge technologies like remote sensing and field data monitoring. Despite the challenges of trait variability, uncertainty, and the complexity of integrating diverse ecological data, functional traits offer a promising path forward. They hold potential for improving conservation planning and adaptive management in response to climate change. By leveraging FTs, we can better navigate the ecological complexities and socioeconomic implications of climate-driven forest changes.

REFERENCES

Anjum, Shakeel Ahmad, Umair Ashraf, Ali Zohaib, Mohsin Tanveer, Muhammad Naeem, Iftikhar Ali, Tahira Tabassum, and Usman Nazir. 2017. 'Growth and Developmental Responses of Crop Plants under Drought Stress: A Review'. *Zemdirbyste-Agriculture* 104 (3): 267–76. doi:10.13080/z-a.2017.104.034

Asner, Gregory P., George V. N. Powell, Joseph Mascaro, David E. Knapp, John K. Clark, James Jacobson, Ty Kennedy-Bowdoin, et al. 2010. 'High-Resolution Forest Carbon Stocks and Emissions in the Amazon'. *Proceedings of the National Academy of Sciences* 107 (38): 16738–42. doi:10.1073/pnas.1004875107

Baraloto, Christopher, C. E. Timothy Paine, Sandra Patino, Damien Bonal, Bruno Herault, and Jerome Chave. 2010. 'Functional Trait Variation and Sampling Strategies in Species-rich Plant Communities'. *Functional Ecology* 24 (1): 208–16. doi:10.1111/j.1365-2435.2009.01600.x

Berchoux, Tristan, Craig W. Hutton, Oliver Hensengerth, Hal E. Voepel, Van P. D. Tri, Pham T. Vu, Nghia N. Hung, Dan Parsons, and Stephen E. Darby. 2023. 'Effect of Planning Policies on Land Use Dynamics and Livelihood Opportunities under Global Environmental Change: Evidence from the Mekong Delta'. *Land Use Policy* 131 (August): 106752. doi:10.1016/j.landusepol.2023.106752

Bond, William J., and Jeremy J. Midgley. 2001. 'Ecology of Sprouting in Woody Plants: The Persistence Niche'. *Trends in Ecology & Evolution* 16 (1): 45–51. doi:10.1016/S0169-5347(00)02033-4

Breiman, Leo. 2001. 'Random Forests'. *Machine Learning* 45: 5–32.

Bulkeley, Harriet, and Peter Newell. 2023. *Governing Climate Change*. 3rd ed. London: Routledge.

Chapin, Francis Stuart, Pamela A. Matson, Harold A. Mooney, and Peter Morrison Vitousek. 2011. *Principles of Terrestrial Ecosystem Ecology*. New York: Springer.

Chave, Jerome, David Coomes, Steven Jansen, Simon L. Lewis, Nathan G. Swenson, and Amy E. Zanne. 2009. 'Towards a Worldwide Wood Economics Spectrum'. *Ecology Letters* 12 (4): 351–66. doi:10.1111/j.1461-0248.2009.01285.x

Chaves, Manuela M., Joao P. Maroco, and Joao S. Pereira. 2003. 'Understanding Plant Responses to Drought — from Genes to the Whole Plant'. *Functional Plant Biology* 30 (3): 239. doi:10.1071/FP02076

Cornelissen, J. H. C., S. Lavorel, E. Garnier, S. Díaz, N. Buchmann, D. E. Gurvich, P. B. Reich, et al. 2003. 'A Handbook of Protocols for Standardised and Easy Measurement of Plant Functional Traits Worldwide'. *Australian Journal of Botany* 51 (4): 335. doi:10.1071/BT02124

Cornwell, William K., Johannes H. C. Cornelissen, Kathryn Amatangelo, Ellen Dorrepaal, Valerie T. Eviner, Oscar Godoy, Sarah E. Hobbie, et al. 2008. 'Plant Species Traits Are the Predominant Control on Litter Decomposition Rates within Biomes Worldwide'. *Ecology Letters* 11 (10): 1065–71. doi:10.1111/j.1461-0248.2008.01219.x

Davis, Margaret B., and Ruth G. Shaw. 2001. 'Range Shifts and Adaptive Responses to Quaternary Climate Change'. *Science* 292 (5517): 673–679. doi:10.1126/science.292.5517.673

Diaz S., Hodgson J. G., Thompson K., Cabido M., Cornelissen J. H. C., Jalili A., Montserrat M., et al. 2004. 'The Plant Traits that Drive Ecosystems: Evidence from Three Continents'. *Journal of Vegetation Science* 15 (3): 295–304. doi:10.1111/j.1654-1103.2004.tb02266.

Dyderski, Marcin K., Sonia Paz, Lee E. Frelich, and Andrzej M. Jagodzinski. 2018. 'How Much Does Climate Change Threaten European Forest Tree Species Distributions?' *Global Change Biology* 24 (3): 150–1163. doi:10.1111/gcb.13925

Feeley, Kenneth J., Yadvinder Malhi, Przemyslaw Zelazowski, and Miles R. Silman. 2012. 'The Relative Importance of Deforestation, Precipitation Change, and Temperature Sensitivity in Determining the Future Distributions and Diversity of AMazonian Plant Species'. *Global Change Biology* 18 (8): 2636–47. doi:10.1111/j.1365-2486.2012.02719.x

Fernandez-Garcia, Victor, Elena Marcos, Peter Z. Fule, Otilia Reyes, Victor M. Santana, and Leonor Calvo. 2020. 'Fire Regimes Shape Diversity and Traits of Vegetation under Different Climatic Conditions'. *Science of the Total Environment* 716: 137137. doi:10.1016/j.scitotenv.2020.137137

Garnier, Eric, Marie-Laure Navas, and Karl Grigulis. 2016. *'Plant Functional Diversity: Organism Traits, Community Structure, and Ecosystem Properties'.* Oxford University Press. doi:10.1093/acprof:oso/9780198757368.001.0001

Gill, S. E, J. F Handley, A. R Ennos, and S. Pauleit. 2007. 'Adapting Cities for Climate Change: The Role of the Green Infrastructure'. *Built Environment* 33 (1): 115–33. doi:10.2148/benv.33.1.115

Kattge, Jens, Sandra Diaz, Sandra Lavorel, Colin Prentice, Paul Leadley, Gerhard Bonisch, Eric Garnier, et al. 2011. 'TRY–A Global Database of Plant Traits'. *Global Change Biology* 17 (9): 2905–2935. doi:10.1111/j.1365-2486.2011.02451.x

Kearney, Michael, and Warren P. Porter. 2009a. 'Mechanistic Niche Modeling: Combining Physiological and Spatial Data to Predict Species' Ranges'. *Ecology Letters* 12 (4): 334–350. doi:10.1111/j.1461-0248.2008.01277

Klein, Tamir, Melanie J. B. Zeppel, William R. L. Anderegg, Jasper Bloemen, Martin G. De Kauwe, Patrick Hudson, Nadine K. Ruehr, Thomas L. Powell, Georg Von Arx, and Andrea Nardini. 2018. 'Xylem Embolism Refilling and Resilience against Drought-induced Mortality in Woody Plants: Processes and Trade-offs'. *Ecological Research* 33 (5): 839–55. doi:10.1007/s11284-018-1588-y

Korner, Christian. 2003. *Alpine Plant Life: Functional Plant Ecology of High Mountain Ecosystems.* 2nd ed. Berlin: Springer. doi:10.1007/978-3-030-59538-8

Kozlowski, Theodore1997. 'Responses of Woody Plants to Flooding and Salinity'. *Tree Physiology* 17 (7): 490–510. doi:10.1093/treephys/17.7.490

Krauss, Ken W., and Marilyn C. Ball. 2013. 'On the Halophytic Nature of Mangroves'. *Trees* 27 (1): 7–11. doi:10.1007/s00468-012-0767-7

Kumar, Amit, Gopal Krishna Dwivedi, Salil Tewari, Jai Paul, Rahul Anand, Narendra Kumar, Parmanand Kumar, Hukum Singh, and Rajesh Kaushal. 2020a. 'Carbon Mineralization and Inorganic Nitrogen Pools under *Terminalia chebula* Retz.-Based Agroforestry System in Himalayan Foothills, India'. *Forest Science* 66 (5): 634–643.

Kumar, Amit, Gopal Krishna Dwivedi, Salil Tewari, Jai Paul, Vinod Kumar Sah, Hukum Singh, Parmanand Kumar, Narendra Kumar, and Rajesh Kaushal. 2020b. 'Soil Organic Carbon Pools under *Terminalia chebula* Retz. Based Agroforestry System in Himalayan Foothills, India. *Current Science* 118 (7): 1098–1103.

Kumar, Amit, Hukum Singh, Garima Kumari, Sarita Bisht, Apurva Malik, Narendra Kumar, Manish Singh, Asha Raturi, Santhan Barthwal, Ajay Thakur, and Rajesh Kaushal. 2022. 'Adaptive Resilience of Roadside Trees to Vehicular Emissions via Leaf Enzymatic, Physiological, and Anatomical Trait Modulations.' *Environmental Pollution* 313: 120191.

Kumar, Amit, Parmanand Kumar, Hukum Singh, and Narendra Kumar. 2021b. 'Modulation of Plant Functional Traits under Essential Plant Nutrients during Seasonal Regime in Natural Forests of Garhwal Himalayas'. *Plant and Soil* 465: 197–212.

Kumar, Amit, Parmanand Kumar, Hukum Singh, and Narendra Kumar. 2021c. 'Impact of Plant Functional Traits on Near Saturated Hydraulic Conductivity of Soil under Different Forests of Kempty Watershed in Garhwal Himalaya'. *Indian Journal of Soil Conservation* 49 (1): 38–44.

Kumar, Amit, P.armanand Kumar, Hukum Singh, Sarita Bisht, and Narendra Kumar. 2021a. 'Relationship of Physiological Plant Functional Traits with Soil Carbon Stock in Temperate Forest of Garhwal Himalaya'. *Current Science* 120 (8): 1368–1373.

Kumar, Narendra, Neha Jeena, Amit Kumar, Rowndel Khairakpam, and Hukum Singh. 2021d. 'Comparative Response of Rice Cultivars to Elevated Air Temperature in Bhabar Region of Indian Himalaya: Status on Yield Attributes'. *Heliyon* 7: e07474.

Kumar, Narendra, Neha Jeena, and Hukum Singh. 2019. 'Elevated Temperature Modulates Rice Pollen Structure: A Study from Foothill Himalayan Agro-Ecosystem in India'. *3 Biotech* 9: 175.

Kumar, Narendra, Deep Chand Suyal, Ishawar Prakash Sharma, Amit Verma, and Hukum Singh. 2017. 'Elucidating Stress Proteins in Rice (*Oryza sativa* L.) Genotype under Elevated Temperature: A Proteomic Approach to Understand Heat Stress Response'. *3 Biotech* 7: 205.

Kumar, Narendra, Sikhar Chand Shankhdhar, and Dipti Shankhdhar. 2016. 'Impact of Elevated Temperature on Antioxidant Activity and Membrane Stability in Different Genotypes of Rice (*Oryza sativa* L.)'. *Indian Journal of Plant Physiology (Now Plant Physiology Report)* 21 (1): 37–43.

Lausch, Angela, Stefan Erasmi, Douglas King, Paul Magdon, and Marco Heurich. 2017. 'Understanding Forest Health with Remote Sensing-Part II—A Review of Approaches and Data Models'. *Remote Sensing* 9 (2): 129. doi:10.3390/rs9020129

Lavorel, Sandra, and Eric Garnier. 2002. 'Predicting Changes in Community Composition and Ecosystem Functioning from Plant Traits: Revisiting the Holy Grail'. *Functional Ecology* 16 (5): 545–556. doi:10.1046/j.1365-2435.2002.00664.x

Laxa, Miriam, Michael Liebthal, WilenaTelman, KamelChibani, and Karl-Josef Dietz. 2019. 'The Role of the Plant Antioxidant System in Drought Tolerance'. *Antioxidants* 8 (4): 94. doi:10.3390/antiox8040094

Lefsky, Michael A., Warren B. Cohen, Geoffrey G. Parker, and David J. Harding. 2002. 'Lidar Remote Sensing for Ecosystem Studies'. *BioScience* 52 (1): 19. doi:10.1641/0006-3568(2002)052[0019:LRSFES]2.0.CO;2

Lloret, Francisco, and Enric Batllori. 2021. 'Climate-Induced Global Forest Shifts Due to Heatwave-Drought'. *Ecosystem Collapse and Climate Change* 155–186. doi:10.1111/gcb.15518

Menzel, Annette, Tim H. Sparks, Nicole Estrella, Elisabeth Koch, AntoAasa, Rein Ahas, Kerstin Alm-Kübler, et al. 2006. 'European Phenological Response to Climate Change Matches the Warming Pattern'. *Global Change Biology* 12 (10): 1969–1976. doi:10.1111/j.1365-2486.2006.01193.x

Milovancevic, Marija, MilanaTrifunovic-Momcilov, Olga Radulovic, Snezana Milosevic, and Angelina Subotic. 2024. 'Drought Stress Effects and Ways for Improving Drought Tolerance in *Impatiens walleriana* Hook.f.—A Review'. *Horticulturae* 10 (9): 903. doi:10.3390/horticulturae10090903

Morin, Xavier, David Viner, and Isabelle Chuine. 2008. 'Tree Species Range Shifts at a Continental Scale: New Predictive Insights from a Process-Based Model'. *Journal of Ecology* 96 (4): 784–794. doi:10.1111/j.1365-2745.2008.01369.x

Nagy, Laszlo, and Georg Grabherr. 2009. *The Biology of Alpine Habitats*. Oxford: Oxford University Press. doi:10.1093/oso/9780198567035.001.0001

Neophytou, Charalambos, KatrinHeer, Pascal Milesi, Martina Peter, TanjaPyhajarvi, MarjanaWestergren, Christian Rellstab, and Felix Gugerli. 2022. 'Genomics and Adaptation in Forest Ecosystems'. *Tree Genetics & Genomes* 18 (2): 12. doi:10.1007/s11295-022-01542-1

Ollinger, Scott V. 2011. 'Sources of Variability in Canopy Reflectance and the Convergent Properties of Plants'. *New Phytologist* 189 (2): 375–394. doi:10.1111/j.1469-8137.2010.03536.x

Pacala, Stephen W., and Mark Rees. 1998. 'Models Suggesting Field Experiments to Test Two Hypotheses Explaining Successional Diversity'. *The American Naturalist* 152 (5): 729–37. doi:10.1086/286203

Parmesan, Camille, and Gary Yohe. 2003. 'A Globally Coherent Fingerprint of Climate Change Impacts across Natural Systems'. *Nature* 421 (6918): 37–42. doi:10.1038/nature01286

Peng, Changhui. 2000. 'Understanding the Role of Forest Simulation Models in Sustainable Forest Management.' *Environmental Impact Assessment Review* 20 (4): 481–501. doi:10.1016/S0195-9255(99)00044-X

Penuelas, Josep, and Marti Boada. 2003. 'A Global Change-Induced Biome Shift in the Montseny Mountains (NE Spain)'. *Global Change Biology* 9 (2): 131–40. doi:10.1046/j.1365-2486.2003.00566.x

Pla, Laura, Fernando Casanoves, and Julio Di Rienzo. 2012. *Functional Diversity Indices*. Springer. doi:10.1007/978-94-007-2648-2_3

Poorter, Lourens, Masha T. Van Der Sande, Eric J. M. M. Arets, NatalyAscarrunz, Brian J. Enquist, Bryan Finegan, Juan Carlos Licona, et al. 2017. 'Biodiversity and Climate Determine the Functioning of Neotropical Forests'. *Global Ecology and Biogeography* 26 (12): 1423–34. doi:10.1111/geb.12668

Richardson, Andrew, Trevor Keenan, Mirco Migliavacca, Youngryel Ryu, Oliver Sonnentag, and Michael Toomey. 2013. 'Climate Change, Phenology, and Phenological Control of Vegetation Feedbacks to the Climate System'. *Agricultural and Forest Meteorology* 169: 156–173. doi:10.1016/j.agrformet.2012.09.012

Schleuning, Matthias, Eike Lena Neuschulz, Jorg Albrecht, Irene Bender, Diana Bowler, Matthias Dehling, Susanne Fritz, Christian Hof, Thomas Mueller, Larissa Nowak, Marjorie Sorensen, KatrinBohning-Gaese, and Wilson Daniel Kissling. 2020. 'Trait-Based Assessments of Climate-Change Impacts on Interacting Species'. *Trends in Ecology & Evolution* 35: 10–19. doi:10.1016/j.tree.2019.12.010

Sheppard, Jonathan P., James Chamberlain, Dolores Agundez, Prodyut Bhattacharya, PaxieWanangwaChirwa, AndreyGontcharov, Willie Cliffie John Sagona, Hai-long Shen, WubalemTadesse, and Sven Mutke. 2020. 'Sustainable Forest Management Beyond the Timber-Oriented Status Quo: Transitioning to Co-Production of Timber and Non-Wood Forest Products—a Global Perspective'. *Current Forestry Reports* 6 (1): 26–40. doi:10.1007/s40725-019-00107-1

Suding, Katharine N., Sandra Lavorel, Franklin Sen Chapin, Johannes H. C. Cornelissen, Sandra Díaz, Eric Garnier, Deborah Goldberg, David U. Hooper, Stephen T. Jackson, and Marie-Laure Navas. 2008. 'Scaling Environmental Change through the Community-level: A Trait-based Response-and-effect Framework for Plants'. *Global Change Biology* 14 (5): 1125–40. doi:10.1111/j.1365-2486.2008.01557.x

Thompson, Ian D., ed. 2009. *Forest Resilience, Biodiversity, and Climate Change: A Synthesis of the Biodiversity / Resiliende / Stability Relationship in Forest Ecosystems*. CBD Technical Series 43. Montreal: Secretariat of the Convention on Biological Diversity.

Thuiller, Wilfried, Sandra Lavorel, Miguel B. Araujo, Martin T. Sykes, and Innowel Colin Prentice. 2005. 'Climate Change Threats to Plant Diversity in Europe'. *Proceedings of the National Academy of Sciences* 102 (23): 8245–50. doi:10.1073/pnas.0409902102

Valladares,Fernando,ErnestoGianoli,andJoseGomez.2007.'EcologicalLimitstoPlantPhenotypic Plasticity'. *New Phytologist* 176: 749–763. doi:10.1111/j.1469-8137.2007.02275.x

Violle, Cyrille, Marie-Laure Navas, Denis Vile, Elena Kazakou, Claire Fortunel, Irène Hummel, and Eric Garnier. 2007. 'Let the Concept of Trait Be Functional'. *Oikos* 116 (5): 882–892. doi:10.1111/j.2007.0030-1299.15559.x

Wardle, David A., Richard D. Bardgett, John N. Klironomos, HeikkiSetala, Wim H. Van Der Putten, and Diana H. Wall. 2004. 'Ecological Linkages between Aboveground and Belowground Biota'. *Science* 304 (5677): 1629–33. doi:10.1126/science.1094875

Woodall, Christopher W., Christopher M. Oswalt, James A. Westfall, Charles H. Perry, Mark D. Nelson, and Andrew O. Finley. 2009. 'An Indicator of Tree Migration in Forests of the Eastern United States'. *Forest Ecology and Management* 257: 1434–1444. doi:10.1016/j.foreco.2008.12.013

Wright, Ian J., Peter B. Reich, Mark Westoby, David D. Ackerly, Zdravko Baruch, FransBongers, Jeannine Cavender-Bares, et al. 2004. 'The Worldwide Leaf Economics Spectrum'. *Nature* 428 (6985): 821–27. doi:10.1038/nature02403

Wright, S. Joseph, Kaoru Kitajima, Nathan J. B. Kraft, Peter B. Reich, Ian J. Wright, Daniel E. Bunker, Richard Condit, et al. 2010. 'Functional Traits and the Growth–Mortality Trade-off in Tropical Trees'. *Ecology* 91 (12): 3664–74. doi:10.1890/09-2335.1

Zakharova, Liubov, Merav Seifan, and Katrin Meyer. 2019. 'Trait-Based Modelling in Ecology: A Review of Two Decades of Research'. *Ecological Modelling* 407: 10–19. doi:10.1016/j.ecolmodel.2019.05.008

Index

Made in the USA
Las Vegas, NV
19 September 2026

27093b3a-65ed-4ddf-818f-d4c44554dd82R02